FREUDIAN
FRAUD

Also by the author

FREUDIAN FRAUD

THE MALIGNANT EFFECT OF FREUD'S THEORY ON AMERICAN THOUGHT AND CULTURE

E. FULLER TORREY, M.D.

HarperCollins*Publishers*

HarperCollins books may be purchased for educational, business, or sales promotional use. For information, please call or write: Special Markets Department, HarperCollins Publishers, Inc., 10 East 53rd Street, New York, NY 10022. Telephone (212) 207-7528; Fax: (212) 207-7222.

Library of Congress Cataloging in Publication Data

Torrey, E. Fuller (Edwin Fuller), 1937–
    Freudian fraud: the malignant effect of Freud's theory on American thought and culture / E. Fuller Torrey.—1st ed.
     p. cm.
    Includes bibliographical references and index.
    ISBN 0-06-016812-9
    1. Psychoanalysis—Social aspects—United States—History. 2. Freud, Sigmund, 1856–1939—Influence. 3. Nature and nurture—History. 4. Science and psychology—History. 5. Psychoanalysis—United States—History. 6. United States—Intellectual life—20th century. I. Title.
    BF175.T68 1992
    150.19'52'0973—dc20   91-58343

92 93 94 95 96 ◆/HC 10 9 8 7 6 5 4 3 2

To Vada and Ted Stanley

*For their extraordinary generosity to the
mentally ill, and their friendship.*

Contents

Acknowledgments

Those of us who do research on ideas are deeply indebted to libraries and librarians. The Library of Congress has been invaluable to my efforts; Chuck Kelly was especially helpful. The staffs of the National Institute of Mental Health Neurosciences Center Library (Laverne Corum and Dera Thompkins), the St. Elizabeth's Hospital Library, the Archives of the American Museum of Natural History in New York (Andrea LaSala), the American Philosophical Society Library in Philadelphia, and the Houghton Library at Harvard University in Cambridge were unfailingly helpful.

Many people directed me to source material of which I was unaware; I wish to thank especially in this regard Peter Swales, Julie Weiss, and Saleem Shah. I am also indebted to the following for responding to my request for information and allowing me to quote them: Daniel Bell, Noam Chomsky, John K. Galbraith, Irving Kristol, Norman Podhoretz, David Riesman, Arthur Schlesinger, Jr., and Robert Silvers. Benjamin Spock, Lester Sontag, and Frank Mankiewicz kindly gave of their time for interviews.

The book was also substantially improved by suggestions of friends who read part or all of earlier drafts: Halsey Beemer, Llewellyn Bigelow, Irving Gottesman, Stephen Hersh, Robert Taylor, Harold Wise, and Sidney Wolfe. Judy Miller both typed and corrected the manuscript, while Camille Callahan typed the notes and clipped pertinent articles for me. Ms. Carol Cohen, Ms. Andrea Sargent, and the staff at HarperCollins provided their usual high quality assistance which makes publishing a book with them a pleasure.

My largest debt is to my wife, Barbara, for her ideas, criticism, and support. Being married to your best reader is not only efficient but vastly enjoyable as well.

In addition to the above, I am grateful to the following for permission to quote from published and unpublished sources:

Annals of the New York Academy of Sciences for permission to quote from an article by Geoffrey Gorer.

Commentary magazine for permission to quote from articles by Milton Klonsky and Irving Howe.

Esquire magazine and the Hearst Corporation for permission to quote from an article by T. B. Morgan.

The Saturday Evening Post for permission to quote an editorial by George Lorimer.

Nation magazine for permission to quote from an article by Franz Boas.

New Republic magazine for permission to quote from articles by Max Eastman and Franz Boas.

American Anthropological Association for permission to quote from an article in the American Anthropological Association Memoirs by A. L. Kroeber. Reproduced by permission of the *American Anthropological Association Memoirs*, no. 61, 1943. Not for sale or further reproduction.

National Association of Social Workers and Social Work for permission to quote from an article by H. J. Karger.

Brain and Behavioral Sciences and the Cambridge University Press for permission to quote from an article by R. D. Bock and M. F. Zimowski.

Pediatrics and the Williams and Wilkins Company for permission to quote from an article by Benjamin Spock and Mary Bergen. Reproduced by permission of *Pediatrics*, vol. 34, 112, copyright 1964.

Professor John C. Burnham for permission to quote from his unpublished doctoral dissertation, "Psychoanalysis in American Civilization Before 1918."

Ms. Carolyn S. Holmes for permission to quote from the unpublished doctoral dissertation of her mother, Catherine L. Covert, "Freud on the Front Page."

Mr. Mervyn Jones for permission to quote from unpublished letters of his father, Ernest Jones.

Oral History Research Office at Columbia University for permission to quote from the unpublished transcripts "The Reminiscences of Franziska Boas."

Houghton Library, Harvard University for permission to quote from two letters of Madison Grant.

The Special Collections of the American Museum of Natural History for permission to quote from letters of Madison Grant and Henry Osborn and from an unpublished manuscript of Henry Osborn.

Dr. Barbara Sicherman for permission to quote from her unpublished doctoral dissertation, "The Quest for Mental Health in America, 1880–1917."

Partisan Review for permission to quote from an editorial by William Phillips.

New York Times for permission to quote from an article by Alfred Kazin in the New York Times Magazine.

Institute for Intercultural Studies, Inc. for permission to quote from two articles by Margaret Mead in Redbook.

Lescher and Lescher, Ltd. and Dr. Benjamin Spock for permission to quote from three articles by Dr. Spock in Redbook and from an interview with him.

Preface

This book concerns Sigmund Freud's theory and the nature-nurture controversy, two subjects about which, individually, volumes have been written, but they have never been considered together as far as I can ascertain. And that is precisely the point of this book. Freud's theory of human behavior postulates early childhood experiences, especially those that are sexual in nature, as being the crucial determinants of adult personality and behavior. The nature-nurture controversy juxtaposes genetic forces (nature) against nongenetic forces (nurture) as being the crucial determinants of adult personality and behavior. Because early childhood experiences are one kind of nongenetic force, it was inevitable that Freudian theory would become part of the nature-nurture controversy and that these two historic streams would become confluent. It is in fact not possible to fully account for Freud's popularity in America without taking into consideration the nature-nurture controversy.

I recognize that some readers may have trouble with aspects of this history. The Freudian paradigm is so intertwined with liberalism and humanism in America that to doubt the former is to implicitly denigrate the latter. Furthermore, people who have invested hundreds of hours and thousands of dollars in therapies arising from Freudian theory are not pleased to learn that the theory is devoid of any scientific foundation. Even worse, the ashes of Auschwitz still permeate our intellectual life and anyone who proclaims genes to be the most important determinant of personality and behavior risks being labeled as a neo-Nazi. Some readers may also disagree with my utilization of personal information from the lives of those I am discussing, but I believe that facts such as Margaret Mead's bisexuality, Ruth Benedict's lesbianism, and Karl Menninger and Benjamin Spock's extraordinarily domineering mothers are directly relevant to understanding their attraction to, and advocacy for, Freud's theory. I ask the reader only to keep an open mind and look at the facts, separating the scientific aspects of the issues from the ideological aspects. There is, after all, no inherent reason why liberalism, humanism, or any other ideology need necessarily accrue to nongenetic determinants of behavior any more than they should accrue to genetic determinants.

The roots of this book go back to my first introduction to Freudian theory in a university psychology course. It seemed to me then, as it seems to me now, a strange theory. When I later began training in psychiatry, Freud was part of the landscape of my profession, more or less prominent depending on one's point of view but never out of sight entirely. I became increasingly puzzled by several aspects of the theory and the movement which grew up around it.

For example, why had Freud's theory taken root and grown so much more vigorously in America than in any other nation? Why, despite its lack of any scientific foundation, had it appealed so strongly to intellectuals who usually pride themselves on thinking critically? Why had it been embraced so enthusiastically by Socialists in the 1930s, then by liberals like Margaret Mead and Benjamin Spock after World War II? Did psychoanalysis—the Freudian method of psychotherapy—offer anything more than other brands of psychotherapy or was it, as Macdonald Critchley claimed, merely "the treatment of the id by the odd"? What about the remarkable similarities between the Freudian movement and religious movements—was Freud's theory mere catechism or was it more cataclysm? Most importantly, how had Freud affected American thought and culture—was the bottom line of the ledger luminescence or rather legerdemain? This book is a record of my search for answers to these questions.

Midway through writing the book I chanced to be in Vienna and went to 19 Berggasse. Visitors—mostly Americans—walked quietly through the rooms where Freud had lived for 47 years. Conversations were in hushed, reverential tones. In the corner of one room three roses had been placed on a bust of Freud and a woman was silently weeping. The milieu was identical to the Cathedral of St. Stephen which I had visited earlier in the day. I recalled Freud's remark that "biographical truth is not to be had," yet this ecclesiastical setting seemed to capture the essence of the true Freud.

FREUDIAN
FRAUD

1

Sigmund Freud, Sexual Freedom, and Social Reform

———

If I had my life to live over again I should devote myself
to psychical research rather than to psychoanalysis.
—SIGMUND FREUD, IN A LETTER TO HEREWARD CARRINGTON, 1921

The marriage of Sigmund Freud's theory to sexual freedom and social reform in America took place in October 1895; the bride wore red. Her name was Emma Goldman, and she was in fact the first visiting American to discover Freud. She was in Vienna to study midwifery at the city's highly rated general hospital, the same hospital in which Freud had completed medical training ten years earlier, when she learned of some lectures being given by "an eminent young professor," as she called him. Goldman had not yet acquired the subversive sobriquet, Red Emma, that was to make her the most infamous woman in America and cause her deportation to Russia. She was sufficiently known to authorities, however, that she chose to travel under an assumed name.

Although only 26 years old, Goldman was already a leader in the fledgling American anarchist movement. She had been converted during the 1886 Haymarket Square massacre when a bomb killed seven policemen in a Chicago labor dispute. Eight anarchists were convicted and sentenced to death; when four of them were hanged on November 11, 1887, Goldman recalled that "something new and wonderful had

1

been born in my soul . . . a determination to dedicate myself to the memory of my martyred comrades, to make their cause my own. . . ."

She did exactly that, taking anarchist Alexander Berkman as a lover and helping him plan the assassination of Henry Clay Frick, chairman of the board of Carnegie Steel, which was locked in a bitter battle with the Amalgamated Association of Iron and Steel Workers in Homestead, Pennsylvania in 1892. Goldman apparently agreed with Berkman's contention that "the killing of a tyrant . . . is in no way to be considered as the taking of a life." She took to the streets as a prostitute to earn money to buy Berkman's gun, then helped him test explosives he intended to use to kill himself after shooting Frick. Berkman did shoot Frick and also stabbed him with a sharpened file, but failed to kill him. Berkman was foiled in his suicide attempt and was subsequently sentenced to 22 years in prison. It was widely suspected that Goldman had assisted in the crime, but conclusive evidence with which to charge her was lacking.

Emma Goldman resumed her anarchist activities in 1893 and, following a rally of workers in New York's Union Square, was arrested and charged with inciting to riot. She was convicted and sentenced to one year in New York's Blackwell Prison. While in prison she was assigned to work as a nurse despite having had no training; it was that experience which led her, after her release, to Vienna for the midwifery course. She realized that she needed to find a means to support herself so that she could continue her efforts to bring about a revolution of the workers.

Goldman's attraction to Freud and his theory was instantaneous and profound. "It was Freud," she recalled, "who gave me my first understanding of homosexuality." More importantly, Freud explained how sexual repression caused neurasthenia, frigidity, depression, and the intellectual inferiority of women. "For the first time I grasped the full significance of sex repression and its effect on human thought and action," she wrote. Freud also "helped me to understand myself, my own needs." Goldman needed such understanding for, despite her vigorous advocacy of free love, she herself suffered intense ambivalence in her physical relationships with men. "I always felt between two fires," she wrote later; "their lure remained strong, but it was always mingled with violent revulsion." Freud's theory provided both solace and an explanation for her feelings.

Freud's influence on Goldman and her subsequent advocacy of free love dated to her stay in Vienna. "His simplicity and earnestness and the brilliance of his mind," she recalled, "combined to give one the feeling of being led out of a dark cellar into broad daylight." "Only people of depraved minds could impugn the motives or find 'impure' so great and fine a personality as Freud." When Freud came to Clark University to lecture in 1909, Emma Goldman was in the front row and described by a reporter from the *Boston Evening Transcript* as "plump, demure, chastely garbed in white" and wearing a red rose at her waist. The article also referred to her as "Satan," for by then her notoriety was widespread as a proponent of free love and anarchy. Goldman's recollection of Freud at Clark University was that "he stood out like a giant among pygmies."

Freudian Theory

The lectures that originally drew Emma Goldman's attention to Freud took place at Vienna's medical college on October 14, 21, and 28, 1895. It is not surprising that she heard of Freud soon after arriving in the city, for he was rapidly acquiring a reputation for his theory of sexuality. Ernest Jones, Freud's official biographer, contended that this reputation was one reason Freud was passed over during the 1890s for the honorary title of Professor. Another of Freud's colleagues recalled that "in those days when one mentioned Freud's name in a Viennese gathering, everyone would begin to laugh as if someone had told a joke. [Freud] was the man who saw sex in everything. It was considered bad taste to bring up Freud's name in the presence of ladies. They would blush when his name was mentioned."

Freud was at the time a 39-year-old physician in the private practice of medicine, specializing in cases of neurosis. He did not actually see "sex in everything," but did believe that sexual abuse of children and sexual repression were the causes of anxiety neurosis, phobias, obsessions, hysteria, and neurasthenia. In 1895 he specifically singled out "voluntary or involuntary abstinence, sexual intercourse with incomplete gratification [and] coitus interruptus" as pathological factors. By 1905 he had broadened his theory to indict sexual problems as the cause of all neuroses and argued that "the unique significance of sexual experiences in the etiology of the psychoneuroses [is] incontestably established." "The energy of the sexual instinct," Freud contended, "supplies the only constant and most important source of energy in the neurosis." In a statement widely quoted by his followers, Freud asserted that "no neurosis is possible with a normal *vita sexualis.*" The logical implication of such a theory was that neurosis would be far less common if people's sexual lives were less inhibited and less repressed. Freud himself was reluctant to make public pronouncements along such lines, but privately told his friends that "I stand for an infinitely freer sexual life. . . ." It was this message Emma Goldman took back to New York and incorporated into her advocacy for a better world.

At the time of the 1895 lectures Freud was engaged in active research on the relationship of sexual function to neurosis with his collaborator and close friend Wilhelm Fliess. Fliess was a Berlin ear, nose, and throat specialist who believed that "the misuse of the sexual function"—especially masturbation, coitus interruptus, and the use of condoms—caused damage to the nervous system and also to tissues in the nose. Fliess had localized specific "genital spots" in the nose and believed that neurosis caused by the misuse of the sexual organs could be treated by applying cocaine to those spots and by electrically cauterizing them. In an 1893 textbook Fliess described 131 cases of neurosis, all of which were treated with cocaine and cauterization.

The recent publication of Freud's voluminous correspondence with Fliess revealed how enthusiastically Freud supported Fliess's theories. One week prior to his initial 1895 lecture, Freud had written Fliess urging him to publish his pamphlet, "The Nose and Female Sexuality." Freud, like Fliess, believed that many physical and mental symptoms were caused by a "nasal reflex" that originated in the genitalia, proceeded to the nose, and then was transmitted to other organs. For example, earlier in 1895 Freud had described a patient with a "one-sided facial spasm," in which Freud believed he had localized the "nasal reflex" to a specific spot on the nasal mucosa; he told Fliess he might send the patient to him for definitive treatment. Similarly, Fliess described to Freud cases of "neuralgic stomach pains" that could be treated by the application of cocaine and cauterization of a specific "stomach ache spot" on the nasal mucosa.

The most significant case on which Freud and Fliess collaborated in 1895 was that of Emma Eckstein, a young woman who had come to Freud with "stomach ailments and menstrual problems." Suspecting that the woman's problems were caused by masturbation, which was mediated through a "nasal reflex" to her stomach and uterus, Freud summoned Fliess to Vienna to operate on Ms. Eckstein's nose. Following the surgery, but after Fliess had returned to Berlin, Ms. Eckstein began hemorrhaging profusely and dangerously from the nose, and a Viennese surgeon had to be called in. The surgeon removed "at least half a meter of gauze" that Fliess had inadvertently left in the nasal cavity. Several additional surgical procedures were necessary to stop the intermittent hemorrhages; as a consequence of the procedures "her face was disfigured—the bone was chiseled away and on one side caved in." Freud initially believed that the hemorrhages were a consequence of the surgery, but later decided that Ms. Eckstein's "hemorrhages were hysterical in nature, the result of sexual longing."

Emma Eckstein's inauspicious outcome did not shake Freud's confidence in Fliess's theories. Five months later Freud went to Berlin, where Fliess operated for the second time on Freud's own nose for swelling and recurring nasal infections that Freud was experiencing. Freud praised Fliess after the Emma Eckstein episode for "holding in your hands the reins of sexuality, which governs all mankind: you could do anything and prevent anything." (Freud also asked Fliess if he could name the baby his wife was expecting after him if it turned out to be a boy; the baby, however, was a girl and was named Anna.)

Although Freud accepted Fliess's theories for treating neurosis, he continued at the same time to look for the specific sexual origins of these disorders. On October 15, 1895, the day after the first of the lectures, Freud excitedly wrote to Fliess that he had discovered the "great clinical secret [that] hysteria is the consequence of a presexual *sexual shock* [and] obsessional neurosis is the consequence of a presexual *sexual pleasure* which is later transformed into [self-]*reproach.* 'Presexual' means actually before puberty, before the release of sexual substances; the relevant events become effective only as *memories.*" So certain was Freud of his theories that he wrote Fliess again the following day, announcing that "I consider the two neuroses essentially conquered."

These letters constitute one of the earliest expressions of Freud's theory of infantile sexuality, which was to become the essence of his theory of human behavior. Freud believed that traumatic events relating to sexual development permanently shape one's personality traits. In subsequent years Freud elaborated his "presexual" period into oral, anal, and genital stages of development and focused especially on the Oedipal conflict in which, Freud claimed, little boys want to incestuously possess their mothers and kill their fathers. Such thoughts were said to overwhelm a child's psyche and to lead to repression and later neurosis.

It should be noted that Freud's theories concerning the unconscious and the use of dreams, theories for which Freud is best known today, were techniques developed by him to bring to light the repressions of childhood. Both concepts were well developed in European thought prior to 1880 and Freud borrowed heavily from his predecessors. Lancelot L. Whyte, in *The Unconscious Before Freud,* showed conclusively that the idea of the unconscious had become "*topical* around 1800 and *fashionable* around 1870–1880. . . . It cannot be disputed that by 1870–1880 the general conception of the unconscious mind was a European commonplace, and that many special applications of this general idea had been vigorously discussed for several decades." By the time of Freud's student years, Friedrich Nietzsche had written regularly about the unconscious (e.g., "Every extension of knowledge arises from making conscious the unconscious"), and while studying at the University of Vienna, Freud belonged to a Reading Society that studied Nietzsche in detail and even corresponded with him. Freud also was indebted to Nietzsche for the concept of the id.

Freud's ideas about dreams were also heavily influenced, if not taken directly, from existing European literature. Henri Ellenberger's *The Discovery of the Unconscious* and Frank J. Sulloway's *Freud, Biologist of the Mind* both cite sources from which Freud adapted his dream theories. As early as 1861 German psychiatrist Wilhelm Griesinger had described dreams as "the imaginary fulfillment of wishes" and the same year Karl Scherner published *The Life of the Dream*, in which the sexual symbols of dreams were discussed in detail. Sulloway concluded that virtually all of Freud's ideas about dreams had been published prior to Freud's writings, including "the claim that dreams have a hidden meaning, that they are wish fulfillments, that they represent disguised expressions of unacceptable thoughts, that they elicit the archaic features of man's psyche, that they involve a regression to the dreamer's childhood experiences and successive personalities. . . ." Freud, then, elaborated and popularized ideas that were already extant about the unconscious and dreams.

By utilizing dreams and other techniques to explore the unconscious, Freud hoped to uncover repressed memories of childhood sexual trauma which he believed were the cause of neuroses. In *Three Essays on the Theory of Sexuality* (1905), Freud clearly described this theory. In the ensuing years Freud's followers have argued about precisely when in childhood the putative traumas occur (e.g., some psychoanalysts have theorized that they even occur *in utero*, before the child is born), what part of the psyche is most damaged (e.g., ego psychology object relations), and what techniques (e.g., various forms of transference) are most effective in psychotherapy to overcome the repression and make the unconscious conscious. Despite such disputes the primary postulate and core of Freudian theory has continued to be the importance of sexual traumas in early childhood. Freud emphasized this in a 1905 essay in which he stated that "my views concerning the etiology of the psychoneuroses have never yet caused me to disavow or abandon two points of view: namely the importance of sexuality and of infantilism."

Fame, Occultism, and Cocaine

During the years Freud was searching for the childhood antecedents of behavior in the 1890s, he was also pursuing three other interests, which have been touched upon only lightly by most of Freud's biographers, for they are difficult to integrate with the portrait that is usually drawn of him. It is important to understand these pursuits, however, in order to fully understand the origins of Freudian theory.

One such interest was fame. As a student at the University of Vienna Freud had lingered in the great hall containing the busts of the university's most famous graduates. "He dreamed of the day when he would be similarly honored [and] knew exactly what inscription he wanted on the pedestal, a line from *Oedipus Tyrannus*: 'Who divined the riddle of the Sphinx and was a man most mighty.' " During his medical internship from 1882 to 1885, according to Sulloway, "Freud was continually preoccupied with the hope of making an important scientific discovery—one that would bring him early fame and the promise of a large private practice, and thus allow him to marry without having to wait another five to ten years."

Monetary gain, however, was not the most important aspect of fame for which Freud longed. Rather he desired fame as the fulfillment of his personal destiny. This can be seen clearly in an 1885 letter that Freud wrote to his fiancée:

> *I have almost finished carrying out an intention which a number of as yet unborn and unfortunate people will one day resent. Since you won't guess what kind of people I am referring to, I will tell you at once: they are my biographers. I have destroyed all my notes of the past fourteen years, as well as letters, scientific excerpts, and the manuscripts of my papers. As for letters, only those from the family have been spared. Yours, my darling, were never in danger. . . . I certainly had accumulated some scribbling. But that stuff settles around me like sand drifts round the Sphinx; soon nothing but my nostrils would have been visible above the paper; I couldn't have matured or died without worrying about who would get hold of those old papers. Everything, moreover, that lies beyond the great turning point in my life, beyond our*

*love and my choice of profession, died long ago, and must not be
deprived of a worthy funeral. As for the biographers, let them worry,
we have no desire to make it too easy for them. Each one of them will
be right in his opinion of "The Development of the Hero," and I am
already looking forward to seeing them go astray.*

This letter was written as Freud was just completing his medical training and before he had accomplished anything of note. In later years,
according to Sulloway, Freud falsified some details about the origins of
psychoanalysis in an effort to portray himself within "the myth of the
hero" and to show "that he was fulfilling a heroic destiny."

One particular aspect of Freud's sense of destiny is noteworthy.
Sometime during the 1890s he became drawn to the historical figure
of Moses. By 1909 he had clearly identified with Moses, as seen in a
letter Freud wrote to Jung in which Freud recounted their recent successes:

*We are certainly getting ahead. If I am Moses, then you are Joshua
and we will take possession of the promised land of psychiatry, which I
shall only be able to glimpse from afar.*

By this time Freud had become intensely interested in a large statue of
Moses sculpted by Michelangelo. He first saw the actual statue in 1901
when he visited Rome, but was almost certainly already familiar with
it since a replica of the statue stood in the Vienna Academy of Art. In
1913, during another visit to Rome, Freud wrote that "every day for
three lonely weeks . . . I stood in the church in front of the statue,
studying it, measuring and drawing it. . . ." The following year, in an
essay entitled "The Moses of Michelangelo," he described this experience:

*Sometimes I have crept cautiously out of the half-gloom of the interior,
as though I myself belonged to the mob upon whom his eye is
turned—the mob which can hold fast no conviction, which has neither
faith nor patience, and which rejoices when it has regained its illusory
idols. But why do I call this statue inscrutable? There is not the slightest doubt that it represents Moses, the Lawgiver of the Jews, holding
the Tables of the Ten Commandments.*

In 1938 his book *Moses and Monotheism* was published, a book that
was widely criticized for Freud's idiosyncratic interpretation of history.

Freud's biographers have theorized about his 40-year obsession with Moses. Ernest Jones, for example, surmised that Freud "had emotional reasons for identifying himself with his mighty predecessor." Reuben Fine contends that "there can be little doubt that Freud had a strong personal identification with Moses . . . ," and Peter Gay added that "one cannot avoid the pretty obvious conclusion that . . . Freud had identified himself with Moses."

An avid interest in the occult was the second of Freud's other major interests in the 1890s, as indeed it was throughout his adult life. This aspect of Freud's personality was well described in Ernest Jones's biography. Freud was convinced that mental telepathy, or "thought transference" as he called it, was possible; he likened it to "a kind of psychical counterpart to wireless telegraphy." He first experienced it in the 1880s when he was engaged to be married but separated from Martha Bernays; while in Paris studying under Jean Martin Charcot, a foremost neurologist, Freud said he "often heard his name being called, unmistakably in her voice." In 1913 Freud arranged a seance at his home with psychoanalytic followers Otto Rank, Hanns Sachs, and members of Freud's family. Ten years later Jones recorded another attempt to prove the validity of telepathy, this time carried out by Freud with his daughter, Anna, and Sandor Ferenczi. Freud's attempts at telepathy with Anna are especially curious since he had also psychoanalyzed her; Jones referred to their relationship as "a quite peculiarly intimate relationship between father and daughter." Freud periodically consulted "soothsayers" who were alleged to have telepathic powers. One notable occasion when Freud, accompanied by Ferenczi, made such a visit was in 1909 en route home from their visit to America. Stopping in Berlin they consulted Frau Seidler, who convinced Freud that she could read his mind.

Freud was also superstitious about numbers and convinced that he was destined to die at age 61 or 62; according to Jones, "He referred to it over and again in his correspondence." The number 31 held a special significance for him, being half of 62. Freud occasionally performed "magical actions," which Jones said were "unconsciously carried out with the aim of averting disaster." In 1905, for example, when his eldest daughter was gravely ill, Freud "found himself skillfully aiming a blow with his slipper at a little marble Venus [statue] which it smashed"; the statue was one of Freud's highly valued antiquities and breaking it "was a sacrificial offering to preserve his child's life."

A belief in the significance of certain numbers was also important to Wilhelm Fliess, especially the number 28, signifying the female menstrual cycle, and the number 23, which Fliess believed represented an analogous cycle in males. The Freud–Fliess correspondence is replete with discussions of such numbers and shows how deeply Freud himself believed in their importance. A shared interest in the occult was fundamental to Freud's initial attraction to Jung. It was Sandor Ferenczi, however, who was Freud's major collaborator on telepathy and other occult experiences. Ferenczi attempted to combine telepathy with psychoanalysis and believed that patients could sometimes read their analysts' minds. This would, as Jones noted, have made "a revolutionary difference to the technique of psychoanalysis." Ferenczi on one occasion brought a telepathist to a meeting of the Vienna Psychoanalytic Society and announced "jestingly but a little proudly his intention of presenting himself in Vienna as the 'Court Astrologist of Psychoanalysis.' "

Freud's interest in the occult was well known to his friends and peers. In 1911 he became a Corresponding Member of the Society for Psychical Research in London and in 1915 was made an Honorary Fellow of the American Society for Psychical Research. He wrote a paper in 1921 on "Psychoanalysis and Telepathy," but was dissuaded from publishing it by Ernest Jones and other followers because of the damage that might be done to the psychoanalytic movement if it were linked to the occult. Freud, according to Jones, was also invited to be coeditor of three different periodicals devoted to the study of occultism. Freud refused the offers, but in his reply to one he confessed that "if I had my life to live over again I should devote myself to psychical research rather than to psychoanalysis."

Freud's use of cocaine in the 1890s was a third important influence on his thought during the time when the foundations of his theory were being laid. His experiments with cocaine between 1884 and 1886 have been well described, including four papers he wrote reporting the effects of the drug on various physical and mental functions. Freud also played an indirect role in introducing cocaine and its derivatives as local anesthetics for use in medicine. Freud was extremely enthusiastic about cocaine before it was proved to be addicting, and he gave quantities of it to his fiancée and sisters as well as taking it regularly himself. He called it "a magical drug" and said that it produced "the most gorgeous excitement [and] exhilaration and lasting euphoria." He found it especially useful to counteract depression and nervousness; for example, while in Paris in 1886, he "fortified himself with a dose of cocaine" prior to going to a dinner party at Charcot's house. Freud apparently believed that cocaine had aphrodisiac properties, as illustrated in an oft-quoted 1884 letter to his fiancée:

Woe to you, my Princess, when I come. I will kiss you quite red and feed you till you are plump. If you are froward you shall see who is the stronger, a gentle little girl who doesn't eat enough or a big wild man who has cocaine in his body. In my last severe depression I took coca again and a small dose lifted me to the heights in a wonderful fashion. I am just now busy collecting the literature for a song of praise to this magical substance.

The important question about Freud and cocaine is not how enthusiastically he promoted it between 1884 and 1886, but rather how frequently he used it after 1886, when it had been clearly shown to be addictive and publicly labeled "the third scourge of humanity" along with alcohol and opium. It is known that Freud's friendship with Wilhelm Fliess began in 1887 and that Fliess used cocaine for many years to treat patients. Publication of the Freud–Fliess letters clearly established that Freud took cocaine, at least irregularly, in the early and mid-1890s:

[May 30, 1893] A short time ago I interrupted (for one hour) a severe migraine of my own with cocaine; the effect set in only after I had cocainized the opposite side [of the nose] as well.

[January 24, 1895] Last time I wrote you, after a good period which immediately succeeded the reaction, that a few viciously bad days had followed during which a cocainization of the left nostril had helped me to an amazing extent. I now continue my report. The next day I kept the nose under cocaine, which one should not really do; that is I repeatedly painted it to prevent the renewed occurrence of swelling.

[April 20, 1895] Today I can write because I have more hope; I pulled myself out of a miserable attack [of nasal problems] with a cocaine application.

[April 26, 1895] I put a noticeable end to the last horrible attack [of nasal problems] with cocaine; since then things have been fine and a great amount of pus is coming out.

[June 12, 1895] I need a lot of cocaine.

[July 24, 1895] This was the occasion of Freud's dream about Irma's injection, the dream that many Freud scholars have identified as the

stimulus for Freud's theory regarding dreams as wish fulfillments. In reporting it in The Interpretation of Dreams *Freud noted: I was making frequent use of cocaine at that time to reduce troublesome nasal swellings.*

There is also evidence that Freud continued to use cocaine after 1895. On October 26, 1896, Freud reported to Fliess that "the cocaine brush has been completely put aside." Although there are no specific references to Freud's further use of cocaine in their ensuing correspondence, there are continuing allusions to Freud's physical symptoms that were probably caused by cocaine, similar to those he complained of during his period of known cocaine use (and which were extensively discussed by E. M. Thornton, who concluded that they were almost certainly due to cocaine).

Most prominent among Freud's symptoms were cardiac problems (especially irregular heart beats), headaches, and nasal problems (especially swelling and stuffiness). According to contemporary studies of cocaine abusers, these are indicators found in almost all individuals who use cocaine steadily, and Freud complained regularly about such symptoms in his correspondence between 1895 and 1899. For example, in a letter to Fliess on December 12, 1897, Freud wrote that he was "now suffering painfully from [nasal] suppuration and occlusion. . . . If this does not improve I shall ask you to cauterize me in Breslau." In another letter to Fliess on September 27, 1899, Freud recorded several episodes of "cardiac weakness with mild headache, . . . cardiac fatigue, . . . [and] headache without cardiac pain" within a period of eight days. Freud had no known heart disease either at that time or in subsequent years.

Additional evidence that Freud continued to use cocaine in the late 1890s was recently unearthed by Freud scholar Peter Swales. Ernest Jones, who had extensive access to Freud's private papers, wrote in an unpublished 1952 letter to another Freud historian: "I don't think he [Freud] gave up interest in cocaine and I guess he took it himself off and on for 15 years." Since the date of Freud's first use of cocaine is clearly established as August 30, 1884, this would mean that Freud used cocaine until approximately 1899. In his authorized biography of Freud, Jones made no mention of Freud having used cocaine except during the initial 1884–1886 period when he first experimented with it. In another unpublished letter Jones noted: "I am afraid Freud took more cocaine than he should—though I am not mentioning that." Jones elsewhere added: "Before he [Freud] knew about its [cocaine's] dangers he must have been a public menace, the way he thrust it on everyone he met!"

In summary, during the 1890s Sigmund Freud was a man with a mystical sense of personal destiny, who believed in telepathy and numerology, and who was using cocaine at least intermittently. During these same years he developed his theory that early childhood experiences, especially those of a sexual nature, are the most important determinants of adult behavior. As Freud biographer Peter Gay noted, "by the time he published *The Interpretation of Dreams* at the end of 1899, the principles of psychoanalysis were in place." Specifically, Freud acknowledged "making frequent use of cocaine" during the period when he analyzed his own dream of Irma's injection in 1895; this dream, labeled by Ernest Jones as "an historic moment," has been called the prototypic "dream specimen" of psychoanalysis. The dream of Irma's injection also inaugurated Freud's own self-analysis, during which he developed his theory of childhood sexual development and the Oedipal complex.

The fact that Freudian theory evolved simultaneously with a sense of destiny, an interest in the occult, and the use of cocaine does not in itself negate the validity of the theory. It does, however, cast a shadow over its scientific foundation. It also makes more comprehensible Freud's own recollections of the development of his theory, published in 1914 as a pamphlet titled "The History of the Psychoanalytic Movement":

> *Since, however, my conviction of the general accuracy of my observations and conclusions grew and grew, and as my confidence in my own judgement was by no means slight, any more than my moral courage, there could be no doubt about the outcome of the situation. I made up my mind that it had been my fortune to discover particularly important connections, and was prepared to accept the fate that sometimes accompanies such discoveries.*

Sexual Freedom

Sigmund Freud's introduction to most Americans in the early twentieth century was as an apostle of sexual freedom. There were, to be certain, other warriors who had been fighting Victorian morality in the early years of the century, but the outcome of the struggle had been questionable. It was only when Freud became commanding general in the second decade that the forces of celibacy and puritanism were put to rout. Freud, as titular leader of the sexual revolution in America, became an appelative symbol of sexual freedom.

Prior to the arrival of Freud's ideas, Havelock Ellis had been the best known advocate of sexual freedom in Europe and the United States. An Englishman trained in medicine and psychology, Ellis established his reputation with the 1897 publication in London of a book on homosexuality, *Sexual Inversion*. The book caused a scandal and was promptly confiscated by English authorities; one bookseller was arrested and convicted of selling "a certain lewd, wicked, bawdy, scandalous and obscene libel." Ellis thereupon arranged for the book's distribution by a Philadelphia publisher of medical books, with the publisher stipulating that sales must be confined "only to physicians and lawyers." Despite this restriction the book sold briskly and was the first of Ellis's six-volume *Studies in the Psychology of Sex*, which became available between 1897 and 1910.

Ellis became aware of the sexual ideas of Sigmund Freud and Wilhelm Fliess early in his research; and Freud was mentioned prominently in *Studies in the Psychology of Sex*, cited ten separate times in volume 6 alone. Ellis was especially interested in Freud's theory that infant children have sexual feelings, and quoted Freud as follows: "In reality the new-born infant brings sexuality with it into the world, [and] sexual sensations accompany it through the days of lactation and childhood . . ." Ellis also cited Freud in support of his contention that sexual abstinence is harmful to individuals and to society. "Immoral," said Ellis, "never means anything but contrary to the *mores* of the time and place." Ellis was especially tolerant of masturbation and homosexuality and was a strong proponent of sex education for children.

The sexual revolution in the early years of the twentieth century was part of a broader social revolution taking place in America. It included birth control, divorce laws, and women's suffrage, and many of the same people were involved in more than one movement. Havelock Ellis, for example, also advocated "semi-detached" marriages in which a husband and wife maintained separate residences and outside friendships. Similarly, Ellen Key, a well-known Swedish feminist, advocated "laissez- faire sexual morality" in her widely read book *The Century of the Child*, published in America in 1909. Margaret Sanger, leader of the birth-control movement, frequently quoted Freud as a supporter of sexual freedom. And Emma Goldman promoted both sexual freedom and birth control in her tempestuous speeches. The importance of this for Freud was that it gave him a broader audience of advocates who spread his message. For example, Dr. William J. Robinson, a leading advocate of birth control (and vocal defender of Emma Goldman), published the first translation of Freud's paper on "Civilized Sexual Morality and Modern Nervousness" in his medical journal in 1915.

It should also be noted that it was Freud's theories of sexuality, not his psychotherapeutic techniques, which led to his invitation to address the historic gathering at Clark University in 1909. Dr. G. Stanley Hall, president of Clark and founder of the American Psychological Association, had been extremely troubled in adolescence by his own sexual impulses. "The chief sin of the world," Hall wrote later, "is in the sphere of sex, and the youth struggle with temptation here in the only field where . . . being in the hands of a power stronger than human will become literally true." At Clark University, Hall pioneered the teaching of sex education, "but he had to discontinue the course because outsiders crowded the lecture hall and even listened at the door." His influential book, *Adolescence: Its Psychology and Its Relations to Physiology, Anthropology, Sociology, Sex, Crime, Religion and Education* (1904), discussed ways for turning the "barbaric and bestial proclivities" of adolescence into more constructive channels. In the 1907 edition, Hall cited Freud's work several times in the section on sexual developments and seemed especially interested in the possibility that individuals may have sexual impulses in early childhood. Inviting Freud to lecture at Clark University, then, was a means of promoting public discussion about sex and sex education.

Hall had no idea how successful this strategy was to be. Freud's trip to the United States exposed him to the American media and his name soon became synonymous with sexual freedom. The *New York Times* referred to Freud as the "Viennese libertine"; Freud was further said to be associated with the "worship of Venus and Priapus," and it was claimed that his teachings were "a direct invitation to masturbation, perversion, illegitimate births, [and] unions out of wedlock." In 1914 Dr. William S. Sadler published a book titled *Worry and Nervousness*, in which he noted: "I am beginning to think that the future will look back upon the present day and generation as having gone sex-mad. . . . This modern sex mania threatens to take possession of psychic [psychiatric] medicine." The association of Freud with sexual freedom continued to grow, so that by September 30, 1926, *Life* magazine featured on its cover a seductive young woman holding a book titled "Psychoanalysis" and surrounded by other books labeled "Freud" and "Havelock Ellis."

Freud anticipated the criticism he would receive in America. In a 1909 letter written shortly before his trip to Clark University, he predicted: "I also think that once they [the American professionals] discover the sexual core of our psychological theories they will drop us. Their prudery and their material dependence on the public are too great." Ridicule of Freud's ideas of sexuality began shortly after his trip. Within two years H. L. Mencken was publishing satires of Freud's dream interpretations in the *Smart Set*:

> *The true meaning of a dream about a murder is not that the dreamer is soon to be married, or that his brother, Fred, in Texas, has been trampled to death by hippopotami, or that the [Philadelphia] Athletics will win the Pennant, but that deep down in the dreamer's innards, somewhere South of the tropic of cancer, the cartilages of last night's lobster are making a powerful resistance to digestion.*

The psychoanalyst's word-association test, developed by Jung, had also become a target by 1914:

> *To Freudian writers the entire language is made up of two groups of symbolic words, half meaning the male, the other the female genitalia. If any words happened to be left over they stand for incest, rape, anus, or foecal associations, or the fornicative, generally speaking.*

By 1916 the association of Freud with sexual freedom was firmly implanted in the public's mind. The *Nation* published frequent criticisms, claiming that Freud had "an ingeniously obscene imagination" that could "discover a sexual motive in a binomial theorem." The New York *Medical Record* noted that Freud had been called a "peddlar of pornography, a sink of salaciousness and in general about three shades worse than the mayor of Gomorrah." In Boston, when one professor heard about Freud's theory he exclaimed: "But this is a matter for the police court!" Religious publications such as the *Catholic World* predictably renounced Freudian theory as being "of such vile nature that the mere mention of it would stain the pages of a decent periodical. . . . If such a philosophy, which glorifies the lowest tendencies and drags the finest products of the mind into the slime, became universally accepted, life would lose its value." According to one historian of this period, "lovers took to urging their companions to forget their 'repressions' " for "to resist [Freud's] doctrines was to arouse the suspicion that one was inhibited or neurotic." Another historian claimed that "people gave freer rein to their latent behavior because they could now explain that 'Freud says it is best not to inhibit.' "

Freud's followers in America noted the criticism. Edwin B. Holt, who published a popular book on Freud in 1915, prefaced his work by observing that "the idea has gone abroad that the term 'Freudian' is somehow synonymous with 'sexual,' and that to read Freud's own works would be fairly to immerse oneself in the licentious and the illicit." Another leading Freudian in America recalled that "psychoanalysis and sex were considered identical," and many laypersons believed that psychoanalysts had sex with their patients as part of therapy. R. S. Woodworth, a respected psychologist who was sympathetic to Freudian theory, asserted in 1917 that "the element of sex gratification is the main factor in the spread of the [Freudian] movement. The books owe their interest principally to the sex element." Woodworth ingenuously confessed that he was certain this was true because sexual interest had been the stimulus for his own study of Freud: "I have devoured many of these writings greedily, and am perfectly aware that my interest has been largely of this sort. . . ." Woodworth also contended that "the psychoanalytic seance is a sort of *coitus sublimatus* (sometimes homosexual) both on the physician's part and on that of the patient."

Although Freud's followers were concerned about his name becoming synonymous with sexual freedom, their responses were often not helpful to their cause. Freud had set the pattern when he argued that those opposed to his ideas were themselves afflicted with repression and sexual inhibition. As early as 1911 Ernest Jones, in a letter to a Philadelphia neurologist, used this reasoning to discredit opponents of Freud's theory:

> *Yes, I am of course familar with the penetrating criticism of Freud's work that you so well describe. It really is only another way of saying that these gentry in question know that they are incapable of dealing with sexual topics in any shape that would not be indeed obscene, and they very naturally concluded that no one else can; to them sexuality is synonymous with obscenity. . . . Such fanaticism is of course at bottom a reaction against repressed desires.*

Another psychoanalyst stated publicly that Puritans were "sexually abnormal" and that those who adhered to such doctrines were afflicted by "nothing but a dignified neurosis."

Freud's hopes of achieving respectability in America received additional setbacks by means of scandals in which Freud's name was linked to sexual freedom. Rumors that psychoanalysts occasionally recommended sexual intercourse as a treatment for their patients proved to be true, and as early as 1910 Freud tried to quiet such accusations with an essay titled "Wild Psychoanalysis." A physician had told a woman who had left her husband, said Freud, "that she could not tolerate the loss of intercourse with her husband and so there were only three ways by which she could recover her health—she must either return to her husband, or take a lover, or obtain satisfaction from herself." In discussing the case Freud acknowleged that "psychoanalysis puts forward absence of sexual satisfaction as the cause of nervous disorders," but he said the physician in question had failed to point out a fourth possible solution—psychoanalysis. Freud did *not* say in this essay, however, that a recommendation to take a lover was necessarily wrong.

Other scandals linking Freud's name to sexual freedom surfaced from time to time in the media or circulated in New York City's salons. In 1916, for example, a New York physician's wife, who had undergone psychoanalysis with Carl Jung in Zurich, left her husband to live with another psychoanalyst. The woman justified her behavior as the pursuit of "higher and nobler knowledge"; the *New York Times* argued that "psychoanalysis was not an excuse for loose morals" or adultery. In 1921 Freud himself was threatened with a lawsuit by the husband of a wealthy New York woman; the woman had been urged by her former psychoanalyst, Horace Frink, and by Freud himself, to divorce her husband and marry Frink. Freud subsequently justified his advice by saying: "I thought it the good right of every human being to strive for sexual gratification and tender love."

Social Reform

At the same time Freud's name was being linked with sexual freedom in America, among some physicians and psychotherapists it was also becoming associated with social reform in a relationship that would prove to be far more important. The introduction of Freud's ideas into reform-minded America was as an adjunct treatment for mental disorders, especially hysteria. In the early years of the century there was much interest in hypnotism, faith healing, Christian Science, and other forms of spiritual healing, such as the Emmanuel Movement, which even had a regular section devoted to it in *Good Housekeeping* magazine. Boris Sidis's *Psychology of Suggestion* (1902) and Morton Prince's *The Dissociation of Personality* (1905) were immensely popular. That same year Paul Dubois's *The Psychic Treatment of Nervous Disorders* was published and was said "to have electrified many physicians." In 1906 Pierre Janet gave a series of 15 lectures on hysteria at Harvard Medical School, and a medical dissertation briefly noted Freud's use of psychoanalysis for such cases, commenting that his work deserved to be better known. At the time Freud's *The Interpretation of Dreams* had neither been translated into English nor had it received a single review in America, and his name was known only to individuals who were promoting sexual freedom, such as Emma Goldman and Havelock Ellis.

The evolution in America of Freud from sexual liberationist to social reformist began as early as 1906, when James J. Putnam, a respected professor of neurology at Harvard Medical School, published the first article specifically on Freud's work in an American professional journal, *The Journal of Abnormal Psychology*. Putnam had experimented with psychoanalysis on a few patients and had become interested in Freud's theory that repression of unconscious memories inhibits the full development of human potential. Following extensive discussion of Freud's theory with Ernest Jones in 1908 and with Freud himself in 1909 subsequent to the Clark University lectures, Putnam became an enthusiastic Freudian convert.

Putnam was totally dedicated to community service and the spiritual improvement of humanity. At the same time he discovered Freud he was also active in the Emmanuel Movement of divine healing, which had originated at Boston's Emmanuel Church. Putnam was interested in psychoanalysis as a method of eliminating weaknesses in the individual and thereby producing "a nobler self." The resolution of a person's inner conflicts was not an end in itself, then, but rather a means for achieving the "idealist ethic of community cooperation in the service of Divine Purpose." At the International Psychoanalytic Congress in Germany in 1911, Putnam urged that psychoanalysis be used to identify "man's spiritual consciousness with the Infinite." His imposition of Christian ethics on Freud's theory must have seemed novel to the participants who were themselves mostly Jewish. Freud politely described Putnam's views as "a decorative centerpiece which all look at but none touch"; another psychoanalyst in attendance recalled the audience "glowing with the noblest sentiments but the copious rush of thoughts left behind a certain bewilderment." In 1915, following publication of Putnam's book *Human Motives*, one reviewer suggested that he had combined psychoanalysis "with an attempt to prove the existence of God." Thus Putnam was the first American, but would not be the last, to try and utilize Freud's theory for achieving the improvement of mankind.

At the same time Putnam was experimenting with psychoanalysis on patients with neurosis at Harvard, Dr. Adolf Meyer was also applying it to patients with schizophrenia and manic-depressive psychosis at Manhattan State Hospital. An immigrant physican from Switzerland, Meyer introduced Jung's word-association test and was cautiously enthusiastic about psychoanalysis as a new therapeutic technique for disturbed patients. Meyer never completely accepted Freud's ideas, but he did introduce them to many colleagues, including August Hoch, Abraham A. Brill, C. Macfie Campbell, and Smith E. Jelliffe, all of whom would play important roles in the dissemination and legitimation of Freud's theory.

Adolf Meyer's most important contribution to promoting Freud's theory was an indirect one. In 1907 Clifford Beers, a Yale University graduate who had been hospitalized for two years because of manic-depressive illness, approached Meyer and asked for assistance in getting published his account of his illness, *A Mind That Found Itself* (1908), and in forming a national association to improve mental hospitals. Beers had named his organization "The National Society for the Improvement of Conditions Among the Insane" and had already secured William James's support. Meyer accepted, but persuaded Beers to change the name of the organization to the National Committee for Mental Hygiene and to adopt a much broader mandate; rather than merely improving mental hospitals, the organization's goal would be to prevent mental illness.

The possibility of preventing mental illness was enormously attractive to psychiatrists like Meyer in the early years of the century. Smallpox, yellow fever, typhus, cholera, typhoid fever, and syphilis were all coming under control through advances in medicine and sanitation. Psychiatrists in America wanted to be part of these medical advances, but they were looked down upon by their medical colleagues. In 1894, for example, a prominent neurologist, Dr. S. Weir Mitchell, had "levied one of the most blistering attacks on mental hospital superintendents [who were all psychiatrists] that had ever been heard" at their annual meeting. Watching the formation of the Association for the Study and Prevention of Tuberculosis in 1904 and the American Association for the Study and Prevention of Infant Mortality in 1909, it must have seemed to Meyer and his colleagues that a National Committee for Mental Hygiene was an idea whose time had come. It also coincided with public interest in social reform, which was widespread at that time, having been ignited by books such as Lincoln Steffens's *Shame of the Cities* (1904) and Upton Sinclair's *The Jungle* (1906).

Almost from the day it was officially founded on February 19, 1909, the National Committee for Mental Hygiene amalgamated Freud's theory of human behavior with ideals of social reform and functioned as a major vehicle for the synthesis of the two in America. That Freud himself had shown virtually no interest in social reform did not appear to matter; the psychiatrists accepted Freud's emphasis on experiential, as opposed to hereditary or organic, causes of behavior, especially emphasizing the crucial childhood years. Consequently, Freud became an unwitting but major player in attempts by American reformers to improve society.

During the first year of the national committee's existence Adolf Meyer dominated it. When he had a falling out with Beers and resigned in 1910, Meyer's close associate, August Hoch, assumed leadership. In subsequent years Freud's followers played major roles in determining the national committee's policies. Because they believed that they understood the antecedents of human behavior, Freudians felt free to prescribe social changes that were intended to ameliorate living and working conditions. For example, Dr. C. Macfie Campbell, an early convert to Freud's theory and a leader of the mental hygiene movement, said that "mental hygiene . . . aims to develop those qualities which give to human life its value and without which the preservation of the most superb physique loses all its meaning. Mental hygiene aims to supplement and thus to justify physical hygiene." Campbell went on to urge specific reforms in working conditions, such as improvement in "the moral atmosphere as well as lighting and ventilation" in workplaces. The following year Dr. William A. White, whose 1909 textbook had been the first to describe psychoanalysis and who had started the first American psychoanalytic journal *The Psychoanalytic Review*, claimed that mental hygiene should be expanded to include "all forms of social maladjustment and even of unhappiness." White previously had urged reforms of factory conditions, tenement inspection, child labor laws, and juvenile courts in the interest of mental hygiene. White was especially interested in improving conditions for children, and *Mental Hygiene of Childhood* (1915) was his attempt to explain psychoanalytic principles to lay persons.

Members of the National Committee for Mental Hygiene who had not received formal training in psychoanalysis were often more aggressive in implementing Freud's theory than those who had been so trained. Dr. Thomas W. Salmon, who became the organization's medical director in 1912 and later its executive director, illustrated this. Salmon's training had been in bacteriology, and he likened the pathological family relationships described by Freud to "psychic infections." Serious mental illnesses, he claimed, were "dependent very largely upon errors of education, unsuitable environment, and the acquisition of injurious habits of thought and the suppression of painful experiences, usually in the sexual field which later in life form the basis for psychoses." In order to correct this, Salmon encouraged mental hygiene to broaden its purview to include such fields as education, immigration, and criminology. Echoing Freud, he said that "practically all the hopeful points of attack in this field exist in early childhood, and if the psychiatrists are to take up such work they must be permitted to enter the schools."

By 1917 Freud's theory had become fused with social reform efforts in America. The foundation for the future child guidance movement, the community mental health movement, and for mental health professionals' involvement in such fields as education and criminology had been well laid. To Sigmund Freud in Vienna it must have all been very flattering if a bit puzzling.

Freudianism in New York

New York City became—and remains—the mecca for Freud's theory in America, and the principal reason it became the psychoanalytic epicenter was Abraham A. Brill. An Austrian immigrant, Brill arrived in New York alone at age 14, initially supporting himself by sweeping floors and giving mandolin lessons. He worked his way through school, university, and medical school, becoming along the way "an ardent admirer and student of Spinoza." "If not for Spinoza," he wrote, "I might have been a Rabbi, or a Methodist preacher, or a Catholic priest." As a young physician he tried psychiatry but found it "a barren, more or less descriptive, science with a poor background, a hopeless prognosis and a haphazard therapy." He was attracted instead to hypnotism and in 1907 went to Paris to study Charcot's method. Disappointed, he moved on to Jung's clinic in Zurich where he had been told "they are doing that Freud stuff." He became an immediate convert and "worked heart and soul in the pioneer work of testing and applying the Freudian mechanisms to psychiatry." The following spring he journeyed to Vienna to meet Freud, then returned to New York to translate Freud's works into English, proselytize, and become America's first private practicing psychoanalyst.

Although Brill was initially attracted to Freud because of his ideas about dreams, it was Freud's sexual theory that sustained his interest. Brill owned "a fine collection of 20 volumes on the history of sex practices" and, according to J. C. Burnham, he "seemed even more preoccupied with sex than his fellow American analysts." Believing that childhood sexual traumas were "invariably" the cause of neuroses, Brill frequently quoted Freud's dictum that "no neurosis is possible with a normal *vita sexualis*" and added that a healthy sex life is as necessary as "pure air and food." "The urge is there," wrote Brill, "and whether the individual desires or no, it always manifests itself." Burnham noted that "Brill conspicuously included in his writings far more of the grossly sexual than his fellow American analysts did. . . . It was obvious that he enjoyed shocking his colleagues and others concerning the 'facts' of human behavior, all of which turned out to be sexual." Another historian noted Brill's "frank insistence on sexual details and his delight in risqué jokes."

For a man who enjoyed shocking colleagues and friends, New York City was the place to live in the years prior to World War I. The Armory Show of 1913 showcased Cubism and Futurism, and paintings such as Marcel Duchamp's *Nude Descending a Staircase* caused a collective raising of eyebrows; supporters called the show "the most important public event that has ever come off since the signing of the Declaration of Independence." Ibsen and Shaw were introducing new frankness into the theater, while dances such as the tango and turkey trot were denounced from pulpits and caused Columbia University to specify that dancing couples must be separated by a minimum of six inches. The revolt against morals was centered in Greenwich Village, with the city above 14th Street said to be "cut off from the Village like the Ego from the Id." The heart of Greenwich Village was Washington Square, where Dadaists occasionally picnicked atop the Washington Square arch and on several occasions drunken revelers declared independence from the United States and proclaimed the square to be a separate republic.

Socially and politically the Village was determined to shock the world. At Polly Halliday's restaurant on MacDougal Street anarchist Hippolyte Havel addressed his customers as "bourgeois pigs." Above the restaurant was the Liberal Club, where Margaret Sanger formulated actions to legalize birth control, Bill Haywood organized the Industrial Workers of the World (IWW) for the Paterson strike, John Reed wrote his account of the Russian Revolution—*Ten Days That Shook the World* (1919), and Emma Goldman argued politics with Lincoln Steffens,

Walter Lippmann, and Theodore Dreiser. The unofficial voice of the Village was the *Masses*, a Marxist publication whose masthead advised that it was "directed against rigidity and dogma wherever it is found . . . a magazine whose final policy is to do as it pleases and conciliate nobody, not even its readers." In this fertile New York ground of liberalism and revolt the seeds of Freudian theory took root and grew.

Max Eastman, the editor of the *Masses*, was one of the most important early converts to Freud's theory. In 1906 Eastman, seeking treatment for fatigue, had spent three "serenely happy" months at Doctor Sahler's New Thought Sanitarium outside New York City. Diagnoses at the sanitarium were arrived at through the "astral body" of one of Sahler's female assistants; when a treatment plan was needed Sahler hypnotized the assistant and she prescribed what should be done. If this failed, an alternative treatment used mild electric shocks "delivered through a serrated gold crown." In his biography years later Eastman recalled how "I wanted to believe in this institution. I wanted to believe in mental healing."

By the time Eastman assumed the editorship of the *Masses* in 1913, he had become a devout Marxist. In 1914 he underwent psychoanalysis with Smith E. Jelliffe, another Freudian who was active in the mental hygiene movement, and later also with Abraham A. Brill. Eastman claimed that he "read Freud and every book on Freud then available in English" and that "I never heard of an infantile fixation of which I could not find traces in my make-up." Eastman was so enthusiastic that he persuaded both his sister and mother to also undertake psychoanalysis with Brill.

Eastman's major contribution to the popularization of Freud came in 1915 when he wrote, for a one-thousand-dollar fee, two long articles for *Everybody's Magazine*, a popular monthly with over 600,000 readers. He described psychoanalysis as a new treatment "which I believe may be of value to hundreds of thousands of people," a method for dissecting out "mental cancers . . . [which will] leave you sound and free and energetic." The "mental cancers" were said to be "desires which dwell in our minds without our knowing they are there, and . . . if we can be made clearly aware of these desires their morbid effects will disappear."

The article included pictures of Freud, Jung, Brill, and Jelliffe and several accounts of miracle-like cures. Eastman described dream analysis and the word-association test, but his main emphasis was on the traditional Freudian theory of infantile sexuality. "The attitude which little children develop toward their parents and immediate family," wrote Eastman, "is after heredity the chief influence in determining the trend of their character, in determining their attitude toward the world."

Max Eastman's associate editor at the *Masses* was Floyd Dell, a young writer and Marxist, who was so enthusiastic about psychoanalytic theory that he was frequently referred to by colleagues as Freud Dell. As early as 1913 Dell and his friends were said to be "busy analysing each other and everyone they met. . . . It was a time when it was well for a man to be somewhat guarded in the remarks he made, what he did with his hands." By 1916, when Dell began an extended psychoanalysis, he recalled that "everyone at that time who knew about psychoanalysis was a sort of missionary on the subject, and nobody could be around Greenwich Village without hearing a lot about it."

Dell's psychoanalyst was Samuel A. Tannenbaum, a controversial physician who believed strongly in the dangers of sexual abstinence; he publicly advocated the legalization of prostitution and urged young men to use prostitutes to avoid "frustrated excitement." Tannenbaum said that a happy sex life could "change a thin, pale, sleepless, irritable and morose wife into a healthy, beautiful and contented mate." Dell later claimed to be "immensely indebted to psychoanalysis" for giving him "a new view of the whole world . . . a new view of history, not supplanting the Marxian but supplementing it." Freudian concepts became prominent in Dell's writings, and in 1930 he addressed the First International Congress on Mental Hygiene on the subject of "Sex and Civilization."

Walter Lippmann, who later would be called the "prophet of the new liberalism," was another influential convert to psychoanalysis in its early years. His friend Alfred Booth Kuttner was one of Brill's first psychoanalytic patients, and Kuttner agreed to assist Brill translate Freud's *The Interpretation of Dreams* into English. In the summer of 1912 Lippmann and Kuttner shared a cabin in Maine, where Lippmann read Freud's work as Kuttner translated it. Lippmann was fascinated, later telling a friend that he had initially reacted to Freud's theory "as men might have felt about *The Origin of the Species*." Young Lippmann was a graduate of Harvard University, where he had founded a chapter of the Intercollegiate Socialist Society and successfully petitioned the administration to introduce socialism into the curriculum. Following graduation Lippmann had worked for Lincoln Steffens and then for George R. Lunn, the Socialist mayor of Schnectady.

During the summer of 1912 Lippmann was writing *A Preface to*

Politics, the book that would establish his reputation as a major liberal thinker. When it was published the following year, reviewers were surprised to note Lippmann's strong endorsement of Freudian theory, which Lippmann called "the greatest advance ever made toward the understanding and control of human character." Lippmann used the concepts of repression and sublimation widely and applied them to political situations. Lippmann also drew parallels between social issues and individual instincts, linking, for example, "the suffrage movement, industrial consolidation, and labor unrest with irrational impulses Freud had discovered in the human psyche." Lippmann was essentially using psychoanalytic constructs to analyze the social problems of America.

In 1914 Lippmann was named associate editor of the *New Republic*. Both Lippmann and his friend, Kuttner, contributed highly flattering accounts of Freudian theory to the fledgling weekly. In 1915, for example, Lippmann wrote of Freud: "I cannot help feeling that for his illumination, for this steadiness and brilliancy of mind, he may rank among the greatest who have contributed to thought." (According to one of Lippmann's biographers, "A few years later in Vienna Lippmann actually met Freud, who invited him to a meeting of the Psychoanalytic Society.") So enthusiastic was Lippmann for Freud's ideas that by 1919 one of his political science colleagues complained: "I wish Walter Lippmann would forget Freud for a little, just a little." It is uncertain whether Lippmann personally undertook a psychoanalysis, although he appears to have shared many ideas with Abraham Brill. Brill, for example, also publicly declared that the women's suffrage movement was a consequence of sexual repression. "In continental Europe," Brill added, "where there is not so much prudery and where women are not forced to be men [women's suffrage] may be dispensed with."

As an evangelist for Freudian theory, Walter Lippmann brought the word to Mabel Dodge's salon at 23 Fifth Avenue, certainly the most influential social gatherings in New York in the prewar years. Dodge was a wealthy patron of the arts, who in 1913 began a series of "evenings" to which she invited the elite of the city's social, political, and intellectual communities. Almost within the shadow of the Washington Square arch, she gathered "Socialists, Trade-Unionists, Anarchists, Suffragists, Poets, Relations, Lawyers, Murderers, 'Old Friends,' Psychoanalysts, I.W.W.'s, Single Taxers, Birth Controlists, Newspapermen, Artists, Modern Artists. . . ." Dodge salon regulars such as Emma Goldman, Max Eastman, and Floyd Dell were early and enthusiastic

Freud followers, but it remained for Walter Lippmann to, in the words of his biographer, "display his own virtuosity by presenting some lectures about the new psychology of Sigmund Freud."

On one noteworthy occasion in 1914 Lippmann invited Brill to address the Dodge gathering. Brill recalled that his talk "aroused a very interesting and lively discussion"; Dodge later claimed, however, that "several guests got up and left, they were so incensed at his assertions about unconscious behavior and its give-aways." Brill specifically recalled that "Big" Bill Haywood was among the guests that evening. Haywood was a huge man with no formal education and one blind eye that fixed a person with its immobile stare, described by one guest as "a large, soft, overripe Buddha with one eye and the smile of an Eminent Man, [who] reclined in the yellow chaise lounge with two or three maidens at his feet." At the time Haywood was the Socialist leader of the IWW Lawrence and Paterson textile workers' strikes; six years earlier he had been tried for the murder of former governor Steunenberg of Idaho in a miners' dispute and had been found innocent with the help of his brilliant lawyer, Clarence Darrow. As Brill later noted, "The questions [about psychoanalysis] I was asked there by such people as . . . Bill Haywood and others equally distinguished were quite different from those posed by medical men."

Mabel Dodge was so intrigued with Freud's theory that she began an immediate psychoanalysis with Brill. Sexual themes were important to Dodge, who was bisexual. According to Christopher Lasch, she was fascinated by women's breasts and was "a pioneer in the cult of the orgasm." At the time of her psychoanalytic soirée, Dodge maintained John Reed as her live-in lover, had had two previous marriages, and two more were to come in the future. Brill was always looking for new methods to spread the Freudian gospel, and he persuaded Dodge to write regular columns about psychoanalysis for the Hearst newspapers. Brill specifically advised Dodge to direct her columns toward working-class women, like "the woman making up the beds in the hotel . . . the shop-girls and the young clerks." Under headlines such as "Mabel Dodge Writes About Mother Love," her columns were carried in the *New York Journal*, which then had the largest circulation of any newspaper in the United States. In later years Dodge continued her advocacy for Freud; she also supported D. H. Lawrence, one of the first English novelists to utilize Freudian ideas in fiction.

If Freudian theory was ever going to spread beyond the confines of Greenwich Village, popularized versions such as those by Mabel Dodge would be essential. Brill was skilled at using his patients to accomplish this task and one of the most effective was Alfred Kuttner, who had helped convert Walter Lippmann. Kuttner contributed articles on Freud's theories to several lay publications. In 1913 he wrote a full-page, highly flattering review of Brill's new book, *Psychoanalysis, Its Theory and Practical Application*, for the *New York Times*. Next to a large picture of Brill he described the "revolution in the last few years in the method of treating the insane" and claimed that psychoanalysis "bears the same relationship to mental and nervous diseases that the microscope does to pathology." Kuttner's claims were not modest: "Freud and his followers literally tear hidden secrets from the bewildered minds of their patients, and the latter do not know that they are revealing them." A year later Kuttner wrote another long article for the *New York Times*, this time on Freudian theories of slips of the tongue, to coincide with the publication of Brill's translation of Freud's *The Psychopathology of Everyday Life*.

Another of Brill's patients, Max Eastman, wrote the adulatory articles on psychoanalysis referred to previously in the June and July 1915, issues of *Everybody's Magazine*. That same year *Good Housekeeping* carried an extended, two-part series on Freud and psychoanalysis, which included pictorial illustrations of miracle cures brought about by this new method of healing. One picture showed a woman with a paralyzed arm who had suddenly been made well when her "mind was opened to the significances of a certain incident that happened before she was five years old." "Every case of nervous invalidism not arising from a physical cause," claimed *Good Housekeeping*, "has its origin in some maladjustment of the sex life. . . . This dangerous maladjustment has its beginning in the experiences of the child before its first five years are past." The article included a version of Freud's dictum frequently quoted by Brill: "With a normal sexual life there is no such thing as a neurotic."

Newspaper and magazine editors were enchanted by the commercial possiblities offered by sexual wishes in childhood. As an editor of the Chicago *Journal* noted when his paper ran an account of Freudian theory: "This . . . is more engrossing than our limerick contest. Our readers will be thrilled to learn they are all potential lunatics who want to stab their fathers or go to bed with their mothers. . . . Can you imagine anybody in 1913 discovering something new about sex?" Best of all, such Freudian titillation could be presented as the "scientific discoveries" of European professors, lending it an aura of respectability and raising it above the Comstock laws which prohibited the publication of salacious material.

Between 1909 and 1917 Freud's ideas spread rapidly among New York's intelligentsia. According to one observer, Freudian theory, which implicitly encouraged sexual freedom, became a wedge used "to liberate American literature from pruderies and other social restrictions. . . . It may well be that the freedom to write about sex, which was linked with other freedoms, would have been won without the intervention of Freud. But the literary exploitation of Freud was a heavy reinforcement at a decisive moment and materially assisted the coming of age of our literature."

Brill and his colleagues played an important role in moving New York's literary community beyond its Victorian heritage. Following an article in the *New York Times Magazine* in 1910, in which Brill's work was favorably described, he recalled that writers "came to me from everywhere wanting to know about it." On one occasion Brill was the guest speaker at a meeting of the Authors' League of America during which, he recalled, "I gave them a number of 'plots' which frequently came to my attention through patients, but of which they had never heard or imagined."

Theodore Dreiser, one of the most acclaimed novelists of the era, utilized many ideas from Brill. Dreiser's *Sister Carrie* had been suppressed by its publisher in 1900 because of public outcry against the book's realistic portrayal of sexual problems. Dreiser was favorably disposed toward Freud's ideas when he began a long, social relationship with Brill in the prewar years. Dreiser read books recommended by Brill, consulted him regarding character motivation while writing *An American Tragedy*, and even "began a series of twelve conversations with Dr. Brill . . . on the subject of life and happiness . . . [which were] taken down by a stenographer" with an intent (never realized) of eventual publication. When Dreiser went to Europe he asked Brill for a letter of introduction to Freud, but Dreiser was not able to meet him.

Dreiser also had a special interest in Freud because of his own sexual problems. According to W. A. Swanberg he was a compulsive womanizer and "was always entangled in intrigue and deceit. While he sometimes admitted that his libido was a burden, he more often felt that sex was the most rewarding of experiences . . . and linked it with courage and adventure." Dreiser also felt "that the conflicts of sex heightened his awareness and creativity." Dreiser praised Freud for his "strong, revealing light thrown on some of the darkest problems that haunted and troubled me and my work." Like Brill, Dreiser believed that sexual repression could be harmful, and he insisted on sexual freedom for himself. On one occasion, according to Swanberg's *Dreiser*, when Helen Richardson, his mistress and eventual wife, objected to one of his affairs, Dreiser purposefully brought the woman home, slept with her, and insisted that Helen serve them both breakfast in bed the next morning, which in fact she did.

Several other American novelists were swept along by the Freudian tide washing into the Village. In Waldo Frank's *Unwelcome Man* (1917), "the novel begins with an extremely detailed account of the hero's emotions as a baby" and was said to draw "heavily on Freud." Frank's *Our America* (1919) explained American history as a consequence of the Puritans having repressed their natural desires, which led to fundamental defects in the American character as it "sickened and shriveled and grew perverse." Frank called Freud "one of the heros of modern thought." Frank was at the time the associate editor of *The Seven Arts*, the most influential of the myriad literary journals being published in New York. The editor of *The Seven Arts* was James Oppenheim, who had also been psychoanalyzed and whose psychoanalyst had in fact contributed to the creation of *The Seven Arts* by persuading one wealthy patient to fund it "as a kind of therapy for Oppenheim," who was also his patient. Oppenheim himself later became a lay psychoanalyst.

The literary journals were an important route for disseminating Freud's ideas and spreading them beyond New York City. According to critic Bernard De Voto, between 1912 and 1920 "three-quarters of the poetry, drama and fiction they [the journals] printed were Freudian in inspiration and at least intended to be Freudian in content." Young writers trying to get started were inevitably influenced by what their New York elders were calling important. Conrad Aiken, for example, "decided very early, I think as early as 1912, that Freud and his co-workers and rivals and followers were making the most important contribution of the century to the understanding of man and his consciousness; accordingly I made it my business to learn as much from them as I could." Aiken continued to use Freudian themes in his writing throughout his long career. Similarly, young F. Scott Fitzgerald, in *This Side of Paradise* (1920), has one woman character petting freely and exclaiming to her young man: "Oh, just one person in fifty has any glimmer of what sex is. I'm hipped on Freud and all that, but it's rotten that every bit of real love in the world is ninety-nine per cent passion and one little soupçon of jealousy."

The Broadway theater also incorporated Freud early and often. Arthur Hopkins, who had been psychoanalyzed by Dr. Smith E. Jelliffe, is credited with writing the first play to use specific Freudian concepts. *The Fatted Calf* opened in February 1912 and depicted the cure of paranoid symptoms by psychotherapy. The following year *The Smouldering Flame* portrayed the consequences of sexual repression, and in 1915 the award-winning *Children of Earth* was built around the same theme, with "a minor character, the old village idiot [being] the symbol of the author's thesis that unless the individual fulfills his sexual life he may become, like this old man, 'love-cracked.' "

Suppressed Desires, which opened in 1916, was the first play specifically about psychoanalysis, "an ingenious and delightful satire on the effects of amateur psychoanalysis in the hands of a giddy fadist." One woman in the play is warned not to reveal her dreams to her sister or "she'll find out that you have an underground desire to kill your father and marry your mother"; another character is said to have been successfully psychoanalyzed in just two weeks. The authors of the play were Susan Glaspell and George C. Cook, prominent members of the Village literary establishment. Glaspell once explained the psychoanalytic ambiance of that period in the Village by saying: "You could not go out to buy a bun without hearing of someone's complex." Another New Yorker recalled a man at a party starting to describe a dream when he was interrupted by a guest: "I beg your pardon, Sir, but since you are not a Freudian, you are unwittingly making the most intimate revelations. I do not wish to be an eavesdropper, even in such a way."

In the early 1920s Freud became even more prominent in the New York theater. In his comprehensive *Freud on Broadway*, W. David Sievers called this period "the psychoanalytic era" and said that it was marked by the Oedipus complex, usually with a "dominating mother and fixated son whose life was warped by her jealousy of his sexual partner." Sexual suppression and frustration were portrayed as harmful in these plays, and Freud was used as a rationale for giving in to one's desires. As Sievers described it: "With flasks on their hips and words like 'inhibited' and 'suppressed' on their lips, the younger generation was depicted as violating the pre-war moral codes in the name of self-expression and individual freedom." Especially noteworthy among the plays was Theodore Dreiser's 1919 *The Hand of the Potter*, in which the behavior of a sexual psychopath was explained as "a great force" which Freud was said to be studying. In Sherwood Anderson's 1925 novel *Dark Laughter*, one character says: "If there is anything you do not understand in human life, consult the works of Dr. Freud." And in Eugene O'Neill's *Strange Interlude*, written in 1928 while O'Neill was in psychoanalysis, incestuous and suppressed sexual desires are intertwined with lines such as, "O Oedipus, O my king? The world is adopting you!" *Strange Interlude* ran for 426 New York performances and won a Pulitzer Prize.

In addition to the New York writers, Dr. Abraham Brill and his small coterie of fellow psychoanalysts also availed themselves of opportunities to carry the Freudian word to other groups. As Brill recalled: "As soon as I became known as an exponent of Freud, I was actually swamped with invitations to speak on problems of sex." Among other groups, he addressed the Child Study Association on the subject of "masturbation" and he taught regular courses in the department of pedagogics at New York University. If mothers and teachers would utilize psychoanalytic principles in their dealings with children, Brill said, "we could reduce nervous and mental diseases as much as we have smallpox and typhoid." Columbia University's student publication encouraged readers to "unshackle our libido," and by 1920 it was said that psychoanalysis was being taught "in extension classes, summer schools, and girl colleges [sic] in the New York area."

Another example of Brill's influence in disseminating Freud's theory was through advertising. A leading New York publicist of the post-war era was Edward L. Bernays, the son of Freud's sister Anna. Bernays had been raised and educated in New York and worked for the American Tobacco Company, which was attempting to promote smoking among women. In 1929 Bernays sought Brill's advice on how to accomplish this goal and was told that "smoking is a sublimation of oral eroticism" and that cigarettes are phallic symbols and thus masculine symbols. Cigarettes could therefore be used, said Brill, as "torches of freedom" by women to demonstrate the fact that they were liberated. Based on Brill's advice Bernays arranged for ten young debutantes to smoke publicly in New York's Easter Parade, causing a national stir with front-page stories in newspapers. Bernays contended that this was "the first instance of its [Freud's theory] application to advertising."

New York City, then, had become the center of the Freudian New World a decade after Freud's 1909 American visit. There were small pockets of psychoanalytic enthusiasm in Boston, Washington and Chicago, but Greenwich Village was Freud's stronghold. Since New York was also the center for writers and publishers, Freud's ideas were disseminated through the media much more quickly than if they had taken initial root elsewhere.

It was also in New York City that Freud's theory became most successfully fused with ideas of sexual freedom and social reform. Birth control, divorce laws, and women's suffrage mingled easily with sexual freedom, and Freud's followers were never far away. Dr. Brill, for example, supported both birth control and the legalization of abortions. Many of Freud's followers were also involved in social reform movements including working conditions, child labor laws, and housing. Artistic experimentation was another facet of social reform, so that Cubism, the new theater, and even the new dances became identified with Freud. Dr. Smith E. Jelliffe, for example, coauthored *Psychoanalysis and the Drama* (1922) and was also a staunch defender of modern painting, which he viewed as "an important mode of expressing hitherto forbidden wishes." Similarly Brill spoke out publicly in favor of the new dances, which he claimed were a good antidote for "Puritan prudery and Anglo-Saxon hypocrisy. . . . Viewed in this light the modern dances must be considered beneficial to our present social system." Brill cited several of his patients who had been specifically helped, including "two timid and shut in persons who were completely changed by the new dances" and "no longer fear to meet persons of the opposite sex and are thinking seriously of matrimony." Jelliffe and Brill thus became prototypes for later American psychiatrists who would endorse social change in the name of promoting mental health.

An important aspect of this fusion of Freudian theory with sexual freedom and social reform was political. Freud's followers in America were, from the outset, drawn from the left of the political spectrum. Psychoanalysts such as Smith E. Jelliffe, James Putnam, and William A. White were said to be "sympathetic to the vague co-operative socialism of the pre-war period." Among prominent nonpsychoanalysts who were followers of Freud were Walter Lippmann, who espoused traditional liberalism; Max Eastman and Floyd Dell, who were Socialists; Theodore Dreiser, who later joined the American Communist Party; and Emma Goldman and her colleagues, who were anarchists. It was said that Trotsky "wrote intelligently about Freud," and Max Eastman claimed that Marx and Freud "both contemplate the same facts and their view of these facts is in complete harmony." William Dean Howells commented that socialism "smells to the average American of petroleum, suggests the red flag, and all manner of sexual novelties. . . ."

All of these threads came together in Greenwich Village at Mabel Dodge's salon, where it was said that one could "determine the ministers of the forthcoming revolutionary government from the guest list." Invitees such as Walter Lippmann, Emma Goldman, and Abraham Brill moved easily with their hostess from discussion of libido to social reform and revolution, all between salad and dessert. Many of these same people could also be found at William and Margaret Sanger's uptown apartment, where talk of birth control and Freudian theory was mixed with revolution; guests included such notables as John Reed, "Big" Bill Haywood, Socialist leader Eugene Debs, Emma Goldman, and anarchist Alexander Berkman, who had been released from prison after serving 14 years for shooting Henry Clay Frick.

Freud had definitely arrived in New York, and many of his followers hoped that "a psychoanalytic revolution would transmute the Victorian Era into a Golden Age."

The Rise of Sex, the Fall of Freud

The sexual revolution that took place in the second and third decades of twentieth-century America was closely linked to Freud's name. Helen Thompson Woolley, a researcher summarizing the literature on the psychology of sex in 1910, had difficulty finding any published material; by 1914 she noted a flood of new studies that she attributed to "the emphasis placed on sex by the Freudian school and the interest in sex education, to say nothing of the whole feminist and woman's suffrage movement." Surveys showed that the number of Americans who approved of extramarital relations, divorce, and birth control "grew quickly after 1918 and reached a peak between 1925 and 1929, especially among intellectuals." Studies by Alfred Kinsey showed that the percentage of women who experienced orgasm and who had premarital intercourse increased sharply "around 1916 among women born around 1900. . . . Kinsey attributed the new pattern partly to a change in attitudes which he associated with the doctrines of Freud and Havelock Ellis." Many factors contributed to the sexual revolution in America, but Freud's name became its shibboleth.

Two technological innovations—condoms and cars—enabled the sexual revolution to occur. Significant advances in the vulcanization of rubber in 1905 and 1906 made it possible to mass-produce reliable condoms for the first time. Margaret Sanger's campaign for birth control began in 1914 when her clinics started dispensing condoms and other birth-control devices at the time Freud's ideas were starting to receive popular press.

Women also became full citizens in August 1920 with the ratification of the Nineteenth Amendment, giving them the right to vote. Women's liberation was further promoted by Prohibition; saloons had always been men's territory, but "the speakeasy usually catered to both men and women. . . . Under the new regime not only the drinks were mixed, but the company [was] as well."

Meanwhile the divorce rate in America was doubling from 8.8 per 100 marriages in 1910 to 16.5 per 100 in 1928, and in urban areas the traditional stigma of being divorced was being replaced by "an air of unconventionality, just enough of a touch of scarlet to be considered rather dashing and desirable."

The availability of automobiles was essential to the sexual revolution, for they provided ladies and gentlemen with mobile venues that could be used to explore new dimensions in their relationships. In 1919 there were 6.6 million automobiles in the United States, but 90 percent of them were open. By 1927 the number of cars had climbed to 20.2 million and, more importantly, 82 percent of them were closed. In *Only Yesterday*, Frederick Allen noted, "The closed car was in effect a room protected from the weather which could be occupied at any time of the day or night and could be moved at will into a darkened byway or a country lane. . . . [It was] an almost universally available means of escaping temporarily from the supervision of parents and chaperones." In *Middletown*, a sociological study published in 1929, a judge called the automobile a "house of prostitution on wheels" and said that two-thirds of the sex offenses being committed by women were taking place in cars.

Preoccupation with sex was reflected in literature and in popular culture of the era. Ernest Hemingway's *The Sun Also Rises* (1926) and Aldous Huxley's *Point Counter Point* (1928) offered some readers a sexual titillation which other readers sought in magazines such as *True Story*, first published in 1919; by 1926 the magazine had almost two million readers, "a record of rapid growth probably unparalleled in magazine publishing." Pulitzer Prize juries had increasing problems finding books which reflected the committee's necessary criteria of "the wholesome atmosphere of American life and the highest standard of American manners of manhood," so officials changed the criteria by substituting "whole" for "wholesome" and dropping the "highest standard" clause altogether. Among young Americans sexual activity had become a topic of conversation; as Frederick Allen noted, "Not only did they believe that [such talk] should be free, but some of them believed it should be continuous." As one popular magazine noted, "sex o'clock" had struck in America.

By the end of the 1920s the major battles in America's sexual revolution had been fought. Public preoccupation with sex was already waning when the October 1929 events on Wall Street shifted public attention to more mundane matters like jobs and money. By 1936 *Fortune* magazine officially proclaimed: "Sex is no longer news. And the fact that it is no longer news is news."

As the sexual revolution began winding down in the 1920s, so too did America's interest in Freudian ideas. Freud was increasingly viewed, according to one observer, as a fad "like mah-jhongg or miniature golf." In places like Greenwich Village it was still *au courant* for young men "to preface their seductions with chapter and verse from Freud," but elsewhere seductions continued to take place without assistance from Vienna. Catherine Covert, in her study of Freudian ideas in American newspapers, noted that references to Freud "crested in the mid-Twenties and subsided at the beginning of the Depression." Similarly, Nathan Hale in *Freud and the Americans* reported that "the number of articles about psychoanalysis . . . reached a peak between 1925 and 1928." (In this author's own study of articles about Freud and psychoanalysis, determined as a yearly rate per total articles listed in the *Readers' Guide to Periodical Literature*, it was found that their number peaked between 1915 and 1922 and thereafter slowly declined; the annual rate of such articles for 1930–1935, in fact, was just one-third what it had been for 1915–1922 and it remained at the lower rate until the 1950s.)

Even more significant than the declining volume of articles about Freud and psychoanalysis was the fact that an increasing percentage of them were derisive or even frankly derogatory. E. E. Cummings suggested "Every Issue an Oedipus Complex" as an official slogan for the New York *Daily News*. The *New York Times* reported the suicide of one Mercy Rogers, "who, after reading 102 books on psychoanalysis and allied subjects, despaired of her life and turned on the gas." Henry Ford's Dearborn *Independent* parodied the fascination of New York newspapers with Freud by reviewing a mythical book titled "Psychic Psarah, a Pstudy in Psychoanalysis" and awarding it the prize of the "American Society for the Promotion and Promulgation of the Abnormal." *Vanity Fair* pictured "Doctor Paul Ehrich, the popular New York psychoanalyst . . . probing the complexes of a young lady who had been dreaming of red lights, boa constrictors, and caviar. Disregarding the fact that she had supped on Welsh rarebit, he will diagnose her case as agoraphobia and acute mania resulting from an Electra Complex." And Morris Fishbein's *New Medical Follies*, a popular 1927 book, characterized psychoanalysis as "a species of voodoo religion characterized by obscene rites and human sacrifices" in which a psychoanalyst can "find some obscene cause for practically any action on which a person may fix his attention, or about which he prefers not to speak."

By the onset of the Great Depression, then, Freud and his theory were increasingly being forgotten by Americans at large. He had been an important force in the sexual revolution, but with increasing openness about sex Freud was falling into decline. In 1928 *Vanity Fair* nominated Freud "as a candidate for oblivion." In 1933 *Commonweal* noted under the heading "Farewell to Freud" that "psychoanalysis has about run its course" and was becoming "an outmoded fad, its credit hopelessly impaired through association with self-indulgence, charlatanry and exploitation." And in 1935 *American Mercury* published an article titled "The Twilight of Psychoanalysis," in which it claimed, "The psychoanalytic jargon has lost its novelty and sunk into esoteric unintelligibility."

There were, to be certain, small groups of Americans for whom Freudian theory continued to be important. The New York intellectual community, especially those in the theater world, utilized psychoanalysis in their personal lives and Freudian concepts in their professional lives. Readers of the *New Republic* and small literary magazines continued to see Freud referenced but, by contrast, a survey of major pulp magazines in the 1920s reported "no trace of psychoanalysis." Social workers and psychiatrists connected with the mental hygiene movement and criminology (see Chapter 7) revered Freud but, according to Nathan Hale, Freud's ideas had broadly "touched only a small minority of psychiatrists." Freud's most popular book, *The Interpretation of Dreams*, had sold by 1932 only 16,250 copies in America and Great Britain during the 19 years since its English translation. As Hale concluded in *Freud and the Americans*, "By 1930 the influence of Freud and psychoanalysis probably still was confined to comparatively small groups of Americans."

Yet three decades later Sigmund Freud had become one of the most widely recognized names in America. In 1959 sociologist Philip Rieff assessed Freud as having had an "intellectual influence . . . greater than that of any other modern thinker," a man who "has changed the course of Western intellectual history," and whose writings were "the most important body of thought committed to paper in the twentieth century." Surely this is one of the most remarkable transformations of a man's ideas in intellectual history.

It is a long road from a beachhead among New York intellectuals to a widespread influence in almost every phase of American life. Literature, drama, anthropology, sociology, child rearing, education, criminology, and many other parts of American thought and culture were to become permeated by Freud. The vehicle that would be needed to carry Freud from Greenwich Village to Utica, Wichita, and Yakima serendipitously appeared just when it was needed. Without the nature-nurture controversy Freud's name would today be merely a footnote in social history, cited as one of the early advocates for sexual reform. But with the nature-nurture controversy Freud achieved a mission and purpose greater than even he had envisioned.

2

Race, Immigration, and the Nature-Nurture Debate

———

The racialists . . . want us to believe that a Nordic idiot is worth more than a Chinese genius.
— FRANZ BOAS

Among the many eminent guests invited to Sigmund Freud's 1909 lectures at Clark University, Franz Boas would ultimately prove to be the most important for the dissemination of Freud's theory in America. Remembered today primarily as an anthropologist and expert on the Native Americans of the northwest coast, Boas was at the time better known as the leading opponent of immigration restriction and racism. Thomas W. Gossett, in *Race: The History of an Idea in America*, suggested that "it is possible that Boas did more to combat race prejudice than any other person in history." As Freud's name became attached to issues surrounding immigration and race, it also became part of the ongoing debate about nature and nurture, code words for those who believed that genetics (nature) or cultural and experiential factors (nurture) were the most important determinants of human behavior. Boas thus unknowingly laid the groundwork for Freud's reification as the messiah of liberalism and humanism in the years following World War II.

Although Boas himself apparently never underwent psychoanalysis, he was an enthusiastic supporter of Freud's theory. When interviewed by the Boston *Evening Transcript* following the 1909 lectures, Boas praised what he referred to as Freud's "epoch-making discovery." Boas himself presented a paper at the conference titled "Psychological Problems in Anthropology," and he was positioned in the front row, along with Freud, Jung, William James, Adolf Meyer, and G. Stanley Hall in the official conference photograph. When Boas returned to Columbia, where he was a professor of anthropology, he taught a seminar on Freud's theory. In later years Boas was sometimes critical of Freud's own anthropological ideas, especially Freud's *Totem and Tabu* (1913) and his assertion that totem poles were sex symbols, but Boas nevertheless continued to be supportive of Freud's theory through correspondence with some of the leading psychoanalysts, support for *Psychoanalytic Review*, America's first psychoanalytic journal, participation in the First International Congress for Mental Hygiene, and encouragement of his daughter to undergo psychoanalysis. Boas admired Freud's emphasis on childhood experiences rather than genes as the crucial determinant of human behavior, and Boas saw Freud as an ally in the monumental fight he had undertaken. Boas would need every ally he could find.

Franz Boas Versus the Anglo Elite

Franz Boas was born in 1858 into a German Jewish family that had participated in the 1848 German democratic uprising. Boas described his home as one "in which the ideals of the Revolution of 1848 were a living force." He was by training both a democrat and a revolutionary and his willingness to stand up for his beliefs resulted in two duels before he had turned 21. At least one of these was precipitated by an anti-Semitic remark to which Boas had taken exception; Boas is said to have thrown the offender out of a café, then insisted on a duel even after the man apologized. Both duels left Boas with facial scars, especially one traversing his left cheek under his shaggy black eyebrows, and these contributed significantly to his reputation for obduracy. As A. L. Kroeber later described Boas: "So decisive were his judgments and so strong his feelings that his character had in it much of the daemonic. . . . His convictions sprang from so deep down, and manifested themselves so powerfully, that to the run of shallower men there was something ultra-human or unnatural about him; he seemed impelled by forces that did not actuate them." Boas's youthful duels, then, were merely a prelude to his life.

After completing his university education with a degree in physics and geography, Boas joined a German expedition to Baffin Island, where he was deeply impressed by the Eskimos. He recalled:

> *I had seen that they enjoyed life, and a hard life, as we do; that nature is also beautiful to them; that feelings of friendship also root in the Eskimo heart; that, although the character of their life is so rude as compared to civilized life, the Eskimo is a man as we are; that his feelings, his virtues and his shortcomings are based in human nature, like ours.*

He left the Arctic with a strong belief in the common humanity of all people as well as an appreciation of how the physical environment can shape both customs and character.

Boas returned to Germany for a year but was discouraged by the anti-Semitism that was prevalent among conservative Germans of the late 19th century. Following anthropological fieldwork among the Northwest Coast Indians in Canada, Boas settled permanently in America, eventually joining his uncle, Dr. Abraham Jacobi, in New York. Jacobi had been imprisoned during the 1848 uprising, and later became a prominent founder of American pediatrics. In New York Boas obtained a position teaching anthropology at Columbia University in 1896 and three years later was appointed a professorship.

If Franz Boas thought that he had left anti-Semitism behind in Germany, he quickly learned differently. The aura of racism that hung over New York City in 1896 was almost palpable. It was not merely a "get off the sidewalk kike" type of anti-Semitism muttered by many illiterate people, the kind of remark that most educated Jews like Boas had learned to disregard. He recounted in a letter to his sister how he had ignored two women in a store who had called him a Jew and said loudly, "These Jews, they all look alike—stupid camels." The new racism of 1896 was a more pernicious type, propagated by intellectuals with Anglo-Saxon surnames and Ivy League degrees, a racism which they claimed to be based on scientific principles and statistical studies. It was a racism emanating from the brain, not merely from the bowels.

The bellwether for the new scientific racism was an article titled "Restriction of Immigration" in the June 1896 issue of the *Atlantic Monthly*. Written by Francis A. Walker, president of the Massachusetts Institute of Technology, it expressed alarm that "within the decade between 1880 and 1890 five and a quarter millions of foreigners entered our ports." Approximately half of them, Walker said, were "Hungarians, Bohemians, Poles, south Italians, and Russian Jews," groups he described as "ignorant and brutalized peasantry . . . degraded below our utmost conceptions," who had "habits repellant to our native people." Unless restrictions on immigration were enacted, "there is no reason why every foul and stagnant pool of population in Europe, which no breath of intellectual life has stirred for ages, should not be decanted upon our soil." Such individuals were said to "have none of the inherited instincts and tendencies which made it comparatively easy to deal with the immigration of the older time. They are beaten men from beaten races, representing the worst failures in the struggle for existence."

Walker's opinions were widely shared by many of his academic colleagues. Such men regarded themselves not as bigoted but rather as scientifically enlightened. Almost invariably they quoted Francis Galton, the wealthy half-cousin of Charles Darwin, whose *Hereditary Genius* (1869) had claimed to show that most eminent men are sired by similarly eminent fathers and mothers and recommended a program called "eugenics" to promote such breeding. Galton was especially disparaging of Jews, calling them "specialized for a *parasitical* existence upon other nations" and questioning whether "they are capable of fulfilling the varied duties of a civilized nation by themselves." Working under Walker at M.I.T. were William Z. Ripley, whose book *The Races of Europe* (1899) divided all Europeans into three types—Nordics, Alpines, and Mediterraneans, and Frederick A. Woods, whose book *Mental and Moral Heredity in Royalty* (1906) purported to demonstrate a causative relationship between intelligence and morality. Woods also claimed that the new immigrants from southern Europe were responsible for America's labor disputes because "all the Nordic peoples have an instinctive horror of anything other than well organized government."

Harvard's contribution to the preservation of patrician genes included three wealthy graduates who in 1894 founded the Immigration Restriction League. Prescott Farnsworth Hall, their leader, was a product of a Brahmin Boston family, had a passion for German philosophy and the occult, and authored *Immigration and Its Effects upon the United States* (1906). The express purpose of the Immigration Restriction League was to stem the immigrant outpour from southern and eastern Europe; to accomplish this the league published data purporting to show how the immigrants coming from these parts of Europe were degrading the American character. Within a year of its formation, the league's literature was being received and reprinted by more than five hundred daily newspapers, and it had established an office for lobbying purposes in Washington, where Senator Henry Cabot Lodge served as its spokesman.

The other end of this Anglo academic axis was in New York City and was led by Charles B. Davenport, Henry F. Osborn, and Madison Grant. A geneticist with A.B. and Ph.D. degrees from Harvard, Davenport believed that virtually all personality characteristics are inherited, including such traits as nomadism, prostitution, pauperism, and an inborn love of the sea (which he believed was a sex-linked recessive gene and therefore found only in males). A zealous eugenicist, Davenport believed that the most urgent need of mankind was to "annihilate the hideous serpent of hopelessly vicious protoplasm." Osborn came from a colonial New England family, had been educated at Princeton University, and was professor of biology at Columbia University when Boas began teaching there in 1896. Osborn referred to "certain races from central Europe" as "a large class of vipers." It is no more possible, he said, for these immigrants to change their nature than "it is for the leopard to change his spots" because environment cannot "fundamentally alter profound and inborn racial differences." Osborn also believed, as he wrote to a fellow eugenicist, "that each race has a distinctive kind of soul."

The most outspoken among the eugenicists forming the Anglo elite was Madison Grant. Raised in an affluent family and educated by tutors at their Park Avenue home before attending Yale University and Columbia Law School, Grant was of Nordic descent and strikingly tall and long-headed; not coincidentally he urged that immigration be restricted to people like himself. His book on Nordic superiority, *The Passing of the Great Race* (1916), became the bible of the eugenics movement and was later widely circulated in Nazi Germany. Grant decried the "wretched outcasts" passing through Ellis Island, "races that have for many thousands of years shown an utter incapacity to appreciate the ideas and ideals that have heretofore dominated this country." He had a special antipathy toward "that half-Asiatic welter of people we call Russians" and toward Jews in particular. Grant argued that virtually all major contributions to civilization had come from individuals of Nordic descent, and he claimed to be able to prove that Jesus Christ also had been Nordic. Anglo-Americans, in short, were the Chosen People.

It is important to emphasize that the Madison Grants of America were not, in the early years of the twentieth century, regarded as a lunatic right-wing fringe. Grant and Osborn were friends with Theodore Roosevelt, with whom they had founded the New York

Zoological Society, and they occasionally dined with the former president when he visited New York. The Boston founders of the Immigration Restriction League were also well-connected in Washington during the Roosevelt, Taft, and Wilson administrations. Beyond New York and Washington, the influence of the eugenicists could be seen in "Fitter Families" contests at state fairs, with awards given to "Grade A Individuals," and in a eugenics sermon contest sponsored by Protestant churches, in which Charles Davenport helped select the winning entry.

The eugenics movement growth was fueled by rapidly increasing immigration from eastern and southern Europe. By 1900 the number of immigrants had swelled to 300,000 per year, yet still they came so that in 1907 almost one million of them entered the United States. As one historian noted, "There was no ignoring the vast Jewish population of New York City, the grimy Slavic settlements that clustered in industrial areas of the East and Midwest, or the troops of Italian laborers hewing and toiling from coast to coast." When economic depression took place, the immigrants were accused of taking American jobs, and ugly racist incidents inevitably followed.

Prior to the 1890s the racial scapegoats in America had been primarily Negroes and Irishmen; indeed, a popular touring professor from England in 1881 suggested that America could solve its racial problems if each Irishman would kill a Negro and then be hanged for it. In the 1890s Italians and Jews also became common, and frequently deadly, targets for the racists. For example, in 1895 in southern Colorado coal miners slaughtered six Italians suspected of crimes, and the following year a mob took three more Italians from jail in a Louisiana town and hanged them. Anti-Semitic incidents became commonplace for the first time in America, especially in the South, where in 1893 "nightriders burned dozens of farmhouses belonging to Jewish landlords in southern Mississippi and open threats drove a substantial number of Jewish businessmen from Louisiana"; in the North "personal taunts and assaults . . . became much more common." By 1897 the persecution of immigrants in general had become more violent; when 150 Polish and Hungarian coal miners in Hazleton, Pennsylvania, attempted to march to a nearby town to organize a strike, the sheriff and his deputies opened fire, killing 21 immigrants and wounding 40 others.

Within this milieu Franz Boas began work in New York City in 1896. He experienced firsthand and saw the effects that the eugenics movement caused, so he set about the task of scientifically refuting the eugenicists' claims, specifically focusing on skull dimensions, which were being used by the eugenicists to divide the races. The measurement most commonly used was the cephalic index, derived by dividing the maximum width of a skull by its maximum length and then multiplying by 100; values of less than 80 were defined as long-headed and more than 80 as short-headed. The prevailing wisdom of the 1890s held that a person's skull dimensions were entirely genetic in origin, fixed at birth, and thus could be used to distinguish the long-headed racial groups of northern Europe from the short-headed races of southern Europe.

Boas's research on the cephalic index of the skulls of 239 Sioux Indians contradicted the prevailing wisdom. Published in 1899 in *American Anthropology*, Boas showed that there was a wide variation in cephalic indices within a single racial group, and he speculated that factors such as height altered the cephalic index. Since height is partly environmentally determined by factors such as nutrition, it follows that the cephalic index is partly environmental in origin. Boas concluded that "while the cephalic index is a convenient practical expression of the form of the head, it does not express any important anatomical relation."

As a Jewish professor in New York at the turn of the century, Franz Boas was ostracized by many of his colleagues. By directly challenging the eugenicists on their own scientific ground, Boas quickly became anathema. In addition to his Columbia post, he also held an appointment at the American Museum of Natural History, where Henry Osborn was a member of the board; following his 1899 paper, Boas found his research budget reduced. According to Boas's colleague A. L. Kroeber, "there was a conscious plan to supersede and oust Boas" from the museum; Boas's irascibility and stubbornness made the task easier. In June of 1906 the president of the board ordered that Boas be fired—and he was; two years later Osborn became president and appointed Madison Grant to be a member of the board. It was the first major skirmish between Boas and the eugenicists in a battle that would continue for more than 30 years.

The early phases of the next major confrontation were taking place in 1909 at the time Boas attended Freud's lectures at Clark University. In 1908 Congress had set up an Immigration Commission to ascertain whether the immigrants arriving in the United States were physically debasing the population; Boas applied to the commission and was given a grant. With these funds he hired 13 assistants and proceeded to measure the height, weight, and head sizes of almost 18,000 immigrants and their children; the largest group studied were Eastern European Jews. The results of his research, published in 1911 as *Changes in Bodily Form of Descendants of Immigrants*, claimed to "demonstrate the existence of a direct influence of environment upon the bodily form of man" by showing that height, weight, and head form were all different among the children of immigrants. Aiming his efforts directly at the eugenicists, Boas maintained that "the American-born Hebrew has a longer and narrower head than the European-born. . . . His face is narrower; stature and weight are increased [over his European-born counterparts]." Furthermore Boas wrote that such changes were progressive and appeared to increase the longer the Jew had been in the United States. That same year Boas published a compilation of his essays, *The Mind of Primitive Man*, which was later labeled as "a Magna Carta of race equality."

The results of Boas's work received wide publicity and caused apoplexy among the eugenicists. Madison Grant wrote to President Taft urging him to disregard Boas's "astonishing conclusions" and "amazing theory that the climate of America, and the subtle influence of American institutions, led to a modification of the skull form in immigrants. . . . This is too silly to discuss were it not for the fact that fallacies of this sort are given wide circulation by the public press." To Senator F. M. Simmons, Grant wrote that Boas

> made a most amazing report for the Immigration Commission, which is absolutely repudiated by scientific men. Dr. Boas, himself a Jew, in this matter represents a large body of Jewish immigrants who resent the suggestion that they do not belong to the white race. . . . He has produced certain alleged measurements, probably from selected examples, to prove that the skull shape, which is unchanged through thousands of years will change in a few years on the transfer of its owner from the Ghetto of some Polish city to the East Side of New York.

To another correspondent Grant warned that "current literature is being swamped by a mass of misleading articles emanating from a group of Jews, financed by those who are trying to encourage immigration. . . . This campaign is especially led by Boas. . . ." Grant further categorized Jews as "that great swamp of human misery" and theorized that they had adapted over time to the "filth and unsanitary surroundings" of the ghettos from which "the survivors perpetuated their superior adaptability to these conditions, and like rats have formed a race able to survive gutter conditions which quickly destroy higher types."

"Lo, the Poor Nordic!"

Despite Franz Boas's efforts, the swell of racism and anti-immigration continued to spread across America in the second decade of the century. In 1910 the existing Immigration Act was amended, making it illegal to admit as immigrants those who were criminals, paupers, and anarchists, and those who had various diseases.

Immigrants laboring in sweatshop conditions played leading roles in the successful factory strikes in 1912 and 1913 in Lawrence, Massachusetts, and Paterson, New Jersey; the latter was led by "Big" Bill Haywood. Members of the Industrial Workers of the World (IWW) spread inexorably from factory to factory, preaching socialism among immigrant workers as patrician families shuddered and poured money into the coffers of the Immigration Restriction League. The issue of immigration restriction became increasingly prominent in each succeeding presidential campaign.

One indication of the respectability of the eugenics movement at this time was the First International Congress of Eugenics held in London in 1912. American vice-presidents of the Congress included former Harvard University president Charles W. Eliot, Stanford University president David Starr Jordan, and Alexander Graham Bell, while one of the English vice-presidents was Winston Churchill. By 1914 many major universities were offering courses in eugenics, including M.I.T., Harvard, Columbia, Cornell, Brown, Wisconsin, and Northwestern. At Princeton an undergraduate named F. Scott Fitzgerald wrote a song titled "Love or Eugenics" for the annual Triangle Club show.

> *Men, which would you like to come and pour your tea?*
> *Kisses that set your heart aflame,*
> *Or love from a prophylactic dame.*

Another indication of increasing support for eugenics was the spread of compulsory sterilization laws. The first such law passed in Indiana in 1907, and by 1928 20 other states had followed. Many of the laws were written broadly, such as Iowa's, which mandated the "prevention of procreation [by] criminals, rapists, idiots, [the] feeble-minded, imbeciles, lunatics, drunkards, drug fiends, epileptics, syphilitics, moral and sexual perverts, and diseased and degenerate persons." Charles Davenport and his colleagues in the Eugenics Records Office drew up a model eugenics law for states to use and recommended including under compulsory sterilization individuals who were blind, deaf, crippled, and dependent ("orphans, ne'er-do-wells, the homeless, tramps, and paupers"). There seemed to be no limit to how far compulsory sterilization might go; in 1929 Missouri proposed legislation to include "chicken stealing [or] theft of automobile" as posssible grounds for sterilization, but the legislation was defeated. It has been estimated that approximately 8,500 individuals (half of them in California) were sterilized under compulsory laws by 1928.

Compulsory sterilization, restriction of immigration, and racism were all ingredients in the American eugenics movement as the nation moved toward participation in World War I. Nothing symbolized this convergence better than Madison Grant's *The Passing of the Great Race*. With a preface by Henry Osborn and a review in *Science*, which called it "a work of solid merit," the book quickly sold more than 16,000 copies and was purported to have marked a turning point in convincing Americans that immigrants in general, and Jews in particular, were indeed threats to their gene pool.

In the book Grant excoriated most immigrant groups, especially "the Polish Jew whose dwarf stature, peculiar mentality and ruthless concentration on self-interest are being engrafted upon the stock of the nation." He expressed grave concern about the effects of marriages between the races for "the cross between a white man and an Indian is an Indian; the cross between a white man and a Negro is a Negro; the cross between a white man and a Hindu is a Hindu; and the cross between any of the three European races and a Jew is a Jew. . . ." Grant specifically ridiculed Boas's work, which he summarized as follows: "A round skull Jew on his way across the Atlantic might and did have a round skull child but a few years later, in the response to the subtle elixir of American institutions as exemplified in an East Side tenement, might and did have a child whose skull was appreciably longer. . . ."

Grant's solution to this putative threat from inferior races was to restrict both their immigration and their procreation. Sterilization, he wrote, should "be applied to an ever widening circle of social discards, beginning always with the criminal, the diseased, and the insane, and extending gradually to types which may be called weaklings rather than defectives, and perhaps ultimately to worthless race types." Beyond the control of immigration and procreation, Grant suggested that other measures might be useful to counteract the threat to the Great Race:

> *Mistaken regard for what are believed to be divine laws and sentimental belief in the sanctity of human life tend to prevent both the elimina-*

tion of defective infants [infanticide] and the sterilization of such adults as are themselves of no value to the community. The laws of nature require the obliteration of the unfit, and human life is valuable only when it is of use to the community or race.

A decade after publication of *The Passing of the Great Race*, it had been translated into German, quoted by Hitler in *Mein Kampf*, and was a favorite reading for Hitler's supporters.

Racist incidents in the United States receded temporarily during World War I, but surged back immediately thereafter. The racism was fueled by postwar economic depression and renewed immigration, which increased so quickly, that by February 1921 the overload at Ellis Island had become too burdensome, prompting immigration authorities to hastily divert New York bound ships to Boston. Prohibition also contributed to racism because immigrants proved to be ready recruits as bootleggers. In New York it was said that "bootlegging was half Jewish, a quarter Italian and one-eighth each Polish and Irish"; immigrants therefore quickly became associated with lawlessness in the public eye.

The postwar racism wore many faces. The Ku Klux Klan, re-established in 1915, was estimated to have reached a peak of between three and five million members in the 1920s; they attacked Catholics, Jews, Negroes, and immigrants of all kinds, but especially immigrant bootleggers. In California the hatred against Japanese-Americans was so intense that a law was passed forbidding their ownership of land. In Alabama antipathy toward Catholics led to an official state convent-inspection commission to look for Protestant women being held against their will. And in West Frankfurt, Illinois, for three days in August 1920, "foreigners of all descriptions were beaten on sight. . . . The crowds burst into the Italian district, dragged cowering residents from their homes, clubbed and stoned them, and set fire to their dwellings." Five hundred state policemen were needed to quell the riot.

Of all the targets of racism in the 1920s, however, Jews were especially conspicuous. Constituting a significant percentage of the newly arriving immigrants, they gathered in ghettos in large eastern cities and provided prominent rationalizations for xenophobia. One of the most vitriolic attacks on them was Henry Ford's serialization of the specious *Protocols of the Elders of Zion* in his newspaper, the Dearborn *Independent*. The *Protocols*, purportedly written by the Czar's secret police to stimulate anti-Semitic outbursts in Russia, supposedly were the secret documents of Jewish elders planning to take over national governments by controlling the world's economy. Ford launched his attack on Jews in May 1920, calling them a "race of people that has no civilization to point to, no aspiring religion, no universal speech, no great achievement in any realm but the realm of 'get', cast out of every land that gave them hospitality. . . ." Ford was widely admired at the time, having been rated with Theodore Roosevelt and Thomas Edison among the five "Great Americans" in a poll by *American Magazine* and mentioned as a possible presidential candidate. The attack on Jews by leaders like Ford was implicitly condoned by official government reports, such as one by the House Committee on Immigration which called Jewish immigrants "abnormally twisted," "unassimilable, [and] filthy, un-American, and often dangerous in their habit."

One of the most striking aspects of racism in the 1920s in America was how widespread and middle class were its roots. It was not merely an American xenophobia confined to red-necked men in blue-collar jobs parading in white sheets; rather, it pervaded the local schools and barber shops on Main Street. Nothing symbolized this fact more clearly than the *Saturday Evening Post*, the most popular periodical of its era. In 1921 it began a series of editorials and articles attacking immigrants in general and Jewish immigrants in particular. The editorials, written by editor George Horace Lorimer, assailed "the mythical magical melting pot warranted to make Americans out of any racial scrap-humanity cast into it. . . ." Madison Grant's *The Passing of the Great Race* was praised as one of the books "every American should read if he wishes to understand the full gravity of our present immigration problem"; liberal portions of Grant's book were quoted, e.g., "New York is becoming a *cloaca gentium* which will produce many amazing racial hybrids and some ethnic horrors that will be beyond the powers of future anthropologists to unravel. . . ."

Lorimer's editorials were accompanied by long articles by Kenneth L. Roberts, a former army intelligence officer whom Lorimer had sent to Europe to investigate the background of immigrants coming to America. Roberts's articles, subsequently published in 1922 as a best-selling book, *Why Europe Leaves Home*, elevated racism to a new stature of respectability. Echoing Madison Grant on almost every page, Roberts described "streams of undersized, peculiar, alien people moving perpetually through consulates and steamship offices" and implied that these new Americans were more closely related to animals than to humans: "The place [the American Consulate Office in Warsaw] was a den of wild beasts, and the stench which rose from the struggling, squirming bodies was sickening." The new immigrants, averred Roberts, included "vermin-ridden women" and carriers of typhus, smallpox, and cholera. Furthermore, "nine-tenths of all the emigrants" coming from Europe "have no political principles or convictions and are entirely without patriotism." The remaining tenth who did have political convictions were said to be "Bolshevik sympathizers."

Roberts made clear that the influx of Jewish immigrants was the source of greatest concern. In 1880, he wrote, there had been just 3,000 Jews in America, "whereas in 1918 there were 1,500,000 Jews in New York City alone." The Polish Jews were said to be especially bad: "Even the most liberal-minded authorities on immigration state that the Jews of Poland are human parasites, living on one another and on their neighbors of other races by means which too often are under-handed, that they continue to exist in the same way after coming to America, and that they are therefore highly undesirable as immigrants." Such immigrants would inevitably mate with Americans of Nordic extraction, producing "mongrels." In passages lifted almost verbatim from Madison Grant, Roberts praised Nordic Americans' "long skulls and blond hair . . . [and their] ability to govern themselves and to govern others." All this was being lost, he warned, by the "mongrelization" of the nation by "parasites" who "never have been successful at governing themselves or at governing anyone else."

A regular reader of the *Saturday Evening Post* was Calvin Coolidge, the Republican governor of Massachusetts, who in 1920 was elected vice-president under Warren G. Harding. Harding had enunciated beliefs about "the dangers which lurk in racial differences" and the need for immigration restriction during his campaign speeches, but he was circumspect. Coolidge, by contrast, was openly racist in his beliefs and in early 1921 published an article in *Good Housekeeping* titled "Whose Country is This?" Coolidge warned that "our country must cease to be regarded as a dumping ground. . . . The Nordics propagate themselves successfully. With other races, the outcome shows deterioration on both sides."

Coolidge's support was an important reason why in May 1921 Congress passed an Immigration Act that established quotas of immigrants by country of origin for the first time in America. Its passage was also due to the persistent efforts of Secretary of Labor James J. Davis, who characterized the new immigrants as "rat-men" bent on tearing down an America which had been built by the "beaver type" Nordic immigrants. The most crucial supporter of the legislation was Republican Representative Albert Johnson, Chairman of the House

Immigration Committee, who corresponded and met regularly with Madison Grant, Charles Davenport, and Prescott Hall. In 1923 the Eugenics Research Association honored Johnson by electing him as its president.

Four months after the passage of the Immigration Act of 1921, racism achieved additional respectability at the Second International Congress of Eugenics. It was held at the American Museum of Natural History. As president of the museum Henry F. Osborn headed the congress and was assisted by Grant and Davenport and their eugenics colleagues. The congress's General Committee was studded with university presidents and community leaders, including Republican Secretary of Commerce Herbert Hoover. Osborn warned the attendees about "certain races from central Europe which have been coming to America in enormous numbers" and which stand "at the very bottom of the list" in intelligence. "We are engaged," he said, "in a serious struggle to maintain our historic republican institutions through barring the entrance of those who are unfit to share the duties and responsibilities of our well-founded government." As the eugenicists had learned to do, Osborn invoked science as the rationalization for racism: "As science has enlightened government in the prevention and spread of disease, it must also enlighten government in the prevention and multiplication of worthless members of society. . . ." Osborn publicly ridiculed Boas's work and his claims that environment could alter head form. In planning the congress, Osborn considered having a public debate on "Rights and Wrongs on the Racial Question" with Boas on one side and Madison Grant on the other; ultimately Osborn rejected the idea. The congress received widespread publicity both in newspapers and in *Science* magazine, and the scientific exhibits concerning immigration which were displayed during the congress were taken to Washington where they were mounted in the Capitol Building and displayed for three years.

Following passage of the Immigration Act of 1921 and the Second International Congress of Eugenics, proponents of terminating immigration to America altogether redoubled their efforts. They were assisted by the publication in 1922 of data on the Intelligence Quotient (IQ) of 1.7 million American soldiers that had been collected during the war and subsequently analyzed. In an article about the IQs in the *Atlantic Monthly* in 1922, Mrs. Cornelia J. Cannon, wife of noted Harvard physiologist Dr. Walter B. Cannon, claimed that "almost half of the white draft, 47.3 percent, would have been classed as morons." The highest percentage of these "inferior men," she said, were immigrants;

70 percent of Polish, 63 percent of Italian, and 60 percent of Russian men who took that test qualified as "morons." These men, warned Mrs. Cannon, "are persons who not only do not think, but are unable to think. . . . Such individuals form the material of unrest, the stuff of which mobs are made, the tools of demagogues. . . ." The following year psychologist Carl C. Brigham published *A Study of American Intelligence*, in which he utilized the army IQ data to show that immigrants from Nordic nations such as England and Scotland had high IQs while those from Russia, Poland, and Italy had an average mental age of approximately 11. An enthusiastic follower of Madison Grant, Brigham concluded that "the army tests indicate clearly the intellectual superiority of the Nordic race group."

The definitive rebuttal to Mrs. Cannon's article and other misuse of the army IQ data came from the pen of Walter Lippmann, the liberal editor of the *New Republic*. In a six-part series beginning on October 25, 1922, Lippmann deftly deflated the "nonsense" of the intelligence testers who had claimed that the average mental age of Americans was "only about fourteen." "The average adult intelligence cannot be less than the average adult intelligence," Lippmann wrote, and to claim otherwise "is precisely as silly as if [the person] had written that the average mile was three quarters of a mile long." Lippmann did not claim that there were no genetic antecedents to intelligence, merely that the IQ tests did not measure them. Aiming his remarks directly at Madison Grant et al., Lippmann said that the IQ test results were "in danger of gross perversion by muddleheaded and prejudiced men . . . [who are] exponents of the New Snobbery."

One can only imagine what Franz Boas, as he continued his fight against racism and eugenics, thought of these postwar developments. He had publicly criticized Grant's *The Passing of the Great Race* as "a dithyrambic praise of the blond, blue-eyed white and of his achievements" and dismissed the book as merely "the attempt to justify a prejudice." He had made extensive appeals for research funds to study the race question, writing to philanthropists like Carnegie and Rockefeller, but they gave financial support to the eugenicists instead. Boas had provided regular technical consultation to New York Congressman Emanuel Celler and other members of Congress who were opposed to further restrictions on immigration. He also testified in court cases involving racial discrimination, including one in which it was contended that Armenians were Mongoloid in origin and therefore could not, like other Asians, legally own property in California; Boas produced data on head measurements that resulted in the case being dismissed. Above all Boas wrote for scientific journals and, increasingly, for lay publications, warning against trying "to raise a race of supermen." Presciently, Boas predicted that "eugenics is not a panacea that will cure human ills; it is rather a dangerous sword that may turn its edge against those who rely on its strength."

Despite occasional victories, Boas's solitary fight against the eugenicists became more difficult with each passing year. In 1914 he was diagnosed as having facial cancer and underwent surgery, during which a major nerve was severed, leaving him with permanent paralysis of the left side of his face. Grant and Osborn followed these developments with great interest, with Grant speculating at one point, "I have rather suspected that the recurrence of the cancer is [a] camouflage." Boas survived but was next in trouble because, like many German-Americans, he publicly urged America to stay out of the war and narrowly avoided being fired by Columbia University for doing so. Following the war Boas wrote an intemperate letter publicly accusing the American government of having used anthropologists as spies against the Germans; the American Anthropological Association, of which he had been a founding member, expelled him from its Council and also forced his resignation from the National Research Council. Another problem was that Boas's own research on changing head form among immigrants was increasingly scrutinized. Some critics maintained that the claimed difference in head form could be explained if the immigrant women in his study had conceived their children illegitimately with long-headed Nordic men after they arrived in New York, a possibility which Boas asserted he had taken into account. Other critics questioned Boas's sampling methods; a Boas supporter in later years acknowledged that "there are some distinct curiosities in it."

Following the 1921 International Congress on Eugenics, held at the American Museum of Natural History, from which Boas also had been dismissed, animus increasingly moved to center stage in the public debate between Boas and the eugenicists. When Osborn gave a paper in 1923 claiming that IQ data had proven that Jews were not as intelligent as previously thought, "despite the misleading teachings of certain anthropologists," Boas responded in the *American Mercury*, accusing Osborn and Grant of being "swayed not by scientific arguments but by prejudice." In a maneuver guaranteed to enrage them, Boas compared the "uncouth barbarian" Nordics of the Middle Ages with the Maya Indians, who at that time were at the height of their civilization: "Would not the Maya have been justified in calling him [the Nordic] an inferior who would never achieve eminence? . . . Lo, the poor

Nordic!" Osborn immediately complained in a letter to the *New York Times* about Boas's "dubious comments about my own race," and Boas responded, also in the *New York Times*, ridiculing Osborn's scientific claims, which he labeled as "entirely in line with the vagaries of Madison Grant."

By early 1924, then, Franz Boas appeared to be winning the war of words but losing all other phases of the battle. Overt racism seemed to be everywhere on the rise, especially anti-Semitism; by the late 1920s, in New York City alone, it was estimated "that Jews were excluded from 90 percent of the jobs" in general office work. When Jews tried to move out of the ghettos they encountered closed doors and restrictive covenants. Compulsory sterilization laws continued to pass and be upheld by the courts, and the pressure to further restrict immigration was relentless. In 1924 the eugenicists achieved their biggest victory of all with passage of a new immigration restriction bill known as the Johnson-Reed Act. It went considerably further than the act of 1921 by banning Japanese immigrants altogether and changing the quota system in such a way that Italians and Jews were sharply discriminated against. In the first year after passage of the Johnson-Reed Act, Italian immigration to the United States fell 89 percent and Polish and Russian immigration decreased 83 percent. It was not a total exclusion of Italians and Jews, as the eugenicists had sought, but such a goal appeared to be within reach. In signing the bill President Coolidge reiterated his earlier pledge that "America must be kept American."

Franz Boas was 66, professionally isolated and deeply discouraged. He was certain that nurture held hegemony over nature but his cause seemed lost. Ironically it was Columbia University that had become one of the first universities to implement a quota system restricting the number of Jewish students; there was even a popular campus song illustrating the prejudice:

> *Oh, Harvard's run by millionaires,*
> *And Yale is run by booze,*
> *Cornell is run by farmer's sons,*
> *Columbia's run by Jews.*

The Freudianization of Nature and Nurture

The debate about "nature and nurture," the "convenient jingle of words" which Francis Galton had adopted as a shorthand for competing theories of human behavior, was on its surface a scientific debate. The eugenicists talked about chromosomes, genes, and how Mendel's laws on heredity made inevitable the transmittal of various physical and personality traits. Those supporting nurture replied with illustrations of how nutrition, culture, poverty, and social conditions could modify these same physical and personality traits. Reading the charges and countercharges in scientific and lay publications in the early years of this century, one might assume that the imbroglio could have been settled by a few well-formulated scientific experiments in a neutral laboratory.

Such an assumption would have been wrong, for the nature–nurture controversy had become much more than a scientific debate by the end of World War I. The hereditarians' belief that genes are the crucial determinants of human behavior had been connected to a belief in the inequality of humankind. This included political inequity, a conclusion that Galton himself had clearly expressed. "The average citizen," said Galton, "is too base for the everyday work of modern civilization" and all citizens are therefore not "equally capable of voting." The inequality also carried with it social consequences which Galton made explicit: Lower-class citizens should be treated with kindness only "so long as they maintained celibacy," but if they "continued to procreate children, inferior in moral, intellectual and physical qualities, it is easy to believe that the time may come when such persons would be considered as enemies to the State and to have forfeited all claims to kindness."

Many of Galton's followers in America shared his political and social biases. Prescott Hall ridiculed the "lust for equality" and the "fatuous belief in universal suffrage"; the archives of his Immigration Restriction League suggest that Hall's campaign to restrict the immigration of "inferior races" was intimately tied to his belief in government by an aristocracy of educated men.

Henry Osborn, in a similar vein, wrote, "The true spirit of American democracy that all men are born with equal rights and duties has been confused with the political sophistry that all men are born with equal character and ability to govern themselves and others. . . ."

But it was left to Madison Grant, as usual, to publicly articulate the political and social prejudices which lay beneath the surface of the nature–nurture debate and that most of his colleagues shared privately. In *The Passing of the Great Race* and in an article published three years later, Grant clearly explicated what he thought of the American inclination "to bend the knee in servile adulation to the great god Demos":

"In the democratic forms of government the operation of universal suffrage tends toward the selection of the average man for public office rather than the man qualified by birth, education and integrity." This leads, wrote Grant, to "the conduct of public affairs by the most incompetent members of the community." "In America," Grant continued, "we have nearly succeeded in destroying the privilege of birth. . . . We are now engaged in destroying the privilege of wealth. . . ." Grant had nothing but contempt for the abilities of the average man: "*Vox populi*, so far from being *Vox Dei*, thus becomes an unending wail for rights and never a chant for duty." Instead of democracy Grant urged a return to aristocracy. "True aristocracy," he claimed, "is government by the wisest and best, always a small minority in population."

On the other side of the debate Franz Boas and his sympathizers were strongly associated with liberal, and in many cases radical, political and social positions. Boas himself was a Socialist, who supported Eugene Debs and later Norman Thomas. In the 1930s Boas labeled Russia as "the greatest experiment that has ever been made in Social Science with a high ideal in view"; the ideal, he said, was "equal rights for every member of humanity." Boas joined organizations such as the American Committee for Democracy and Intellectual Freedom, which were later alleged to have been Communist fronts, and the FBI opened a file on Boas as a suspected Communist. Boas's daughter later staunchly rejected such a possibility, saying that her father "was no more a Communist than [his dog] Muffin."

Boas's openly pacifist stance in World War I also linked him to leftist groups. Following the passage of the Selective Service Act by Congress in May 1917, those opposing entry of the United States into the war exhorted others to refuse the draft. Publication of the *Masses* was stopped by the government, and its editors—Max Eastman, Floyd Dell, and John Reed—were brought to trial on charges of sedition. (Even a writer for the *Masses* had been included in the indictment on the basis of a poem she had written praising Emma Goldman. At the opening of the trial a defense attorney "handed the poem to the judge claiming it contained nothing illegal. His Honor read it and handing it back to [the defense attorney] demanded: 'Do you call that a poem?' 'It's so called in the indictment, your Honor,' replied [the defense attorney]. 'Indictment quashed,' declared the judge.")

If there were light moments in the fight between the government and political leftists during World War I, it was also a serious—sometimes deadly—business. Senator Thomas R. Hardwick, who strongly supported restricting immigration to keep out Bolshevists in his capacity as Chairman of the Senate Immigration Committee, was sent a package bomb in April 1919, which blew the hands off a maid who opened it. Similar bombs were sent to Commissioner of Immigration Caminetti, to J. P. Morgan, to John D. Rockefeller, and to other prominent capitalists and government officials. The following month a bomb exploded outside the Washington, D.C., house of A. Mitchell Palmer, the United States Attorney-General. This triggered the wholesale arrest of over six thousand suspected Socialists, Communists, and anarchists in the early days of 1920, a crackdown which was coordinated by a newly hired lawyer in the Department of Justice, J. Edgar Hoover. In Detroit more than one hundred of them "were herded into a bull-pen measuring twenty-four by thirty feet and kept there for a week." In Hartford, Connecticut, "authorities took the further precaution of arresting and incarcerating all visitors who came to see [those who had been arrested], a friendly call being regarded as *prima facie* evidence of affiliation with the Communist party." Three months later Nicola Sacco and Bartolomeo Vanzetti, reputed anarchists, were arrested for two murders committed during a payroll robbery.

Franz Boas was not arrested, but many individuals who had been sympathetic to his fight were. Emma Goldman was imprisoned for two years following her conviction for sedition, then deported to Russia. She was joined there by "Big" Bill Haywood and many leaders of the labor movement as well as by many liberal intellectuals. The images of being foreign-born, Jewish, pacifist, or politically to the left became merged in popular thought as threatening; this did not help Franz Boas, who was all of these things. Boas had championed the immigrants while the eugenicists had called them a genetic menace; political events seemed to be proving them correct. The magnitude of this immigrant, Jewish, pacifist, leftist menace was further demonstrated when Eugene Debs, the Socialist Party candidate for president, was arrested for publicly praising pacifists, labor leaders, and Bolsheviks, and for calling American patriotism "the last refuge of a scoundrel"; campaigning against Warren Harding and James M. Cox from his prison cell, Debs received almost one million votes (3.5 percent of the total) for president in November 1920.

The nature-nurture debate, then, had become highly politicized. The real argument was not about what is inherited but rather who is superior. It was not about dominant and recessive genes but rather dominant and recessive races. In its starkest form the debate pitted the aristocracy against the proletariat, tradition against new customs, and wealth against poverty. Politically, advocates for nature tended to be nationalistic, conservative, and in favor of aristocratic forms of government; the far right supported fascism while those more moderate were Republicans, the party which was leading the fight for a restriction of immigration. Advocates for nurture were more likely to be pacifists, liberal, and supporters of democratic forms of government; the far left supported socialism or communism, while those more moderate were Democrats, the party of the immigrants.

Although it was not often publicly discussed, there was also a sexual aspect to the nature-nurture debate that would prove in the long run to be important for the disseminaton of Freud's theory in America. Many of the leading eugenicists were associated with conservative sexual mores. Henry Osborn, for example, opposed the dissemination of birth control devices as "fundamentally unnatural" and "connected with a great deal of sexual promiscuity." Charles Davenport also "deplored birth control"; he had been "sexually continent before his marriage" and "bridled at the merest hint of sexual indulgence." Madison Grant left no record of his beliefs about sex, but it is known that he was a lifelong bachelor and apparently sired no children despite his rhetoric about the importance of breeding by members of the aristrocracy. Thus the forces of nature in the nature-nurture debate were associated with sexual continence as well as with conservative politics.

Supporters of nurture, however, were associated with sexual as well as political liberalism. Boas was linked to Freud's theory, while his uncle, Dr. Abraham Jacobi, with whom he continued to spend summer vacations, was one of Margaret Sanger's earliest supporters. Walter Lippmann alternately praised Freud and criticized the eugenicists in the pages of the *New Republic*. By contrast, the *Saturday Evening Post*, which held the eugenicists in such high esteem, ran features ridiculing Freud. Prominent supporters of Freud, such as Max Eastman, Floyd Dell, and Emma Goldman, preached and practiced a public brand of sexual freedom as well as left-wing politics. In short the nature-nurture debate became quietly "Freudianized" as it also became politicized. Freud's theory about the importance of childhood experiences merged easily with other theories about nurture. In 1924, as interest in Freud was starting to dwindle in America, Freud's involvement in the nature-nurture debate was of little consequence since he was then associated with the losing side. But that would not always be the case.

3

The Sexual Politics of
Ruth Benedict and Margaret Mead

———

There is only one problem in life: that fire upon our flesh
shall burn as a knife that cuts to the bone, and joy strip
us like a naked blade. There are no other problems.
We live instead, our passions we fling at the sure lies
and trumperies if perhaps by these we may buy one
burning moment.
—RUTH BENEDICT, FROM A PERSONAL JOURNAL

*R*uth Benedict and Margaret Mead were not the first students of
Franz Boas to combine an interest in Freud with advocacy efforts on
behalf of nurture in the nature-nature debate. Alfred Kroeber, Boas's
first doctoral student, underwent psychoanalysis and practiced as a lay
analyst for two years at the same time as he was promoting culture as
the most important antecedent of human behavior and attacking the
eugenicists' views. Indeed, as Marvin Harris noted in *The Rise of
Anthropological Theory*, "During the 1920s anthropologists and psychoanalysts were natural allies in the intellectual revolt against the con-
straints of sexual and other forms of provincialism." Freud focused on
early childhood experiences, while anthropologists promoted cultural
influences more broadly defined; they united in opposition to the
eugenicists' scientific, social, and political beliefs.

It remained for Benedict and Mead, however, to demonstrate just how effective Boas's students could be in this fight. Benedict was slight of stature, prematurely graying, partially deaf, and affected by a stammer when she began graduate studies under Boas in the spring of 1921. She was 34, married to a research chemist, had taught school, and was already verbalizing a strong conviction that culture, not genes, was the molding force of human behavior. Perhaps for this reason Boas waived his course entrance requirements and accepted her as a doctoral student in anthropology.

Benedict was at this time already interested in Freud as well as in anthropology. She had been introduced to both by Elsie Clews Parsons, from whom Benedict had taken a course, "Sex in Ethnology," at the New School for Social Research in 1919. Parsons, one of Boas's former students, had contributed articles to the *Psychoanalytic Review* as early as 1915, and the following year had been the first anthropologist to discuss Freud in the *American Anthropologist*. As the wife of a former congressman and a friend of Margaret Sanger, Mabel Dodge, and Walter Lippmann, Parsons was "well known for her shockingly unconventional ideas," her impromptu episodes of nude swimming at decorous New England resorts, and her argument that "sex relations should be entirely free from any regulations by society." Then, said Parsons, "passionate love will forget its shameful centuries of degradation to spread its wings into those spaces whereof its poets sing."

There is no evidence that Ruth Benedict ever undertook a personal psychoanalysis, but she utilized Freudian concepts liberally in her writings. In later years she co-led a seminar at the New York Psychoanalytic Institute with analyst Abraham Kardiner, and she also taught at the Washington School of Psychiatry. Benedict was good friends with Harry Stack Sullivan and Karen Horney and influenced both in their amalgamation of cultural and Freudian theories in what would emerge as neo-Freudian psychoanalysis. Culture was the paramount force in Ruth Benedict's worldview, but culture included the effects of early child rearing as had been described by Freud.

Benedict's conviction that culture, not genes, determines behavior arose in part from her year's work with a social services agency in Buffalo following her graduation from Vassar College. She had visited Polish immigrant families in their homes and had seen for herself that their problems were economic, social, and linguistic in origin, not genetic. In her first year as a graduate student under Boas, she argued so forcefully for culture's influence that a colleague asked her whether she had become "a kind of cultural fatalist" in believing that culture was the sole determinant of behavior. Two years later Benedict wrote that man's "responses have been conditioned from birth by the character of the culture into which he was born." "The fundamental question . . . to which the labors of anthropology are directed," added Benedict, "is in how far the forces at work in civilization are cultural, and in how far organic or due to heredity; what is due to nurture, in the rhyming phrase, and what to nature." This was the question on which, above all, Boas wished for definitive answers.

Ruth Benedict shared her mentor's social agenda and opposition to racism, although she did not enter into the public fray until years later. Benedict was, in the words of an anthropologist colleague, a "world improver." In 1940, at Boas's suggestion, she published a strong attack on racism, *Race: Science and Politics*, in which she asked, "How can we stop this epidemic of racism?" Her answer was that racism was a product of inequitable social conditions and would decrease only when "all men shall have the basic opportunity to work and to earn a living wage, that education and health and decent shelter shall be available to all. . . . Until the regulation of industry has enforced the practice of social responsibility, there will be exploitation of the most helpless racial groups." Both Boas and Benedict believed that government should play a major role in ameliorating social and environmental factors which produce racism. Whereas Boas supported Socialist political candidates, Benedict was a strong supporter of Roosevelt's New Deal. Benedict joined Boas in the 1930s in supporting organizations that were later accused of being Communist fronts. Margaret Mead said that Benedict "knew that some of the people with whom she came in contact were Communists, but she refused to believe that what these doctrinaires advocated needed to be taken seriously."

The influence of Franz Boas on Ruth Benedict was profound. Eighteen years after their initial meeting Benedict wrote him: "As for me, there has never been a time since I've known you that I have not thanked God all the time that you existed and that I knew you. I can't tell you what a place you fill in my life." She was not put off, as many were, by Boas's formidable demeanor, his "coal black and piercing" eyes beneath shaggy eyebrows, his dueling scars accentuated by partial facial paralysis from his surgery, or by his autocratic methods. Benedict instead admired Boas as the leader of society's oppressed and as a fighter against racism. According to Judith Modell, one of Benedict's biographers, "For Ruth, above all Franz Boas represented the scientist motivated by a lasting faith in human capacity to create a better world."

Ruth Benedict's interest in Freudian theory and cultural anthropology, then, were closely tied to her social and political agendas, which she shared with Franz Boas. It is also clear that Benedict's interests in these areas were tied to a personal agenda that she did not share with Boas. As early as her college years at Vassar she "apparently had a quiet crush on at least one woman faculty member." Benedict had married, but it had been a barren and loveless marriage, at least in part because of her own frigidity. Benedict's personal diaries and writings reflect a lyrical and intense preoccupation with "that fire upon our flesh," a persistent sexual passion that seemed unattainable:

> *We cast about always searching with what coin we shall purchase it.*
> *Some by incest have broken the crust and travelled to this earth-core of*
> *aliveness; some have made pain the bedfellow of their imagination and*
> *licked blood from the welts they have laid upon the flesh of their most*
> *loved. Some have laid hold on fear; in other generations men feared*
> *hell and quivered, and they feared death too before scales fell from their*
> *eyes and they saw that within is death a bogey. But we, caught*
> *between the accident of birth and desirability of death, how shall we set*
> *stars in heaven? We shall wither to our death as in a dream, knowing*
> *all things are folly, phantoms, shadows.*

Because of these longings she was an outsider, a misfit to both her body and her culture. According to Modell, who reviewed her personal diaries during the years of her marriage, "The more she thought about her lack of fit to the female role, the more she seemed drawn to considering the maleness in herself." Ruth Benedict was determined to prove that her lack of conformity had been imposed on her by her culture and was not due to some genetic or biological defect inherent in herself.

Ruth Benedict's sexual agenda recurred repeatedly in her anthropological writings. The course she had taken from Elsie Clews Parsons on sex in ethnology had emphasized that homosexuality is determined by cultural and childhood experiences, not biological factors. The Plains Indians, for example, were said to honor homosexual and transvestite men with special roles as healers. To define homosexuality as aberrant, Benedict contended, was culturally arbitrary, and she enlarged upon this theme in later writings:

> *The relativity of normality is not an academic issue. In the first place,*
> *it suggests that the apparent weakness of the aberrant is most often and*
> *in great measure illusory. It springs not from the fact that he is lacking*
> *in necessary vigor, but that he is an individual upon whom that culture*

has put more than the usual strain. His inability to adapt himself to society is a reflection of the fact that that adaptation involves a conflict in him that it does not in the so-called normal.

"The etiology of homosexuality," concluded Benedict, "is overwhelmingly social." Homosexuality is merely one of "our culturally discarded traits," which other cultures may value. "A tendency toward this trait [homosexuality] in our culture exposes an individual to all the conflicts to which all aberrants are always exposed, and we tend to identify the consequences of this conflict with homosexuality. But these consequences are obviously local and cultural."

Benedict's preoccupation with the cultural role of aberrant individuals in general and homosexuals in particular was a major recurring theme in her writings. As early as 1925 her friend Edward Sapir, a noted linguist and anthropologist, chided her for her obsession: "There is something cruel, Ruth, in your mad love of psychic irregularities. Do you not feel that you extract your liveliness from a mutely resisting Nature who will have her terrible revenge?" And Margaret Mead observed of Benedict: "She used to wonder whether she would have been happier had she been born in a different time or place, perhaps ancient Egypt. In our discussions we adopted the word 'deviant' for the individual who is a cultural misfit. . . ." Ruth Benedict felt herself to be a cultural misfit throughout her life, and she held her culture to blame.

After she began a homosexual relationship with Margaret Mead, Benedict eventually divorced her husband and then lived in exclusively homosexual relationships for the remainder of her life, most prominently with research chemist Natalie Raymond and later with psychologist Ruth Valentine.

Margaret Mead

Ruth Benedict had been working with Franz Boas for just over a year when she met Margaret Mead in September of 1922. At the time Mead was a senior at Barnard College and had enrolled in Boas's anthropology course in which Benedict was a teaching assistant.

Despite a 14-year difference in their ages, the two women had an extraordinary range of common interests. Both had small town, Protestant upbringings. Both were eldest children with younger sisters who were considered to be much prettier than themselves. Both had grandmothers whom they worshiped, mothers to whom they were indifferent, and fathers who were either distant (Mead's) or dead (Benedict's). Both women were aspiring writers, who exchanged poetry and stories; years later Mead cited Benedict's "exquisite responsiveness to literature" as one of her finest traits. And both revered Franz Boas, who, Mead recalled, "spoke with an authority and a distinction greater than I had ever met in a teacher." A colleague later described Boas as "very much a father figure" to both Mead and Benedict who "were his little girls until the end of their lives. . . . Their activity was messianic; their duty was to transmit the message." Boas was in fact the same age as Benedict's father would have been had he not died when she was a small child.

More than either Franz Boas or Ruth Benedict, Mead was an adherent of Sigmund Freud's theory. At Barnard one of Mead's roommates had undergone psychoanalysis, and Mead had become familiar with both Freudian and Jungian concepts. Mead had an intense interest in dream interpretation, which began in college and lasted throughout her life; as biographer Jane Howard noted, "Mead considered a night without a dream a total waste of time." Although, like Benedict, she did not undergo psychoanalysis herself, she contributed many articles to psychoanalytic journals, and in her book *Growing Up in New Guinea* (1930) she cited psychoanalytic theory as providing "perhaps the most fruitful attacks" upon traditional ideas regarding children's development. In *Male and Female* (1949) Mead wrote that "the solution to the Oedipus situation will depend [upon] a great deal of the way in which a boy or a girl accepts primary sex membership." In later years Mead taught at the Menninger Clinic and other bastions of Freudian orthodoxy and was described by one psychoanalyst as having a "command of psychoanalytic concepts . . . the equal of any professional's I ever knew." Mead did not say that psychoanalysis was the *only* way to achieve insight into human behavior, but it was high on her list, a list which also included doing fieldwork among primitive people.

Margaret Mead also shared the social and political agendas of Franz Boas and Ruth Benedict. At Barnard Mead belonged to a group which called itself the Ash Can Cats after the Ash Can school of painters, a school which combined realism with Socialist and anarchist causes. Mead participated in a mass meeting in support for Sacco and Vanzetti, stuffed envelopes for the Amalgated Clothing Workers Union, walked on picket lines, took a course on the labor movement, and was intrigued with the economic theory of dialectical materialism. On the fifth anniversary of the Russian Revolution Mead went to the college dining hall with her friends "wearing red dresses, sat down among red

flags, red flowers and red candles, and burst into the 'Internationale.' "
So pronounced were her leftist proclivities that anthropologist Melville
Herskovits, in a 1923 letter, expressed surprise that Mead had been
allowed to join Phi Beta Kappa despite her public support of the Communists. Like Benedict, Mead admired Franz Boas for his dedication to
social causes: "He was not working for personal power," she observed,
"but for the good of mankind."

Of all the bonds between Margaret Mead and Ruth Benedict,
however, one of the strongest was certainly sexual. Mead, like Benedict, felt herself to be "special and different" as a schoolgirl: "I had
wondered occasionally what was the basis of my own apparent
deviance from the accepted style of the career-minded women I had
met." Mead also believed that "I had my father's mind, but he had his
mother's mind"; in later years friends sometimes noted her masculine
qualities, but "she abhorred being called 'manlike' and prided herself
on her femininity."

Evidence is suggestive that Margaret Mead had multiple affairs
with women while she was still at Barnard. Marie Eichelberger, seven
years her senior, was one such woman; in later years Mead is recorded
as having said to Eichelberger, "You just looked at me across the room
and fell in love with me." The relationship continued throughout their
lives, with brief periods of cohabitation and with Eichelberger acting as
Mead's personal secretary. According to Jane Howard, "Once a man
wanted to marry Marie, but compared with Margaret no man had a
chance with her." Love letters to Mead from Leah Hanna, another
Barnard student, also suggest that during Mead's senior year theirs was
a sexual relationship. Mead appears to have been frequently attracted to
older women, with one classmate recalling that "Margaret fell for my
mother, and my mother in turn admired her very much. . . ."

Although Margaret Mead appears to have been bisexual, many
friends noted her comparatively lesser interest in men. And yet despite
her affairs with women, during her senior year at Barnard Mead
became engaged to Luther Cressman, a seminary student. Cressman
was active in social causes and in later years left the ministry to become
an anthropologist. Mead was characteristically attracted to men because
of their ideas; letters in Mead's private correspondence in the Library
of Congress, which recently became available to the public, suggest
that her later marriages to Reo Fortune and Gregory Bateson were
viewed by her primarily as professional liaisons to accomplish field
work and to promote anthropology.

Within six months of their initial meeting, Benedict and Mead were seeing each other regularly. Benedict, still married, rented a room near the Columbia campus and began commuting to her husband in Westchester only on weekends. In March 1923 Benedict wrote in her diary that Mead "rests me like a padded chair and a fireplace." Other entries in Benedict's diary during this period included reflections such as, "When touch seems such a sweet and natural human delight I resent sorting it out even in favor of my dearest dream of achieving some sort of dignity in living." According to Modell, Margaret Mead's was a "perfect friendship . . . free of the 'force of custom,' a friendship between women. . . . She had begun expanding the expression of her own sensuality when Mead came, a confident, enthusiastic, open-minded woman. The resulting intimacy was not legislated and therefore not limited. . . . She had achieved her valued 'permanency' with Margaret Mead. The permanency rested on all manner of exchange, from caresses to gossip, from recipes to psychological theories . . . [and] buying a new dress."

The relationship between Ruth Benedict and Margaret Mead continued from 1922 until Benedict's death in 1948. Correspondence between them, both published and unpublished, suggests that it was a tender, loving, and extremely important relationship for them both. They lived together only once, in the summer of 1928, but continued intermittently as lovers throughout the years of Mead's marriages to Luther Cressman, Reo Fortune, and Gregory Bateson. Such an arrangement entailed complications. For example, when Mead and Luther Cressman married in 1923 the newlyweds went directly to visit Benedict and her husband at their summer house in New Hampshire. According to Cressman, Mead was frigid for the period of their visit and the two slept in separate bedrooms for the duration of their honeymoon. Many years later Mary C. Bateson, the daughter of Mead and Gregory Bateson, reflected on her mother's bisexuality:

> *Ruth and Gregory were the two people she loved most fully and abidingly, exploring all the possibilities of personal and intellectual closeness. . . . After Margaret's death I asked my father how he had felt about the idea of Margaret and Ruth as lovers, a relationship that had begun before Margaret and Gregory met, and continued into the years of their marriage. He spoke of Ruth as his senior, someone for whom he had great respect and always a sense of distance, and of her remote beauty.*

Like Benedict, Margaret Mead blamed her culture for causing her to feel like a deviant. Although she never publicly acknowledged her personal situation, Mead wrote in later years that "bisexual potentialities are normal" and urged that "we must recognize bisexuality as a normal form of human behavior." Mead argued that "a very large number of human beings—probably a majority—are bisexual in their potential capacity for love" and added that "the individual who is wholly incapable of a homosexual response has failed to develop one human potentiality." But, she said, culture shapes behavior, and "whether they will become exclusively heterosexual or exclusively homosexual for all their lives and in all circumstances or whether they will be able to enter into sexual and love relationships with members of both sexes is, in fact, a consequence of the way they have been brought up, of the particular beliefs and prejudices of the society they live in and, to some extent, of their own life history." Mead's sexual agenda was never far from the surface in her writings, and she regularly expressed concern about "the process by which an identification with a parent of the wrong sex can produce inverted sex attitudes."

In the closing years of the 20th century, when homosexuals demonstrate publicly for rights and identify themselves candidly, it may be unclear as to how homosexuals were treated in the opening years of the century. Religiously, homosexuality was considered to be an abomination against nature, a mortal sin. Medically, it was thought to be a kind of insanity, and in some places castration or clitoridectomy were still used for "treatment." Legally, homosexuality was a felony and one could be sentenced to life imprisonment for sodomy; police regularly raided Turkish baths and other common gathering places for gays in major cities. And socially, it was regarded as an indescribable perversion, with suicide a common outcome if one was exposed. New York City was liberal by national standards and at Barnard the dean "and several prominent professors were generally assumed to be homosexual." But outside the college gates self-righteous morality reigned; when *The Captive*, a play hinting at a lesbian relationship, opened in 1926 on Broadway, the *New York Times* called it "a revolting theme," and the police closed it down.

Ruth Benedict and Margaret Mead, then, had been implicitly defined as deviants by their culture. Just as Benedict wrote about "the conflicts to which aberrants are always exposed," Mead wrote that "the burden of nonconformity with the attendant sense of sin or guilt, puts a very heavy pressure on the very large proportion of Americans who deviate from the recognized patterns of temperament and behavior." The women shared the bond of being cultural misfits, and the magnitude of their guilt was interpreted by them as a reflection of the power of culture to shape human behavior. Being a misfit also carried with it responsibilities: Mead wrote that "upon the gifted among the misfits lies the burden of building new worlds." Benedict and Mead would spend the remainder of their lives documenting the strength of culture, including its Freudian aspects, and anthropology would be their workshop.

There is one additional aspect of Margaret Mead's thought that is important to understand her. Mead, like Freud, was a firm believer in the occult, and she wrote sympathetically of auras, communication with plants, telepathy, ESP, clairvoyance, and horoscopes; in 1943 she became a trustee of the American Society for Psychical Research. She even stated categorically that "there *are* unidentified flying objects" and speculated that "they are simply watching what we are up to—that a responsible society outside our solar system is keeping an eye on us to see that we don't set in motion a chain reaction that might have repercussions for outside our solar system." Mead believed that she herself was personally accompanied by two "spirit guides," and she visited mediums for advice. For example, immediately prior to Mead's marriage to Gregory Bateson, Ruth Benedict arranged for Mead to visit a medium; the woman advised Mead to "feed [Bateson] chicken." And although Mead's interest in the occult aligned her more closely to Freud, such thinking was antithetical to Franz Boas, whose research methodology was firmly grounded in the physical sciences. Mead described her own approach to science as one of "disciplined subjectivity," and she was regularly criticized in later years for her failure to use scientific methodology in her work.

Benedict and Mead Take Up the Fight

Beginning in the fall of 1923 Margaret Mead joined Ruth Benedict as a graduate student working under Franz Boas. Henry Osborn and Boas were trading verbal swings with increasing intensity. The level of Mead's interest in these events can be gauged by the fact that she told Luther Cressman, whom she had just married, that she needed a separate bedroom on their honeymoon in order to study Carl C. Brigham's *A Study of American Intelligence*, which had just been published. A perusal of Mead's correspondence from this period shows occasional references to the nature-nurture debate.

In the spring of 1923, with Benedict's encouragement, Mead undertook her first study in support of Boas even before beginning her graduate work. In order to complete her college requirements for a degree in psychology, Mead had to carry out field research. She chose to analyze the IQ scores of children of Italian immigrants to determine how much their IQ was dependent on whether English or Italian was the predominant language spoken at home. Years later she acknowledged that "the problem on which I was working had been suggested by Boas." "Those were the days," she added, "when [Congress was] passing the immigration act saying that Italians could never learn—look at their I.Q.'s—and all the things that we say about black people today."

Mead's fieldwork and thesis, "Intelligence Tests of Italian and American Children," did indeed show that "the home language was related to the I.Q. level attained; the scores of those speaking Italian were lower." Mead concluded that the lower scores were due to the children having been exposed to less English and their lack of experience with examinations and urged "extreme caution in any attempt to draw conclusions concerning the relative intelligence of different racial or nationality groups on the basis of tests, unless a careful consideration is given the factors of language, education, and social status, and a further allowance is made for an unknown amount of influence which may be logically attributed to different attitudes and different habits of thought." Nurture, concluded Mead, not nature, was responsible for the differences in IQ. Boas referred to Mead's findings in a 1925 article, entitled "This Nordic Nonsense," in which he argued that lower IQ scores among immigrant groups could be explained by their lesser "degree of assimilation to American conditions, and particularly in regard to the acquisition of the English language."

Passage of the immigration restriction bill in 1924 was a major victory for the eugenicists as Benedict and Mead pursued their graduate studies. Boas was desperately in need of data with which to refute the hereditarian point of view. When Mead proposed going to Polynesia for her graduate fieldwork to study "new and old elements of culture," Boas asked her instead to study adolescence among American Indians. Boas was specifically interested in how much adolescent turmoil was due to biological factors, as claimed by the eugenicists, and how much it was due to cultural factors. Boas, said Mead, was "always tailoring a particular piece of research to the exigencies of theoretical priorities . . ."

In the end Boas and Mead compromised: she would go to Polynesia but study the problem he wanted studied. American Samoa was selected, and in the spring and summer of 1925 Mead prepared for a field stay of up to one year. It was a busy time, and she was not able to begin Samoan language lessons prior to her departure.

As she prepared to leave for Samoa in the summer of 1925, Margaret Mead's bisexuality continued to complicate her personal life. She was married to Luther Cressman but was also carrying on a serious affair with Edward Sapir. Sapir had been one of the original anthropological converts to psychoanalysis, noting as early as 1917 "the far-reaching importance of infantile psychic experiences in adult life" and calling Freud's work "perhaps the greatest fructification that the study of the mind has yet experienced." Sapir pressured Mead to divorce Cressman and marry him, even writing to Boas urging him not to allow Mead to go to Samoa.

At the same time Mead apparently was continuing her friendship with Marie Eichelberger. And there was also Ruth Benedict, with whom Mead traveled to the Grand Canyon on her way to the Pacific. According to Mead's daughter, "Margaret described to Marie [Eichelberger] the way in which they [Mead and Benedict] talked, sitting and overlooking the Grand Canyon" and "decided that neither of them would choose further intimacy with Sapir, but rather preferred each other." Mead was apparently faced with many choices about her personal life, but they could be temporarily deferred during her stay in Samoa.

The product of Margaret Mead's nine-month fieldwork was *Coming of Age in Samoa* (1928), the best-known study of another culture ever carried out and one which has been read by virtually every college student who has taken a course in anthropology. Its idyllic descriptions of Samoan adolescent courtship and sex brought it immediate attention. According to Mead:

> [A]dolescence represented no period of crisis or stress, but was instead an orderly developing of a set of slowly maturing interests and activities. The girls' minds were perplexed by no conflicts, troubled by no philosophical queries, beset by no remote ambitions. To live as a girl with many lovers as long as possible and then to marry in one's own village, near one's own relatives and to have many children, these were uniform and satisfying ambitions.

Mead concluded "that adolescence is not necessarily a time of stress and strain, but that cultural conditions make it so."

This conclusion was exactly what Franz Boas wished to hear and he applauded her efforts. In his foreword to the book Boas noted, "The results of her painstaking investigation confirm the suspicion long held by anthropologists, that much of what we ascribe to human nature is no more than a reaction to the restraints put upon us by our civilisation." In an essay published that same year Boas claimed that "with the freedom of sexual life . . . the adolescent crisis disappears" and "where sexual life is practically free sexual crimes do not occur." By 1934 Boas was citing Mead in support of his contention that "the study of cultural forms" had proven "genetic elements" to be "altogether irrelevant" in determining personality. Margaret Mead had faithfully served her mentor toward whom she felt a "heavy sense of responsibility." In Samoa, Mead even kept a picture of Boas hanging on the wall of her room, occasionally decorating it with "a huge red hibiscus."

But Margaret Mead also served a personal agenda with her field-work in Samoa. Homosexual relations among girls did occur, she asserted, but were of little concern because they were culturally sanctioned:

> These casual homosexual relations between girls never assumed any long-time importance. On the part of growing girls or women who were working together they were regarded as a pleasant and natural diversion, just tinged with the salacious.

Both heterosexual and homosexual relationships were so easily accepted, wrote Mead, that neither sexual problems nor neuroses occurred in Samoa.

> Familiarity with sex, and the recognition of a need of a technique to deal with sex as an art, have produced a scheme of personal relations in which there are no neurotic pictures, no frigidity, no impotence, except as the temporary result of severe illness, and the capacity for intercourse only once in a night is counted as senility.

Mead's conclusions from her Samoan work, as indeed from the work she did throughout her life, were compatible with Freud's theory. There were no sexual problems and no neuroses in Samoa, said Mead, because there was no sexual repression there. Repeatedly in *Coming of Age in Samoa*, Mead turned her anthropological observations back onto American culture, noting, "The present problem of the sex experimentation of young people would be greatly simplified if it were conceived of as experimentation instead of as rebellion, if no Puritan self-accusations vexed their consciences." Especially important, continued Mead, were the experiences of early childhood: "The findings of the behaviourists and of the psychoanalysts alike lay great emphasis upon the enormous role which is played by the environment of the first few years. Children who have been given a bad start are often found to function badly later on when they are faced with important choices. And we know that the more severe the choice, the more conflicts." As if to underline the Freudian message, a review comment by Dr. Abraham A. Brill was included on the back cover of the book: "Illuminating and interesting . . . Corroborates, through practical demonstration, the psychosexual theories promulgated by Freud and his pupils."

Margaret Mead's description of Samoan culture has been severely criticized as inaccurate, most thoroughly and publicly by anthropologist Derek Freeman in his 1983 book *Margaret Mead and Samoa: The Making and Unmaking of An Anthropological Myth*. In an attack on Mead which reached the cover of *Time* magazine, Freeman contended that "the young Margaret Mead was, as a kind of joke, deliberately misled by her adolescent informants." This theory was also put forth by Nicholas von Hoffman in his *Tales from the Margaret Mead Taproom*, in which he wrote:

> *There are supposed to be a bunch of old ladies on the island who claim to be the little girls in Mead's book and who say that they just made up every kind of sexy story for the funny palagi lady because she dug dirt.*

Von Hoffman also alleged that Samoan men are in fact such poor lovers that "I know where the missionaries learnt the position."

Lowell D. Holmes was another anthropologist who raised questions about the accuracy of Mead's work, despite being one of her strongest supporters. Holmes, who spent five months in the same village studied by Mead 28 years after she had been there, found that the sexual freedom of young people was "probably no more than is characteristic of their counterparts in the United States"; that frigidity and other sexual problems existed; and that adultery was "the most common ground for divorce." Holmes did confirm Mead's observation that "homosexuals are accepted without stigma or ridicule," although he also found "Samoans very conservative in regard to sex—at least in regard to talking about it."

Another question about Mead's work is whether she could have learned enough of the linguistically difficult Samoan language to have gathered the information which she claimed. Her formal language training consisted of working with a Samoan nurse for one hour a day for six weeks, yet, as Jane Howard politely phrased it, "Even her most fervent champions have not claimed that her many remarkable talents included a flair for languages." Furthermore, Mead lived with a Samoan family for only ten days, residing the remainder of her time with the family of the chief pharmacist at the United States Navy Dispensary, hardly the ideal milieu for mastering the language or eliciting the innermost secrets of adolescent girls. Moreover, in her fieldwork Mead was not known for her steady patience. Anthropologist Gregory Bateson, her third husband, described being "shocked" at the tactics used by Mead and Reo Fortune in their work in New Guinea: "They bully and chivvy their informants and interpreters and *hurry* them till they don't know whether they are on head or heels."

In looking back at *Coming of Age in Samoa*, one wonders how conscious Mead was of fitting her observations into the political and personal agendas that she brought with her to Samoa. Lowell Holmes summarized Mead's work by saying, "Margaret finds pretty much what she wants to find"; and Derek Freeman argued that Mead's principle goal was to please Boas. It should be remembered that Mead was not trained as a scientist but instead relied on "spirit guides" and other intuitive means of gathering information, in addition to her formal data collection—thus it seems unlikely that she consciously deceived anybody except herself.

Margaret Mead's personal agenda should not be underestimated, since it shaped her work. The burden of her bisexuality was illustrated again by events on her way home from Samoa in 1926. She went by way of Europe, and on board ship she fell in love with Reo Fortune, a handsome New Zealander going to England to study psychology. When the boat landed in Marseilles at the end of their six-week journey, Reo nearly succeeded in persuading Mead to continue on with him to England rather than disembarking to meet her husband, Luther Cressman, waiting on the dock. Common sense prevailed and Mead rejoined Cressman, from whom she had been separated for ten months, for what was to be a two-month journey through southern France. For part of the trip Louise Rosenblatt, a Barnard roommate, joined them. The trip did not go well, and at the end of one month the two women went to Paris and Cressman prepared to return alone to New York.

Paris was at the time witnessing the late stages of Dadaism, an urban crucible of artistic, social, and sexual experimentation. Natalie Barney ("the Pope of Lesbos") and Gertrude Stein ran salons extolling homosexuality, while Tristan Tzara and André Breton concocted visions of artistic rebellion against the bourgeoisie. Just two weeks before Mead and Cressman arrived, George Antheil's *Ballet mécanique* had premiered at the Theatre des Champs Elysées, with T. S. Eliot, James Joyce, Sylvia Beach, Ezra Pound, and Lincoln Steffens in attendance; the production was scored for nine pianos, automobile horns, electric fans, and a huge airplane propeller, and the cacophony had rapidly evolved into a riot.

Shortly after Mead and Cressman went to Paris, Reo Fortune arrived from England. Cressman invited him to a dinner he was giving for Margaret's friends. The following day Cressman returned from a walk to find "Margaret locked in such a tight embrace with Reo that neither heard him come in"; ever the gentleman, Cressman quietly left and returned later. Mead assured Cressman that Fortune was impotent, but Cressman, as he departed for New York, advised Mead where she might purchase contraceptives in London.

Mead and Fortune did go to London for ten days, but then Mead had to leave to meet Ruth Benedict in Rome, where the two women had agreed to rendezvous at an anthropology congress. They spent a week there, then returned together to Paris. There is no evidence that they met Natalie Barney, Gertrude Stein or their Parisian friends, although Mead wrote admiringly of this group in later years. She called them "creative, innovative men and women [who] were privately but quite frankly bisexual in their relationships. . . . It took courage to break with Victorian prejudices and risk encounters with the savage laws against the practice of homosexuality."

In Paris, Mead and Benedict were joined by Reo Fortune, who again came from England to try and persuade Mead to marry him; she declined and instead sailed with Benedict for New York. There Mead set up an apartment separate from Luther Cressman and close to Ruth Benedict, Marie Eichelberger, and other friends from her Barnard days. Her new job was as assistant curator of the American Museum of Natural History, whose board chairman continued to be Henry F. Osborn, senior statesman for the eugenicists. Settled into her sixth-floor office in the autumn of 1926, Margaret Mead contemplated her political and personal agendas and wrote *Coming of Age in Samoa*.

Freudian Theory with a Grass Skirt

In later years Margaret Mead contended that it had been merely coincidence that she had found in Samoa "a culture that made the point so clearly" in support of Boas's position. The results of her next fieldwork, however, strained the credulity of even her strongest supporters.

In 1928, buoyed by the reception of her book on Samoa, Mead returned to the South Pacific, where over the following five years she studied four different New Guinea cultures. She was accompanied in this work by her second husband, Reo Fortune, who had finally succeeded in persuading her to divorce Luther Cressman and marry him. The first group they studied was on the island of Manus. Mead described it as "a puritan society" in which "sex is conceived as something bad, inherently shameful" and homosexual relations between women were rare and frowned upon. Mead said that "the Manus emphasized anality," and she compared the Manus culture in *Growing Up in New Guinea* to her own Puritan culture in America.

The results of her studies with the other three cultures—the Arapesh, Mundugumor, and Tchambuli—were published together in 1935 as *Sex and Temperament in Three Primitive Societies*. Mead described the three societies, located within one hundred miles of each other, as having radically different culturally assigned roles for men and women:

> In one, both men and women act as we expect women to act—in a mild parental responsive way; in the second, both act as we expect men to act—in a fierce initiating fashion; and in the third, the men act according to our stereotype for women—are catty, wear curls and go shopping, while the women are energetic, managerial, unadorned partners.

Mead concluded that sex roles are culturally determined and have virtually nothing to do with biology or genetics. "We may say," Mead claimed, "that many, if not all, of the personality traits which we have called masculine or feminine are as lightly linked to sex as are clothing, the manners, and the form of head-dress that a society at a given period assigns to either sex." She contended that one's temperament—which she defined as inborn personality traits—is a given at birth just as one's sex is but that both temperament and sex are overridden by the expectations imposed by one's culture:

> The differences between individuals who are members of different cultures, like the differences between individuals within a culture, are almost entirely to be laid to differences in conditioning, especially during early childhood, and the form of this conditioning is culturally determined.

It was the same message of Freudian cultural relativity that Mead had brought back from Samoa, one which was immediately useful to Franz Boas and ultimately useful to followers of Freud.

Like her work in Samoa, Margaret Mead's work on sex and temperament in New Guinea was harshly criticized by her colleagues. As early as 1945 anthropologist Jessie Bernard wrote: "I for one found myself constantly confused between the facts Miss Mead reported and the interpretations she made of them . . . even accepting Miss Mead's observations as valid, one can come to exactly opposite conclusions to those she arrives at."

Especially disturbing to her colleagues was Mead's contention that she just happened upon three cultures which fit perfectly into a scheme proving that sexual roles are culturally determined. Mead herself summarized what she was asking readers to believe in a preface to the 1950 edition of her book:

> *According to some readers, my results make "too beautiful" a pattern. . . . This, many readers felt, was too much. It was too pretty. I must have found what I was looking for. But this misconception comes from a lack of understanding of what anthropology means, of the open-mindedness with which one must look and listen, record in astonishment and wonder, that which one would not have been able to guess.*

As she had done in Samoa, Mead also spent time focusing on the role of deviants in the New Guinea cultures. The entire final chapter of *Sex and Temperament in Three Primitive Societies* (1935) is devoted to an analysis of "the individual who is at variance with the values of his society" because his inborn temperament does not fit the prescribed pattern of acceptable behavior. Such individuals may have a "temperamental affinity for a type of behavior that is regarded as unnatural for their own sex and natural for the opposite sex," and Mead cited homosexuals as a case in point. This may produce, she wrote, "the pain of being born into a culture whose acknowledged ends he can never make his own" as well as "the added misery of being disturbed in his psycho-sexual life." With the exception of the pronoun used, the entire chapter reads as an intensely personal document.

Mead's own personal life continued to be complicated, even chaotic, during the period of her synthesis of material on sex and temperament. By the time she and Reo Fortune arrived at Tchambuli their marriage was strained. There they met and stayed with anthropologist Gregory Bateson, working with him for several weeks "in the tiny eight-foot-by-eight-foot mosquito room . . . [moving] back and forth between analyzing ourselves and each other. . . ." Mead and Bateson decided that they were similar in temperament, leaving Fortune as the odd man out. Mead later wrote that "Gregory and I were falling in love," but Reo was not as understanding as Luther Cressman had been. According to Mead's daughter, "Reo repudiated the whole way of working and thinking in a kind of panic, and accused Margaret of simply using it [the work on temperament] to justify a new romance." Mead did in fact divorce Fortune and marry Bateson two years later. On one occasion at Tchambuli, Fortune knocked Mead down, causing her to suffer a miscarriage; he later said that "Gregory ate our baby." To complicate matters still further Mead continued to spend time with Ruth Benedict, with whom she had lived just prior to marrying Fortune, during her periods of return to the United States.

Benedict was at that time writing *Patterns of Culture* (1934), the second most influential anthropological book yet written and one which would later be called "the most important single source of recruitment for anthropology as a profession." Its genesis lay in discussions which Mead and Benedict had had for many years regarding the question of whether a culture, like an individual, could be characterized by one or two overriding psychological traits which are so strong that the traits set the pattern for that culture. Individuals born into that culture would then "become the willing or unwilling heirs to that view of the world."

Patterns of Culture is a description of three markedly different cultures—the Zuni Indians of New Mexico, the Kwakiutl Indians of British Columbia, and the Dobuans of Papua New Guinea—with an analysis of why the cultures are so different. The Zuni were selected because Benedict had spent two summers working in a Zuni pueblo, although, according to Mead, she did not learn the language and "always had to work through interpreters." The Zuni, wrote Benedict, were a culture in which "moderation is the first virtue." She appropriated the term "Apollonian," to indicate that they "are a ceremonious people, a people who value sobriety and inoffensiveness above all other virtues." According to Benedict, the Zuni have "a distrust of individualism" and denigrate leadership qualitities, and "the ideal man in Zuni is a person of dignity and affability who has never tried to lead." The Zuni also strive for moderation in one's emotions. Even in marriage there are no deep emotions, and "marital jealousy is similarly soft-pedalled." When confronted with adultery Zuni husbands and wives "do not contemplate violence." Because there are no extreme emotions the

Zuni are said to have virtually no crime, drunkenness, or suicide. "Suicide is too violent an act, even in its most casual forms, for the [Zuni] to contemplate. They have no idea what it could be."

The second culture selected by Benedict was that of the Kwakiutl Indians. She had never worked with them, but based her profile on published and unpublished work of her mentor, Franz Boas. In direct contrast to the Zuni, Benedict said that the cultural pattern of the Kwakiutl was "Dionysian" because of its inclination toward behavioral excesses and "megalomaniac paranoid trend." Benedict claimed that "the object of all Kwakiutl enterprises was to show oneself superior to one's rivals." Behavior was dominated "at every point by the need to demonstrate the greatness of the individual and the inferiority of his rivals." Cannibalism was institutionalized in Kwakiutl rituals, and an intense "preoccupation with shame" led to frequent suicides. "The gamut of the emotions which they recognized, from triumph to shame, was magnified to its utmost proportions."

The cornerstone of *Patterns of Culture*, however, was the third culture, the Dobuans. Benedict based her description of that culture solely on the work of Reo Fortune, who had studied the Dobuans for six months in 1928 while waiting for Mead to divorce Luther Cressman and join him. Fortune's book, *Sorcerers of Dobu*, was published in 1932, but Benedict reviewed early drafts of his manuscript; Fortune, in turn, reviewed and approved Benedict's chapter on Dobu prior to the publication of *Patterns of Culture*.

The Dobuans, wrote Benedict, were "dour, prudish, and passionate, consumed with jealousy and suspicion and resentment. . . . The good man, the successful man, is he who has cheated another of his place. . . . It is taken for granted that he has thieved, killed children and his close associates by sorcery, cheated whenever he dared. . . . A man normally cares more for his dog than for mere acquaintances. . . . Suspicion and cruelty are his trusted weapons in the strife and he gives no mercy, as he asks none. . . ." The ethical ideal of the culture was "treacherous conflict"; "ill-will and treachery" are "the recognized virtues of their society" and "the concepts good and bad in the purely moral sense do not exist." In this cultural context a man "who enjoys work and likes to be helpful is their neurotic and regarded as silly."

Ruth Benedict drew two conclusions from her book. First she emphasized culture as a powerful determinant of human behavior. In words directly echoing Mead's formulation, Benedict wrote: "Most people are shaped to the form of their culture. . . . They are plastic to the moulding force of the society into which they are born." The corollary of this, Benedict continued, is that "the biological bases of cultural behavior in mankind are for the most part irrelevant." "Man is not committed in detail by his biological constitution to any particular variety of behavior. . . . There is no basis for the argument that we can trust our spiritual and cultural achievements to any selected hereditary germ plasms." Benedict went on to explicitly address the contention of the racial purist by arguing that "racial heredity . . . applied to groups distributed over a wide area, let us say to Nordics, has no basis in reality." Prior to publication of the book, Benedict submitted promotional material to the publisher, which she herself wrote, to insure that readers would not misunderstand her message: "The author demonstrates how the manners and morals of these tribes, and our own as well, are not piecemeal items of behavior, but consistent ways of life. They are not racial nor the necessary consequence of human nature, but have grown up historically in the life history of the community."

Like Mead, Ruth Benedict also incorporated her personal agenda in anthropological writings. The final chapter of *Patterns of Culture* included an extended discussion of homosexuality and its cultural relativity, citing ancient Greece and the Plains Indians as societies in which homosexuals were valued. Benedict said that "when the homosexual response is regarded as a perversion, however, the invert is immediately exposed to all the conflicts to which aberrants are always exposed." Since defining homosexuality as deviancy was simply a culturally relative value, Benedict made a plea at the end of the book for "an increased tolerance in society toward its less usual types."

In 1934 Benedict was living with Natalie Raymond, with whom she had a long relationship. Five years later Benedict would be faced with the possible loss of her teaching position at Columbia by a woman who threatened to "spread word publicly that Benedict was a lesbian." In America in the 1930s, exposure of one's homosexuality still carried enormous social consequences.

Patterns of Culture "was from the outset the object of intense criticism," according to anthropologist Marvin Harris. Colleagues who knew the cultures Benedict had described accused her of caricaturing them rather than characterizing them. As early as 1937 Li An-Che, an anthropologist who had studied the Zuni, called Benedict's description "over-simplified" and "very misleading." "Below the calm surface of Zuni life," asserted An-Che, "lies the same tensions which exist in other societies." Others pointed out that marital friction and suicide were not uncommon among the Zuni and that a rather non-Apollonian alcoholism was remarkabaly prevalent. Similarly, Benedict's depiction of the "Dionysian" Kwakiutl cuture was labeled as "highly colored." Anthropologist Victor Barnouw probably reflected a majority opinion among his colleagues when he concluded: "At any rate, there were Apollonian features in Kwakiutl life, just as there were Dionysian ones in the [Zuni] Pueblos. . . . A difficulty in both cases is Ruth Benedict's tendency to overstatement."

Of the three cultures in *Patterns of Cultures*, however, it was the Dobuans who were described as most anomalous. When A. R. Radcliffe Brown, an eminent English anthropologist, first read Reo Fortune's description of the Dobuans he simply said, "I don't believe it." Radcliffe Brown had abundant company, and in 1964 another anthropologist, Ann Chowning, noted of Fortune's work: "I know of no other case in which so many colleagues have expressed private doubts about a colleague's work without attacking it publicly."

There was solid ground for doubt. Fortune had had no anthropological training at the time he undertook his six-month fieldwork. He also apparently had had no Dobuan language training, yet claimed that he "used no English whatever after my second day there. I had no interpreter but acquired the language by contagion. At the end of three months nothing said passed over me. . . ." Five of his six months were spent on an island inhabited by only 40 people, who excluded him from the village for six weeks and on one occasion threatened to spear him. In descriptions of Fortune in her autobiography, Margaret Mead pictured him as being jealous, prudish, and occasionally paranoid, traits similar to those which he had ascribed to the Dobuans.

The most serious criticism of Fortune's description of the Dobuans, and thus of Benedict's *Patterns of Culture*, is that Fortune's view of them has never been corroborated by anybody. A missionary who lived among them from 1891 to 1908, and who said it took him four years to really know them, described them as "cheerful, laughter-loving folk." Margaret Mead was her ex-husband's strongest defender over the years, and in her autobiography, *Blackberry Winter* (1972), wrote: "There are others who think that [Fortune's] account of Dobu is exaggerated, but fieldworkers like Géza Roheim, Ian Hogbin and Ann Chowning, who have had an opportunity to check the accuracy of his work at first hand, do not." Roheim was a psychoanalyst who spent ten months in 1930 on Normanby Island, close to Dobu, trying to validate cross-culturally the Oedipus complex; he described no behavior remotely resembling Fortune's Dobuans and said that the ideal Normanby man was "a man who [gives] food and other presents to all the members of his clan and also to strangers without expecting a counter-gift." Hogbin worked elsewhere in Papua New Guinea but never near Dobu; when Mead was asked about this discrepancy she said that Hobgin had told her that he had based his assessment on Dobuans whom he had met elsewhere. That leaves only Chowning, an anthropologist who worked on an island approximately 50 miles from Dobu in 1957 and 1958 and who stated in a review that "the Dobuans may not be lovable, but they are not imaginary." Chowning was a personal friend of both Fortune and Mead and was asked to write the review by Mead. (Attempts by this author to obtain additional data from Dr. Chowning to support her contention that Fortune was correct have been singularly unsuccessful.) Fortune's Dobuans, then, appear to have been an anthropological chimera, a most unusual spectacle seen by one observer but apparently by nobody else.

Patterns of Culture thus joined *Coming of Age in Samoa* as a cornerstone of 20th-century teaching that culture is an important—indeed a crucial—determinant of human behavior. This was useful data for Franz Boas and others trying to counter the arguments of eugenicists. It was also useful data for supporters of Freud who were promulgating his theory about the importance of childhood experiences. Mead's and Benedict's sensual Samoans and paranoid Dobuans themselves became part of American culture, literary exemplifications of the dominance of nurture over nature.

It is apparent, however, that as objective and scientific contributions the anthropological studies of Margaret Mead and Ruth Benedict have little value. Both women viewed the cultures they studied through prisms tinged with political and personal concerns; as a result, certain features of the cultures became distorted, producing a panoply of peoples who are fundamentally fictitious. Utilized as political treatises, however, their books were effective for they helped shift the nature-nurture debate toward nurture. As Mead summarized her work in a reflection in 1939: "It was a simple—a very simple—point to which our materials were organized in the 1920s, merely the documentation over and over of the fact that human nature is not rigid and unyielding." And in 1970 Mead added: "Anthropologists had to spend the first four decades of this century showing that the various 'races' of mankind are specializations without any measurable differences."

And what, then, can be said about Franz Boas, who had been trained as a scientist and prided himself on the use of objective and scientific methods in his work? He had supported the work of Mead and Benedict, allowed Benedict to use his Kwakiutl studies in her book, endorsed both *Coming of Age in Samoa* and *Patterns of Culture*, and referred to both books approvingly in his own writings.

Boas was in fact posthumously censured by some of his anthropological colleagues for having allowed Ruth Benedict to distort his Kwakiutl data. In 1955, for example, Verne Ray chided Boas for "his failure to speak out in correction of the errors of his students, such as Benedict. . . ." Melvin Herskovits, a Boas student, was more explicit in his criticism: "Certainly the almost paranoid nature of the behavior of this [Kwakiutl] people as portrayed by Benedict is scarcely in line with the patterns of humility Boas sketches as prevailing within the family." Anthropologist Marvin Harris added: "It is simply impossible to reconcile the Boasian self-image of methodological rigor with the highly impressionistic and scientifically unreliable procedures which characterize the early phases of the culture and personality movement."

Such criticisms are valid insofar as the publications of Mead and Benedict are viewed as anthropological works. If they are instead viewed as political efforts to combat racism and promote tolerance for homosexuals, lesbians, and others who are different, then *Coming of Age in Samoa* and *Patterns of Culture* are seen in a different light. Boas, assisted by Mead and Benedict, had undertaken a lifelong battle against racist and elitist thinking in America. It seems from his point of view that the works of Mead and Benedict were ammunition to be used in the battle. When a man is engaged in a fight to the death and boxes of bullets are dropped into his bunker, he does not stop to read the warranty on the boxes.

4

Hitler's Resolution of the Nature-Nurture Debate

———

[Hitler] believed in biological determinism, just as Lenin believed in historical determinism. He thought race, not class, was the true revolutionary principle of the twentieth century.
—PAUL JOHNSON, *MODERN TIMES*

*D*uring the same years in which Ruth Benedict and Margaret Mead were carrying out their studies to demonstrate the importance of childhood experiences and culture, the balance of power in the dispute between nature and nurture was slowly shifting. The 1924 passage of the Johnson–Reed immigration restriction bill proved to be the high-water mark for racism and eugenicist thinking in the United States. As immigrants from Italy and Eastern Europe dwindled to a trickle, the genetic fears of the populace also diminished. By the early 1930s, when the Great Depression took hold, the number of individuals emigrating *from* the United States had actually surpassed those coming *to* this country, and the bogey of infestation by degenerate genes seemed remote indeed.

Henry Osborn, Madison Grant, and other eugenicists watched the waning of their cause with increasing concern. By 1927 Grant was complaining to Osborn about "the kind of propaganda which is put out by the Jews to break down the [immigration] restriction act. This kind of sob stuff, unfortunately, is very effective among our sentimentalists." It was also in 1927 that Henry Ford, the crown prince of American racism, shocked his friends by deserting their cause. Pressured by President Wilson, and faced with a libel suit against his Dearborn *Independent*, Ford sent two emissaries to the American Jewish Committee to explore a cessation of hostilities. The committee demanded "complete retraction, full apology, and a pledge that Ford would never again indulge in such activities"; Ford signed the statement and released it to the press.

At the same time that Henry Ford was having second thoughts about the fiscal wisdom of anti-Semitism, the leaders of the Carnegie Institution began to question their support for research being done under Charles Davenport in the Eugenics Record Office at Cold Spring Harbor, Long Island. The veneer of science that had covered such research until World War I had worn increasingly thin; Cold Spring Harbor had instead become a proving ground for the shrill cries of self-appointed propagandists like Davenport and Madison Grant. As the study of genetics became increasingly well organized in the 1920s, geneticists in the universities were embarrassed by their eugenics brethren and attempted to disassociate themselves. One consequence of this was that in 1929 the president of the Carnegie Institution asked a committee of scientists to evaluate the Eugenics Record Office. Davenport was still powerful enough to block any substantive change in his operations, but when he retired in 1934, another blue-ribbon commission recommended the closing of the record office and this was carried out.

The most dramatic setback for the eugenicists, however, took place in January 1930 with the publication of an article by Professor Carl Brigham in *Psychological Review*. Brigham had been an enthusiastic supporter of Madison Grant, and Brigham's *A Study of American Intelligence* had been widely quoted for its conclusion that "the army [IQ] tests indicate clearly the intellectual superiority of the Nordic race group." In what has been called "one of the most agonizing retractions in the history of the behavioral sciences," Brigham completely reversed himself in the 1930 article. He contended that IQ "tests in the vernacular must be used only with individuals having equal opportunities to acquire the vernacular of the test. . . . The last condition is frequently violated here in studies of children born in this country whose parents speak another tongue." Proceeding to dissect and devastate the IQ data that he himself had previously used, Brigham concluded with a classic *mea culpa*:

> *This review has summarized some of the more recent test findings which show that comparative studies of various national and racial*

*groups may not be made with existing tests, and which show, in
particular, that one of the most pretentious of these comparative
racial studies—the writer's own—was without foundation.*

By the time the Third International Congress of Eugenics was
convened in 1932 the eugenics movement was becoming anemic. Held
again at the American Museum of Natural History under the direction
of Henry Osborn, the congress attracted fewer than one hundred participants, with major supporters of past years conspicuously absent.
Herbert Hoover, a member of the Organizing Committee for the 1921
Congress, declined to contribute at all. John D. Rockefeller, previously
a reliable source of funds for eugenics causes, regretted through a
spokesperson that "the project does not seem to be one towards which
[Rockefeller] as an individual could contribute."

Most papers given at the 1932 congress predictably advocated further restrictions on immigration and stricter sterilization laws, especially for the chronically unemployed and the poor. An English
eugenicist at the meeting categorized such individuals as a "definite
race of chronic paupers, a race parasitic upon the community, breeding
in and through successive generations. . . ." A few papers took issue
with the assumptions of eugenics; one of these—given by the respected
geneticist Hermann J. Muller—strongly argued for "the dominance of
economics over eugenics" and claimed that poverty was a product of
social and economic factors, not genes. Three years later Muller further disassociated himself from eugenics, which he called "hopelessly
perverted" and its supporters "advocates of race and class prejudice,
defenders of vested interests of church and state, Fascists, Hitlerites, and
reactionaries generally."

At the same time as the Third International Congress of Eugenics
was meeting in New York City, bread lines and shantytowns—
"Hoovervilles"—were growing increasingly prominent as the Great
Depression bottomed out. By the end of 1932 more than 13 million
individuals were unemployed, industry was operating at half its 1929
capacity, banks were closing, and the economy descended toward sub-
terranean levels. In Washington, "hunger marchers" descended on the
White House, while 17,000 World War I veterans set up camp on the
Mall, eventually to be driven from Washington by army troops led by
General Douglas MacArthur and assisted by his young aide, Major
Dwight D. Eisenhower, and by Major George S. Patton of the U.S.
Cavalry.

Bread lines proved to be poor recruiting grounds for eugenics enthusiasts. When speakers at the Eugenics Congress spoke about the necessity for sterilization and restricted immigration for people who were poor, millions of unemployed men and women suddenly realized that they qualified. Nordics were noted to be fully represented in unemployment lines, around urban bonfires built to keep warm, and in old cars wending their way west as dust storms added insult to the injury of the depression. Perhaps, many mused, genes were not as important a determinant of human worth as the Madison Grants of America had claimed.

For opponents of racism and eugenics the increasing emphasis being given to economic and social causes of poverty, and the defections of men such as Henry Ford and Carl Brigham, were very encouraging. For Franz Boas personally, the late 1920s and early 1930s were, however, a difficult time. One daughter died of polio, a son was killed in a train accident, and in 1930 Boas's wife was killed by an automobile on the streets of New York. Finally, in 1932, at the age of 73, Boas suffered a serious heart attack.

Despite personal tragedies and his illness, Boas continued advocating for racial equality. He was encouraged by the widespread interest in Mead's *Coming of Age in Samoa* when it was published in 1928; the same year Melville J. Herskovits, another of Boas's students, published *The American Negro: A Study of Race Crossing*, which argued that racial interbreeding does not produce a deterioration of genes. In 1931, as president of the American Association for the Advancement of Science, Boas delivered an inaugural address, "Race and Progress" and used the work of Mead and Herskovits to reinforce his points.

Boas also continued attacking the eugenicists in popular periodicals. In 1931, for example, he published a letter in the *New York Times* titled "The Rich Are Taller." In it he argued that "outer conditions" such as better nutrition lead to more rapid growth in childhood and that such children are therefore taller as adults. Height, Boas said, "cannot be ascribed solely to the genetic character of the group." Tall, stately Madison Grant, who strongly equated height with genetic worth, could not have missed the message. When Grant published a new book in 1934, *The Conquest of a Continent*, with an introduction by Henry Osborn, Boas ridiculed it in the *New Republic*: "I do not believe that the serious scientific questions involved have ever presented themselves to either Professor Osborn or to Mr. Madison Grant. . . . Contradictions, when convenient, do not trouble the author."

Nazi Eugenics

Franz Boas was apparently one of the first Americans to realize the full implications of Adolf Hitler's appointment as chancellor of Germany by President Paul von Hindenburg on January 30, 1933. Boas promptly sent von Hindenberg an open letter, which was read over German radio, strongly protesting the appointment. Later in the year Boas complained to a friend about "the crazy conditions in Germany" and wrote an article, "Aryans and Non-Aryans," for *American Mercury*. The article was distributed in Germany by anti-Hitler forces and "of everything [Boas] wrote [this] achieved perhaps the widest circulation." In it Boas systematically picked apart the myth of an aryan race and concluded that "the attempt that is being made by those who are in power in Germany to justify on scientific grounds their attitude toward the Jews is built on a pseudo-science." More pointedly, Boas added in 1934 that "Herr Hitler has stated that . . . national problems are, at bottom, race problems."

Boas realized Adolf Hitler was a man about whom the civilized world should be worried. A failed artist who had lived in Vienna from 1906 to 1913 during the years when Sigmund Freud was establishing a reputation, Hitler was known to be both strongly anti-Semitic and anti-democratic. His *Mein Kampf* (1925) outlined a return to greatness for the German people. The Germans were the best educated people in the world at the time, and students in German universities were among the earliest and strongest supporters of Hitler. He preached not politics but Valhalla, a Third Reich in which only Nordics would qualify for admission. They bought his promise even as they ignored his fanaticism.

The idea of German racial purity did not originate with Hitler. As early as 1895 Dr. Alfred Ploetz published a book titled *The Excellence of Our Race and the Protection of the Weak* in which he argued that social programs in Germany were weakening the race by allowing genetically inferior individuals to survive. Among the questions which Ploetz posed was, "Why is the white man more perfect than the Negro, and the Negro more perfect than the gorilla?" Ploetz combined his interest in eugenics with a commitment to anti-Semitism and was a member of a secret Nordic club called the Mittgartbund. Like other German eugenicists of that era, Ploetz looked to the United States for leadership on this issue, and Ploetz even traveled to a small utopian community in Iowa to try and persuade those living there to implement his eugenicist ideas. In 1905 Ploetz helped found the Society for Racial Hygiene, which by 1930 had grown to 20 branches and more than 1,300 members. In 1933, when Hitler came to power, racial hygiene was being taught as a separate course in most German medical schools.

Six months after assuming the chancellorship, Hitler persuaded the Reichstag (the lower chamber of the federal parliament) to pass a Law for the Prevention of Genetically Diseased Offspring. The law mandated compulsory sterilization for presumed genetic conditions including schizophrenia, mental retardation, epilepsy, and severe alcoholism, and it was to be administered through seventeen hundred genetic health courts. The German law was apparently closely patterned after a 1922 model statute for use by American states, which had been drawn up by Charles Davenport and the staff of the Eugenics Record Office in Cold Spring Harbor; after passage of the German law one employee of the record office suggested that Hitler "should be made [an] honorary member of the Eugenics Record Office."

It is important to note that sterilization was illegal in Germany until 1933. German eugenicists in fact held America up as a model; in 1924 a leading German enthusiast for sterilization had publicly argued: "What we racial hygienists promote is by no means new or unheard of. In a cultured nation of the first order—the United States of America— that [legislation] which we strive toward was introduced and tested long ago." By the time the German law was passed in 1933, the United States had carried out approximately 20,000 sterilizations, most of them involuntarily. In Virginia, for example, according to Daniel J. Kevles's book *In the Name of Eugenics*, in the 1930s "state sterilization authorities raided whole families of 'misfit' mountaineers. . . . 'Everybody who was drawing welfare then was scared they were going to have it done to them. They were hiding all through these mountains and the sheriff and his men had to go up after them. . . . The sheriff went up there and loaded all of them in a couple of cars and ran them down to Staunton [the state mental hospital] so they could sterilize them.' " Because of American sterilization laws, "after the war, allied authorities were unable to classify the [German] sterilizations as war crimes, because similar laws had only recently been upheld in the United States."

Once the sterilization law had been passed in Germany, it was carried out with both enthusiasm and efficiency. Doctors were required to report all individuals who might meet criteria for sterilization and were fined if they failed to do so. Individuals who were ordered to be sterilized had the legal right to appeal the decision, but less than ten percent of such decisions were reversed. In 1934, the first year of the law's operation, a total of 56,244 sterilizations were carried out; the total number during the Nazi regime eventually reached approximately 400,000, and in some regions of Germany sterilizations were performed on one percent of the entire population. The most common methods used were vasectomy and tubal ligation (although in later years German doctors experimented with the use of sterilizing X-rays that could be directed at a person's genitals without his knowledge as he sat at a special desk filling out a form).

Sterilization was not the only area of common interest between Nazi eugenicists and their American counterparts. There was much concern in Germany about the falling birthrate, which had declined by almost one-third between 1876 and 1911. To counteract this, "the government provided loans to biologically sound couples whose fecundity would likely be a credit to the *Volk*, [and] a number of German cities established special subsidies for third and fourth children born to the fitter families." Beginning in the 1920s, there was also much public propaganda encouraging "a quality marriage," and by the time Hitler took power approximately half of all German cities had counseling centers to promote marriages based on the principles of racial hygiene. In 1935, under Hitler's direction, the Law for the Protection of German Blood and German Honor—one of the Nuremberg Laws—was enacted and essentially changed the standards promoted by the counseling centers from voluntary to mandatory. A certificate indicating that a person was "fit to marry" was henceforth needed, and individuals cohabiting without a certificate could be sentenced to prison.

It is important to realize how closely the events in early 1930s Germany were associated with the American eugenics movement. The Germans looked to America both to justify their sterilization law and for a model on which to base the law. Henry Osborn made "an enthusiastic trip" to Nazi Germany, where he was awarded an honorary degree by the University of Frankfurt in 1934. Madison Grant's *The Passing of the Great Race* had been translated into German in 1925 and was being widely read. When Grant's later book, *Conquest of a Continent*, was also translated into German in 1937, it included a special foreword by Dr. Eugen Fischer, Hitler's adviser on racial hygiene and the foremost German anthropologist:

> *No one will be surprised that this work met with the most vehement opposition in the land of its origin, where politicians and scholars, led above all by the Jewish anthropologist and ethnologist, Franz Boas,*

dominated all public opinion with the notion that racial differences were determined by environment and changeable with it. . . .

German and American eugenicists also shared a common bond of anti-Semitism. Hitler in 1928 had called for a "cleansing process" to counteract race-poisoning brought about by the "corrupt blood" of the Jews. In 1932 Charles Davenport publicly defended the Nazi pledge to exterminate the Jews. There was also a sexual aspect to the anti-Semitism that tied Freud's name to the struggle. In 1927 Fritz Lenz, the first professor of racial hygiene in Germany, had claimed that "Jews are especially interested in the sexual life" and had cited Freud as a typical example. Increasingly Jewish doctors in Germany were said to be the cause of sexual perversions, homosexuality, "sexual degeneration, a breakdown of the family, and all that is decent." High-ranking Nazis such as Heinrich Himmler "associated sexual relations with the loose immorality" that he blamed on Jews. In 1933 psychoanalysis was banned from the Congress of Psychology in Leipzig as a "Jewish science," and the following year all "non-Aryan physicians and physicians with non-Aryan wives were no longer allowed to begin practice as part of Germany's state-supported health insurance scheme."

On both sides of the Atlantic, then, eugenicists linked Jews, including Freud, to sexual freedom and liberal politics. These were the forces of nurture, the antithesis of those which represented Nordic racial purity, sexual continence, and conservative or Fascist politics. Thus it was not surprising that when the Nazis began publicly burning "subversive" books in May 1933, the books of Boas, Freud, and Marx would be among the first chosen for the bonfires. Freud, with his usual sardonic sense of humor, observed: "What progress we are making. In the Middle Ages they would have burnt me; nowadays they are content with burning my books."

The Marriage of Freud and Marx

For many Jewish professionals living in Europe in the 1930s their future under the Nazis was painfully clear long before they were required to wear a Star of David. Their exodus had no precedent at the time. One out of every five university scientists had left or been forced out of their jobs by 1935; at the University of Berlin the figure was almost one in three. Albert Einstein, Hans Krebs, and others who had won or would win Nobel Prizes were among those who left. Physicians exited in especially large numbers because Jews constituted a disproportionately large percentage of that profession; by September 1933 more than 200 German physicians had emigrated to England alone, and by late 1935 more than one-fifth of all German doctors had departed.

In 1933 German and East European followers of Freud included both psychiatrists and nonmedical (lay) analysts. Among the latter were psychologists Theodore Reik and Bruno Bettleheim, anthropologist Géza Roheim, social scientist Erik Erikson, and lawyer Hanns Sachs. The vast majority of European psychoanalysts were Jews, and as a result, according to Laura Fermi's *Illustrious Immigrants* (1971), "European psychoanalysis found itself in the unenviable position of being the only discipline . . . that Hitler virtually exterminated in continental Europe." In Vienna, for example, the Psychoanalytic Society had 69 members in 1935 but only three remaining ten years later.

It has been estimated that the majority of German physicians who left under Hitler remained in Western Europe or went to Palestine and that only 13 percent of them emigrated to the United States. For psychoanalysts there is no comparable data, but a higher percentage of them almost certainly ended up in America, often passing through other countries en route. Fermi, who has done the most extensive study of this question, estimated that approximately 190 European psychoanalysts emigrated to the United States under Hitler. These included virtually the entire staff of the Berlin Psychoanalytic Institute: Hanns Sachs, Franz Alexander, Karen Horney, Sandor Rado, Otto Fenichel, Theodor Reik, Therese Benedek, Siegfried Bernfeld, and Ernst Simmel. Other psychoanalysts who moved to America to practice included Otto Rank, Sandor Ferenczi, Felix and Helene Deutsch, René Spitz, Heinz Hartmann, Fritz Redl, Erik Erikson, Erich Fromm, Wilhelm Reich, Frieda Fromm-Reichmann, Margaret Mahler, Ernst Kris, David Rapaport, Edith Weigert, Kurt Eissler, Heinz Kohut, and (after spending a year in a concentration camp) Bruno Bettleheim. A few others, such as Sandor Lorand, Paul Schilder, and Fritz Wittels, had emigrated to the United States before Hitler came to power.

The psychoanalytic influx to America in the 1930s was a major reason why Freudian theory became so prominent in this country. Included among the analysts were three members of Freud's original inner circle (Sachs, Rank, and Ferenczi) as well as most of the leaders of European psychoanalysis; Freud himself emigrated to England in 1938. By that time New York had 76 members in the International Psychoanalytic Association and Boston, Washington, D.C., and Chicago together had another 85 members out of 564 members worldwide. When the war ended the United States had more psychoanalysts that the rest of the world combined.

The newly arrived psychoanalysts conferred an aura of respectability on the psychiatric profession in general and on psychoanalysis in particular. Americans, especially members of the media, were at that time still in awe of European scholarship, and the pronouncements of the German professors commanded immediate attention. The increased number of psychoanalysts, most of whom had private practices, also made psychoanalysis available to many more Americans.

The immigrant psychoanalysts also took control of the training institutes and formalized the curriculum. Prior to 1929, Americans had not even been required to undergo training analysis to qualify as psychoanalysts. By the late 1930s courses were in place and training institutes were proliferating. The European psychoanalysts also enabled psychoanalysis to spread west from its East Coast base; the Topeka Institute for Psychoanalysis opened in 1938 with seven American and six immigrant members, and the San Francisco Psychoanalytic Institute was founded in 1942 with five Americans and five Europeans. Most importantly, though, was the founding in 1946 of the Los Angeles Institute for Psychoanalysis by three immigrant psychoanalysts; this institute would provide a crucial nexus for the spread of Freud's ideas through the postwar movie industry.

One of the most important effects of the influx of European psychoanalysts to America in the 1930s was to strengthen the association between Freudian theory and liberal politics. Many of the psychoanalytic refugees had strong Marxist leanings, dating to Alfred Adler's presentation of a paper, "The Psychology of Marxism," as early as 1909. Siegfried Bernfeld, Otto Fenichel, Erich Fromm, and Bruno Bettleheim all made significant attempts to fuse Freudian theory with their Marxist beliefs. Perhaps the best-known Marxist psychoanalyst was Wilhelm Reich, who joined the Communist Party in 1928 and whose extreme views "in the service of Communist ideology" contributed to his suspension from the International Psychoanalytic Association in 1934.

As a result of the increasingly close association between Freud and leftist politics, what had been mere propinquity between Freud and Marx in America developed into consanguinity. Among New York intellectuals, the formalization of this union was embodied by the *Partisan Review*, a bimonthly publication that began in 1934 as the voice of the communist John Reed Club. Following the Stalinist purges in 1936 through 1938, the *Partisan Review* denounced Stalin and aligned itself with Trotsky, becoming, according to its masthead, "A Quarterly of Literature and Marxism"; it was derisively labeled by Edmund Wilson as the "Partisansky Review." Trotsky was especially attractive to the New York intellectuals because he had met Alfred Adler in 1909 while living in Vienna and had become interested in psychoanalysis (Adolph Joffe, one of Trotsky's closest revolutionary collaborators, had been in psychoanalysis with Adler). Trotsky had later urged his colleagues to "keep an open mind to what was new and revealing in Freud." Equally important to the staff of the *Partisan Review* during the 1930s was Freudian theory; one staff member recalled, "We were all more or less saturated with psychoanalytic jargon. Psychoanalysis was at that time very much in the air, and everybody seemed to be in it or contemplating it."

The *Partisan Review* was a remarkable publication in its early years; its array of contributions has probably never been equaled in a literary journal. The issue of Winter 1939, for example, included selections by John Dos Passos, André Gide, Allen Tate, Delmore Schwartz, Lionel Trilling, Franz Kafka, and Gertrude Stein. Even political reactionaries like T. S. Eliot contributed two of his *Four Quartets* to the magazine, in what was, according to Christopher Lasch, "the most ambitious attempt since prewar Village days to fuse radical politics and cultural modernism." The journal was required reading for the aspiring intelligentsia in the cafés around Washington Square; one historian said of the period that "a New York intellectual was one who wrote for, edited, or read *Partisan Review*."

The leaders of the *Partisan Review* worshiped two icons—Marx and Freud. Foremost among these were Philip Rahv and William Phillips, coeditors for more than a decade. Rahv had emigrated from the Ukraine at age 14 and was "a Marxist purist" who scorned capitalism and retained a passionate commitment to socialism until his death. He had read Freud extensively and showed "a heavier and heavier reliance on the insights of Freud." According to a close friend, "Rahv had great respect for Freud and wrote a first-rate and too-little-known essay on Freud and Literature." Viewed by his colleagues as a rude and power-grasping man, Rahv apparently never undertook a personal psychoanalysis. William Phillips referred to him privately as "a manic-impressive" and said of Rahv: "Most of us, under [psycho]analysis, break down and admit our shortcomings. Philip would break down and confess he was a great man."

William Phillips, whose father had changed his name from Litvinsky, had been raised in New York and attended the City College of

New York. City College was at the time the cradle for New York's intellectuals; in addition to Phillips, it produced Sidney Hook, Alfred Kazin, Irving Howe, Irving Kristol, and Daniel Bell. They were intellectual Marxists and congregated in alcove number 1 in the college cafeteria, arguing with the more political Marxists, whose members included Julius Rosenberg, in alcove number 2. Following his graduation Phillips joined the John Reed Club, where he met Rahv and began reading Freud. He was devoutly respectful of classical Freudian theory and in later years edited books on *Art and Psychoanalysis* (1957) and *Literature and Psychoanalysis* (1983) in which he included the writings of Freud, Rank, Kris, Alexander, Reik, Jones, Fromm, and other psychoanalysts.

Dwight Macdonald and Delmore Schwartz also worked for the *Partisan Review*. Macdonald was one of the few *goyim* among the New York intelligentsia, a Protestant product of Exeter and Yale. In the 1930s Macdonald was an admirer of the Soviet Union and an active member of the American Committee for the Defense of Leon Trotsky. "With what must be considered a classic example of chutzpah," according to biographer Stephen Whitfield, Macdonald "told Leon Trotsky how to make a revolution"; Trotsky later assessed Macdonald as "not a snob but a bit stupid." Macdonald was not known for intellectual consistency; one critic later said that his "political career looked as if it had been painted by Jackson Pollack." One area in which Macdonald was consistent was in his use of psychoanalytic concepts; "his articles in *Politics*, which he later edited, abounded with psychological and psychiatric terminology," and he was known to be especially interested in the theories of Wilhelm Reich.

Delmore Schwartz, considered one of the finest poets of his generation, also "flirted with radical politics" before becoming engrossed with Freudian theory and his own psychoanalysis. This was followed by the onset of manic-depressive psychosis and an early death. Schwartz observed that the "vocabulary of psychoanalysis" totally dominated the thinking of the New York intelligentsia.

Another important member of the *Partisan Review* community was Mary McCarthy, who, as a member of the Committee for the Defense of Leon Trotsky, had met Dwight Macdonald and Philip Rahv. She had been orphaned in 1919 at age seven when both parents died in the influenza pandemic and later graduated from Vassar College. She was initially interested in communism, then Trotskyism, and eventually evolved, according to Paul Johnson's *Intellectuals* (1988), into being "nothing at all but mild, all-purpose left." She lived with Philip Rahv for several months and became a regular contributor to *Partisan Review*. Tiring of Rahv, McCarthy began surreptitiously seeing Edmund Wilson, who was also a member of the Trotsky committee and a regular contributor to the journal. In February 1938 McCarthy eloped and married Wilson in a move that "astonished everyone [including] Philip Rahv [who] had no idea that Mary McCarthy had been seeing Wilson." Four months after the marriage, McCarthy began psychoanalysis with Sandor Rado, an analysis that stretched over several years and two more analysts, including Abraham Kardiner, Ruth Benedict's collaborator and friend. McCarthy used psychoanalysis prominently in many of her novels, such as *The Group* (1963), although, according to a biographer, she slowly came to the conclusion "that psychoanalysis was based on a series of myths."

When Edmund Wilson married Mary McCarthy in 1938 he was one of the leading members of the New York intelligentsia. Seventeen years her senior, he had published both fiction (e.g., *I Thought of Daisy*, 1929) and literary criticism (e.g., *Axel's Castle*, 1931) and was described by Irving Howe as a "blend of avant-garde culture and social radicalism . . . we thought of his as the kind of intellectual we too should like to become." Wilson had supported Communist Party candidate William Foster's bid for the presidency in 1932, including holding a fund-raising cocktail party for him, and had been one of the prominent signers of a pamphlet, "Culture and the Crisis," which had urged intellectuals to support the Communists. The integrity of his allegiance to Socialist principles became suspect in later years, however, when he failed to file income tax returns for ten years and was heavily fined by the Internal Revenue Service.

Wilson was equally enthusiastic about Freud as he was about Marx. He recalled being introduced to Freudian theory through Max Eastman's 1915 essay in *Everybody's Magazine*. Wilson was especially interested in Freud's sexual ideas, an interest which he reflected in two essays in *The New Yorker* on the Marquis de Sade, whom Wilson cited as having anticipated Freud's theory of infant sexuality and who stood as "a reminder that the lust for cruelty, the appetite for destruction, are powerful motivations that must be recognized for what they are." Wilson himself was known as an inveterate womanizer and heavy drinker who regularly beat Mary McCarthy during the seven years they lived together. Wilson had also suffered from "a terrifying nervous breakdown" in 1929 and had been psychiatrically hospitalized for three weeks, which may have also accounted for some of his interest in Freud.

Wilson utilized psychoanalytic themes liberally in his literary criticism. He implied that Henry James's *The Turn of the Screw* betrayed the author's own attraction to little girls and attacked Ben Jonson's plays for reflecting anal erotic tendencies. Wilson's widely praised *The Wound and the Bow* (1941) was based on the traditional Freudian premise that artistic creativity is a product of a psychological "wound," especially one arising from early childhood experiences. Wilson's tripartite allegiance to literature, Freud, and Marx was reflected in a 1941 letter in which he wrote: "Yeats, Freud, Trotsky, and Joyce have all gone in so short a time—it is almost like the death of one's father."

Edmund Wilson's chief rival for the leadership of New York's intellectuals was Lionel Trilling, a member of the *Partisan Review* advisory board. A New Yorker who attended and taught at Columbia University, Trilling was eulogized with a front-page obituary in the *New York Times* and a two-page article in *Time* magazine when he died in 1975. Irving Howe in the *New Republic* assessed him as follows: "With the exception of Edmund Wilson, Lionel Trilling was the most influential literary critic in America these past few decades."

Trilling was strongly committed to Marx in his early years and, like Wilson, supported William Foster for president in 1932. Like most of his colleagues he switched allegiance from Stalin to Trotsky following the purges and joined Wilson, Macdonald, and McCarthy as members of the American Committee for the Defense of Leon Trotsky. Trilling was also friends with Whittaker Chambers during the 1930s and used Chambers as a model for one of the characters in his only novel, *The Middle of the Journey* (1947), before Chambers's name became linked to that of Alger Hiss as part of an American household hyphen.

Trilling's ultimate allegiance was not to Marx, however, but to Freud. During the 1930s, "He had had a 'successful' psychoanalysis, which seemed to have come at a crucial period, and evidently remained one of the central and transforming experiences in his life. . . . When Freud's name is invoked by Trilling it is nearly always bathed in something of a numinous glow." In a 1940 essay, "The Legacy of Sigmund Freud," Trilling praised Freud for his "brilliant" methodology and said:

> *The Freudian psychology is, I think, the only systematic account of human nature which, in point of subtlety and complexity, of interest and tragic power, deserved to stand beside the chaotic accumulation of insights which literature has made over the centuries.*

Trilling's literary criticism was heavily laced with psychoanalytic constructs as he found Freudian forces lurking behind the most innocuous-looking pentameter. William Wordsworth's "Intimations of Immortality," for example, was said by Trilling to be not about immortality but rather a reflection of Wordsworth's infantile narcissism. Trilling's *The Liberal Imagination* (1949), included essays on "Freud and Literature" and "Art and Neurosis"; Norman Podhoretz called the book "the single most important expression of the new liberalism." In 1955 Trilling published *Freud and the Crisis of Our Culture*, based on a lecture given to the New York Psychoanalytic Society, and in 1970 he published *The Life and Work of Sigmund Freud*, an abridgment of Ernest Jones's three-volume biography. In *Psychoanalysis and American Literary Criticism*, Louis Fraiberg says of Trilling: "[He] has a grasp of what psychoanalysis tells us about the mind and its functioning that is rarely met with outside professional psychoanalytic circles. . . . No other critic has shown a comparable grasp of the significance of psychoanalysis; no other critic has so well incorporated it into his criticism."

There were of course occasional dissenters to the marriage of Freud and Marx among the New York intelligentsia. Foremost among them was philosopher Sidney Hook. Member of the advisory board of the *Partisan Review*, supporter of William Foster in 1932, and an acquaintance of Whittaker Chambers, Hook was also among the first to criticize Stalin in the 1930s. Hook was one of the best known and most respected of the New York intellectuals, but on Freudian theory he parted company. He called it "a scientific mythology," and in his memoirs, *Out of Step*, said that "Freud's doctrines . . . seemed to be then (as now) projections of poetic fancies unscientifically related to his remarkable observations of human behavior." In this view Hook was indeed out of step with almost every other New York intellectual of his generation.

By the end of World War II, then, Freud and Marx were firmly linked in the literary world around Washington Square. Rahv, Phillips, Macdonald, Schwartz, McCarthy, Wilson, and especially Trilling elevated an Oedipal fixation to the same level of importance as a classless society. The staff of the *Partisan Review* represented the union of Freud and Marx most conspicuously, but this liaison could also be seen in other publications. For intellectuals aspiring to the inner sanctum of the *Partisan Review*, a reading of Freud's *Civilization and Its Discontents* was considered equally as important as *The Communist Manifesto*. The role models for young intellectuals were men like Edmund Wilson and Lionel Trilling, who could discuss economic and Oedipal forces in the same paragraph and frequently did so.

In addition to anthropologists like Benedict and Mead and the

New York intellectuals, there were during the 1930s two other American citadels of psychoanalysis in which Freud's name continued to be invoked as something more than a euphemism for sexual conquest. One of these was social work, especially in child guidance clinics (*see* chapter 7). The other was the New York theater community where, according to Sievers's *Freud on Broadway*, "the influence of psychoanalysis did not diminish." Most New York playwrights were at least reading Freud and many were also undergoing psychoanalysis.

One example was Leopold Atlas, who acknowledged having read works of Freud, Jung, Reik, and Brill. His 1934 play, *Wednesday's Child*, was described by Sievers as being about "the conflicting emotions in a disturbed child growing out of Oedipal jealousy centered on the primal scene." Clifford Odets, whom Sievers assessed as "perhaps the most original . . . playwriting talent which the American drama produced in the thirties," was also a fervent believer who commended Freud especially for his discovery of "the analytic situation and its resultant therapy and the big matter of 'transference.' " Odets was also an example of a playwright who combined Freudian ideas with social reform; as one example, his 1935 play *Waiting for Lefty* prominently featured workers' rights. Even a play as apparently innocent as Thornton Wilder's *Our Town* (1938) consciously incorporated Oedipal themes into the wedding scene; Wilder was friends with Freud and visited him in Vienna.

Broadway musicals were also influenced by Freud's ideas. Moss Hart, who underwent psychoanalysis in 1934 and had read "almost all of Freud," utilized psychoanalytic themes; his 1941 *Lady in the Dark* was said to be "the first musical drama based upon psychoanalytic therapy itself." It was "reportedly written in tribute to Hart's own analyst," contained prominent Oedipal themes, and included lines such as, "What a really great man Freud was." According to Sievers the play "ran for 467 performances and was considerably effective in popularizing psychoanalysis to a musical comedy public."

Other groups of intellectuals in the 1930s were less affected by Freudian ideas. Bernard DeVoto, in a 1939 article, "Freud's Influence on Literature," claimed that "there is no department of literature . . . that psychoanalysis has not affected," but as often as not the depiction of psychoanalysts in fiction was less than flattering. One example was the neurotic analyst in F. Scott Fitzgerald's *Tender Is the Night* (1934). Many writers ignored Freud altogether, and some, such as Joseph Conrad, simply dismissed Freudian theory as "a kind of magic show."

One political scientist who emerged from the 1930s as an enthusiastic advocate of Freud was Harold D. Lasswell. After undergoing psychoanalysis with Theodor Reik while studying in Berlin in 1928 and 1929, Lasswell himself became a lay analyst and even undertook research to prove that psychoanalysis was effective. According to Bruce L. Smith, Lasswell believed "that human motives (including political and economic motives) are generated very largely in the nursery, in the bedroom, in childhood sexual and excretory experiences and reveries, and in a whole network of interpersonal contacts generally thought of as 'private.' " His widely read *Psychopathology and Politics* (1930) discussed society as a patient and attempted to apply the principles of mental hygiene to politics. Like many other admirers of Freud's theories during this period, Lasswell was equally enthusiastic about Marx as he was about Freud.

Although interest in Freud had begun to wane in the 1920s in most segments of the American public, small pockets of Freudian orthodoxy continued to preserve the faith. Foremost among these were anthropologists, New York intellectuals, New York playwrights, and social workers who were working in child guidance and criminal justice (*see* chapter 7). Most followers of Freud were also supporters of social reform and leftist politics, thereby fusing Freud and Marx in American thought. By 1937, when the Lincoln Brigade marched off to fight Franco's Nationalist forces in the Spanish Civil War, New York psychoanalysts who went to watch the parade down Broadway were likely to see most of their patients in attendance, as well as intellectuals, playwrights, social workers, and anthropologists, such as Franz Boas, who was deeply committed to the cause of the Spanish Loyalists.

The Final Solution

As Americans continued to struggle with the depression in the late 1930s, historical forces were set into motion that would lead to a resolution of the nature-nurture debate. It was an era of Clark Gable and Greta Garbo, the Marx Brothers and Mae West, the Lone Ranger, and Charlie McCarthy, Tommy Dorsey and Glenn Miller, *Gone with the Wind* and *The Wizard of Oz*. It was as if an age of innocence was being used as a prelude to events which were to follow to render them all the more horrifying.

Franz Boas, approaching age 80, threw himself into the nature-nurture debate one last time. He tried unsuccessfully to persuade the National Academy of Science and other organizations to pass resolutions condemning the Nazi regime, and in 1937 he retired from his

Columbia professorship so that he could, as he explained to *Time* magazine, "carry on my research, particularly on race questions. . . . With the present condition of the world I consider the race question a most important one . . . for this country too, as here also people are going crazy." That same year he published an article in *Forum* decrying the logic of Nazi pseudoscience, which Boas claimed had been driven "to its fanatical extreme. . . . There is not the slightest scientific proof that 'race' determines mentality, but there is overwhelming evidence that mentality is influenced by traditional culture." In his "Credo," published in 1938, Boas wrote, "The hysterical claims of the 'Aryan' enthusiasts for the superiority of the 'Nordics' have never had any scientific background." Boas had been writing about nature and nurture for over 40 years.

An official change in German policy regarding individuals with mental retardation and mental illness was announced in September 1939. On the same day that Hitler invaded Poland he ordered the murder of the incurably ill in German hospitals. Such individuals, it was argued, were "useless eaters" and "lives devoid of value." Euthanasia was best both for such individuals and for the state. It was not a new idea and had been promoted by Karl Binding and Alfred Hoche, a lawyer and a psychiatrist, as early as 1920 in *The Release and Destruction of Lives Devoid of Value*. At the Nazi Party Congress in 1935 the idea had been further promoted as official government policy; Dr. Gerhard Wagner, a physician, argued that such a program was consistent with "the natural and God-given inequality of men." Nor was euthanasia for the mentally retarded and mentally ill a peculiarly German idea. As late as 1942 Foster Kennedy, a professor of neurology at Cornell University, writing in the *American Journal of Psychiatry*, urged a program of euthanasia "for those hopeless ones who should never have been born—Nature's mistakes. . . . It is a merciful and kindly thing to relieve that defective—often tortured and convulsed, grotesque and absurd, useless and foolish, and entirely undesirable—of the agony of living."

As they had done with sterilization, the Germans set up an extensive and efficient program to carry out their objectives. Twenty-eight hospitals were officially designated as transfer institutions for mentally retarded children; once there the children were killed by injections, exposure, or starvation. In January 1940 the first adult mental patients were killed by shooting, but that method was abandoned as inefficient, so medical authorities sought better means. At a hospital near Berlin a special shower room was set up for "disinfections"; after the mental patients had been herded into the room, carbon monoxide was released, killing them. This hospital was used as a model for five others that carried out similar programs; in 12 months at one hospital alone there were 18,269 patients killed. By the end of the war it is estimated that approximately 275,000 mentally retarded and mentally ill individuals had been killed.

The evolution of Germany's program of mercy killings for the mentally retarded and the mentally ill to "the destruction of unworthy life," more broadly defined, took place at Wannsee, a suburb of Berlin, on January 20, 1942. Adolf Eichmann took minutes of the conference for the 14 assembled officials, half of whom possessed doctoral degrees. Much planning had preceded the meeting at which a "final solution" to the Jewish problem was agreed upon, a solution that would eventually include Slavs, Gypsies, Marxists, prostitutes, homosexuals, and other individuals deemed unworthy of life. According to Robert N. Protor's *Racial Hygiene*, "Gas chambers at psychiatric institutions in southern and eastern Germany were dismantled and shipped east where they where they were reinstalled at Belzec, Majdancek, Auschwitz, Treblinka, and Sobibór. The same doctors and technicians and nurses often followed the equipment." In March 1942, the first people arrived at Belzec and Auschwitz for what was referred to as "resettlement." Minister for Propaganda Joseph Goebbels noted in his diary that "not much will remain of the Jews."

In America a resolution of the nature–nurture question was under way. Eugenicists and others favoring a genetic view of human behavior were inextricably and logically linked in the mind of the public with events taking place in Germany. Hitler's program was represented as applied eugenics, yet Americans willing to publicly defend such programs had become few in number. Frederick Osborn died in 1935 and Madison Grant followed two years later. The Eugenics Record Office closed; Charles Davenport lived until 1944, but he was conspicuously silent about eugenics in Germany. One side of the nature–nurture debate had lapsed into virtual silence.

Those favoring a cultural view of human behavior, on the other hand, grew progressively stronger. The writings of Boas, Benedict, and Mead circulated increasingly broadly among the educated elite. Such writings merged with those of Freud and his followers, who argued that early childhood experiences were crucial determinants of behavior. The name "Freud," once merely a euphemism for sex with overtones of social reform and liberal political belief, slowly became reified into a broader symbol of nongenetic approaches to human behavior, liberalism, and humanism.

Sigmund Freud would probably have been amused by this reification, but he had died. Even in his declining years Freud never lost his sense of irony about life. When the Nazis finally agreed to let him leave Vienna in 1939, they demanded that he sign a document testifying that he had been well treated; Freud asked for and was given permission to append a statement at the end which said: "I can heartily recommend the Gestapo to anyone." The following year Freud lay in his London home listening to the air raid sirens, his facial cancer so far advanced that his pet chow crouched far in a corner of the room to avoid the smell. "It is my last war," Freud told his doctor shortly before he died.

Freud drew his last breath on September 23, 1939. On that day German planes conducted eleven separate raids on Warsaw and newspapers reported that half the city was in flames. The order to kill incurably ill patients in German hospitals had been issued and the program of mass genocide was officially under way. Ironically, the Holocaust would play a crucial role in the resurrection of Freud, but the risen Freud would be substantially different from the one who had lived. Once Freud's body was gone his American followers were free to transfigure his soul into whatever shape they chose.

Margaret Mead was the first to proclaim the victory of nurture over nature. In a 1939 preface to the reissue of her Samoan and New Guinea work, Mead declared: "The battle which we once had to fight with the whole battery at our command, with the most fantastic and startling examples that we could muster, is now won."

For Franz Boas it must have been an enormously satisfying victory after almost half a century of fighting. Characteristically, he did not dwell on the present but rather talked of "a new duty" for the future. "We must do our share," he said, "in trying to spread the art, and to engender the habit of clear thinking. . . . We must do our share in the task of weaning the people from a complacent yielding to prejudice, and help them to the power of clear thought, so that they may be able to understand the problems that confront all of us."

On December 29, 1942, Boas gave a luncheon at Columbia University for a French colleague who had fled the Nazis. The conversation turned inevitably to Germany and to the subject of race. With a glass of wine in his hand Boas stood up, turned to his guest, and said: "I have a new theory about race. . . ." He paused momentarily, then fell backward, dead.

5

Postwar Propagation of the Freudian Faith

The concentration camp had become the
representative institution as the Gothic cathedral had
been of an earlier century. . . .
—STEPHEN WHITFIELD, *A CRITICAL AMERICAN*

*I*n the 1930s Sigmund Freud's theory was like an exotic plant being grown in hothouses in New York and a few other cities. Unless one lived in the environs of such hothouses the theory was of no more consequence than a rare orchid—occasionally depicted in magazines or books but merely a botanical curiosity. Following the war Freud's theory began to spread, first sending roots westward under the Hudson River and eventually extending tendrils into every American city and town. The transformation of Freud's theory from an exotic New York plant to an American cultural kudzu is one of the strangest events in the history of ideas.

No single event contributed more to the dissemination of Freud's ideas than World War II itself. At a superficial level it led to the migration of a large number of Freud's European followers to American shores, thereby making psychoanalysis both more respectable and more available to people who wished to try it. At a deeper level it led to a turning inward of Americans in general and intellectuals in particular as they tried to solve "the crime without a name," as Winston Churchill

*vaguely on the ground, next to an enormous pile of bones, piled up
like cordwood, from which protruded legs, arms, heads. . . . It was
unbearable. People coughed in embarrassment, and in embarrassment
many laughed.*

Kazin "could imagine my father and mother, my sister and myself,
our original tenement family of 'small Jews,' all too clearly—fuel for
the flames, dying by a single flame that burned us all up at once."

Magazines carried stories of the atrocities. The *Catholic World*
claimed that among "the horrible details of the Buchenwald camp
[were] stories of lampshades made of the tattooed human skin of prisoners." The *New Republic* told of a woman who had arrived at a concentration camp one evening with 36,000 other Jews but "the next day
only 2,000 were alive." As she was marched to the work camp "both
sides of the road were [lined with] hundreds of corpses. Most had no
heads. Those with heads had blood gushing from eyes and mouths."
Ladies Home Journal noted: "The Germans are in many ways *like us.*
That is what is terrifying about the concentration camps." Lionel
Trilling, in his influential book *The Liberal Imagination* (1953), recorded
his own reactions: "Before what we now know the mind stops; the
great psychological fact of our time which we all observe with baffled
wonder and shame is that there is no possible way of responding to
Belsen and Buchenwald. The activity of mind fails before the incom-
municability of man's suffering."

The worst was at Auschwitz where a single oven could burn two
thousand bodies every twelve hours and "at night the red sky . . .
could be seen for miles." Dr. Josef Mengele, "an elegant figure . . .
handsome, well groomed, extremely upright in posture," would stand

on the railway platform whistling music by Verdi or Wagner as he worked, directing the incoming Jews and Gypsies either "links" to the gas chamber or *rechts* to the work camps. Approximately two million people were murdered at Auschwitz, their gold teeth removed for the Reichbank, their hair collected for mattress stuffing, and their fat boiled down for soap. Irving Howe spoke for other Jewish intellectuals when he wrote, "We are living after one of the greatest and least explicable catastrophes in human history. . . . We know that but for an accident of geography we might also now be bars of soap."

The Holocaust precipitated a profound crisis of consciousness among American Jews in general and among intellectuals in particular. In Howe's works it elicited "a new rush of feelings, mostly unarticulated and hidden behind the scrim of consciousness. It brought a low-charged but nagging guilt, a quiet remorse." Many Jewish intellectuals sought expiation of their guilt and remorse in psychoanalysis. Freud had been a persecuted Jew himself; his postwar credibility as a symbol for such Jews was further elevated by the fact that four of his elder sisters—all aged eighty or older—had been killed in the death camps at Theresienstadt, Treblinka, and Auschwitz.

At the same time, as the aftermath of World War II was pushing American intellectuals into psychoanalysis, the final dénouement of the nature-nurture debate was taking place. The rise of Hitler had brought political aspects of the debate to center stage, so that by 1941 one professional observer noted, "The controversy regarding nature vs. nurture is being transformed into a conflict between rival social, political, and economic ideologies." The debate, which had raged for half a century, was finally resolved at Nuremberg, where nature was put on trial alongside the Nazi leaders. Nature had been indicted for promoting the belief that some races are genetically more equal than others. The Nazi extermination of millions of people had followed directly from such a belief, just as legislation to restrict immigration, laws restricting marriages, laws mandating sterilization for individuals thought to be genetically unfit, and euthanasia for those who were defective had all followed directly from that belief. Eugenics had evolved into genocide, a Nordic Nunc Dimittis, and nature was sentenced to be hanged along with 12 Nazi defendants. The sentence was carried out on September 30, 1946; henceforth, theories extolling nurture as the primary antecedent of human behavior, including the theory of Sigmund Freud, would be considered politically correct and encounter little intellectual opposition as they moved across America in the postwar years.

increasingly abundant evidence concerning the Great Terror of 1936 to 1938, during which, in Josef Stalin's words, "ten millions" of peasants had been "dealt with." One historian called the period "probably the most massive warlike operation ever conducted by a state against its own citizens." The fact that Stalin's victims included an estimated one million members of the Communist Party, including virtually all of Lenin's colleagues in the Russian Revolution, made it even more difficult for American Marxists to defend. The 1940 murder of Leon Trotsky in Mexico, the one Russian leader whom most intellectuals could unequivocally support, was the final act of disillusionment.

As Marxism waned in America and the "Red Menace" rose in its place, the previous support of Stalin by Western intellectuals looked increasingly fatuous. Joseph E. Davies, American ambassador to the Soviet Union, had written in the early 1930s that Stalin's eyes were "exceedingly wise and gentle. . . . A child would like to sit on his lap and a dog would sidle up to him." George Bernard Shaw, following a 1931 visit to the Soviet Union, had contrasted prisons in England, where "a man enters prison a human being and emerges a criminal type," with those in Russia, where a man enters "as a criminal type and would come out an ordinary man but for the difficulty of inducing him to come out at all." A decade later it had become clear that the most common exit from Soviet prisons was as a corpse. In the words of one observer: "By the end of the 30's Lenin's mummy had begun to stink."

At the same time as Soviet Marxism was being discredited, its American supporters were being increasingly viewed as threats to the state. On August 3, 1948, Whittaker Chambers testified before the House Committee on Un-American Activities, admitting that he had been a Communist in the 1930s and had been given documents by former State Department employee Alger Hiss. Chambers had been a friend of Lionel Trilling and other New York intellectuals and was the editor of *Time* magazine. Two years later Julius and Ethel Rosenberg were indicted for passing atomic secrets to the Soviet Union. Julius Rosenberg was known by many of the New York intellectuals from their days of arguing Marxist principles in the City College cafeteria. According to Alexander Bloom's assessment in *Prodigal Sons*, "The New York Intellectuals exhibited few doubts about the guilt of either Hiss or the Rosenbergs, then or later. . . . Among these intellectuals each case touched nerves or personal issues. . . . As with much of their other thinking, they read their own lives and positions into the Hiss and Rosenberg discussions."

The great American Inquisition was on, with Senator Joseph McCarthy playing Torquemada. The intellectuals, many of whom had been Communist supporters if not members, rapidly repudiated their former Marxism and dove for cover behind the American flag. The bellwether of this new found patriotism was a three-part symposium titled "Our Country and Our Culture," published in the 1952 *Partisan Review*. Issued at the height of McCarthyism and while the Rosenbergs were appealing their death sentences, it included contributions from almost every New York intellectual of note. An accompanying editorial observed:

> *We have obviously come a long way from the earlier rejection of America as spiritually barren, from the attacks of Mencken on the "booboisie," and the Marxist picture of America in the thirties as a land of capitalist reaction. . . . The wheel has come full circle, and now America has become the protector of Western civilization, at least in a military and economic sense. . . . More and more writers have ceased to think of themselves as rebels and exiles. They now believe that their values, if they are to be realized at all, must be realized in America and in relation to the actuality of American life.*

In the symposium Lionel Trilling wrote: "The ideal of the workers' fatherland systematically destroyed itself some time back. Even the dullest intellectual now knows better than to look for a foster-father in Thyestes," a reference to a Greek legend in which Stalin was being compared with Thyestes who had eaten the flesh of his three sons at a banquet. Sidney Hook added: "In the West non-conformists, no matter how alienated, can always win a hearing. . . . In the land of Purges and Brainwashing, the only thing a non-conformist can earn is a bullet

cially dissolved by ,

Driven toward ps,
the same time forced to .
an emerging cold war, ma.
moved closer to Freud. For this g.
for Marxism. This was noted in a study of patients undergoing psychotherapy in New York in 1960, in which it was reported that "Freud
began to replace Marx as the prophet" among such patients following
World War II. Similarly, Richard King, a historian of this period,
observed: "The demise of Marxism as a persuasive radical ideology was
paralleled by the rise of Sigmund Freud's theories to prominence
among postwar intellectuals. After World War II Freudian terminology
became the common coin of the intellectual realm." An eloquent
description of the intellectual substitute of Freud for Marx was playwright Arthur Miller's reflections on his own brief psychoanalysis in
the 1950s:

> My difficulties were surely personal, but I could not help suspecting
> that psychoanalysis was a form of alienation that was being used as a
> substitute not only for Marxism but for social activism of any kind.
> New York, that riverbed through which so many subterranean cultures
> are always flowing, was swollen with rivulets of dispossessed liberals
> and leftists in chaotic flight from the bombarded old castle of self-denial,
> with its infinite confidence in social progress and its authentication-
> through-political-correctness of their positions at the leading edge of his-
> tory. As always, the American self, a puritanical item, needed a scheme
> of morals to administer, and once Marx's was declared beyond the pale,
> Freud's offered a similar smugness of the saved. Only this time the
> challenge handed the lost ones like me was not to join a picket line or a
> Spanish brigade but to confess to having been a selfish bastard who had
> never known how to love.

Miller's assessment of the postwar substitute of Freud for Marx and
social activism was also echoed by Melitta Schmideberg, daughter of
Melanie Klein and herself a psychoanalyst, who observed of the New

York scene in 1961: " 'Being analyzed' was regarded as [a] hallmark of a progressive and liberal attitude (without the stress and possible dangers inherent in being involved in social or racial issues)."

The postwar shift in focus from Marx to Freud among New York's intellectuals was immediately perceptible in the pages of the *Partisan Review*. From 1934 to 1944, despite the fact that virtually everyone on the staff was knowledgeable about Freud and many were in psychoanalysis, there had not been a single article about Freud or psychoanalysis in the journal. In 1945 that abruptly changed with a posthumous essay by Freud ("Dostoevski and Parricide"), which the editors said "opens a series of texts for the times." Also published during 1945 was Randall Jarrell's "Freud to Paul: The Stages of W. H. Auden's Ideology," Saul Rosenzweig's "Freud, Master and Friend— Hanns Sachs," Lionel Trilling's "A Note on Art and Neurosis," and Robert G. Davis's "Art and Anxiety." This preoccupation with Freudian concepts continued in subsequent issues of *Partisan Review*, so that by 1948 there were articles on "Sullivan's Interpersonal Psychology," a book review of Ernest Jones's psychoanalytic study of Hamlet, and Lionel Trilling's review of *Sexual Behavior in the Human Male* (1948), better known as the "Kinsey Report," in which it was said that "the way for the Report was prepared by Freud . . . it often makes use of Freudian concepts in a very direct and sensible way." The *Partisan Review* had come a long way from its days as the Marxist voice of the John Reed Club.

As Freud was substituted for Marx among New York intellectuals in the 1950s, Freud acquired many of the ideological accoutrements which had previously been worn by Marx. Foremost among these was the mantle of the humanist, the benefactor of mankind and hope for its future. By 1952 a prominent psychoanalyst was claiming that his profession was "one of the most powerful humanizing forces in the twentieth century." Three years later Lionel Trilling assessed Freud's work as the "noblest and most generous achievement of Western culture." Joel Kovel, a young Jewish intellectual growing up in New York recalled "those dialectical disciplines, psychoanalysis and Marxism, where the Enlightenment has made a step forward, many of us see as an attempt to overcome the past of the human race." The language of Freud was to be the new futhark with which to write a better future.

The transformation of Freud from a plaything of intellectuals to the savior of mankind entailed a new respect for Freudian theory in the postwar years. No longer was psychoanalysis merely for the amusement of the Mabel Dodges of the world. It had acquired a numina which

importance o̶.
became a quasi-reli̶g̶
sion and transference. This ̶n̶e̶w̶ ̶s̶t̶a̶t̶u̶s̶ ̶w̶a̶s̶ ̶r̶e̶̶
of William Barrett, one of the postwar editors of the ̶P̶a̶r̶t̶i̶s̶a̶n̶ ̶R̶e̶v̶i̶e̶w̶.

> *We were perhaps the first generation in America to take Freud and psychoanalysis seriously. . . . [We] faced the matter of psychoanalysis with intellectual solemnity. If you could dig up the money at all, you had a moral and an intellectual duty to face yourself in psychoanalysis, and you were shirking if you didn't.*

The new Freud was also reflected by Milton Klonsky's 1948 portrayal of Greenwich Village in *Commentary* in which he observed: "When the political cliques of the 30's lost their passion and died, they never really died but rose to the bosom of the Father and were strangely transmogrified. Psychoanalysis is the new look, Sartor Resartus, but the body underneath is the same."

The Rise and Fall of "Diaperology"

With the death of nature at Nuremberg, the nature–nurture struggle was left with but a single contestant. The idea of genetic differences between individuals or groups of people had acquired distinctly Nazi overtones, and anyone voicing such opinions was suspected of being a member of some fascist fringe group. Nurture ruled; rather than a competition against nature there was merely to be a debate among supporters of nurture as to whether early childhood experiences or cultural factors were more important determinants of human behavior.

This preeminence of nurture in the postwar period affected all the social sciences but most especially anthropology. Margaret Mead and Ruth Benedict (until her death in 1948) assumed the role of Franz Boas's heirs in opposition to theories of genetic predetermination. Such theories had led, in Mead's words, to "the full terror of Nazism, with its emphasis on 'blood' and 'race.' " Thus theories emphasizing genetic aspects of behavior were nowhere to be seen in the postwar period, and with a clear field it was easy for theories of culture and early childhood experiences to establish themselves as truth.

For Margaret Mead the postwar period offered unprecedented opportunities for her ideas. In 1939 she had written:

> *The anthropologist is concerned with the interrelations between man's human nature, his natural environment, his technological inventions, his social organization and his symbolic structures of religion, art and philosophy by which he endows life with value and meaning. As the comparative student of many human cultures the anthropologist is in a position to discuss total systems. . . .*

In 1942 in *And Keep Your Powder Dry,* a book about American culture that Mead had written in three weeks, she had noted:

> *We must see this war as a prelude to a greater job—the restructuring of the culture of the world—which we will want to do, and for which, because we are also a practical people, we must realize there are already tools half forged.*

The postwar arena for Mead's "restructuring of the culture of the world" was a project entitled "Research in Contemporary Cultures," established at Columbia University by Mead and Benedict with a $100,000 grant from the Office of Naval Research. Between 1947 and 1953 this project involved the collaboration of 120 persons doing analyses of other cultures. It arose directly from work that Benedict and other anthropologists did for the Office of War Information during World War II, in which they analyzed the cultures of America's enemies and allies and, on the basis of these analyses, made recommendations for American conduct of the war.

The Research in Contemporary Cultures project aimed, according to Mead, to "develop a series of systematic understandings of the great contemporary cultures so that the value of each may be orchestrated in a world built new." It was to be an attempt to fuse "two scientific approaches to human behavior . . . cultural anthropology and the study of child development within the general schema of Freudian psychology. . . . Attention was focused upon the series of stages by which the child learned to perceive and deal with the world through his realization of his own developing body." The project would be, said Mead, "the most fruitful relationship between cultural anthropology and psychoanalysis."

The seminal importance of Freudian theory was clearly outlined by Mead in a 1953 essay describing the project. Psychoanalytic theory, Mead wrote, teaches us "that the child learns to relate itself to other people and to the world around it through the use of its body. . . . As different cultures emphasize different zones or different modes—intake,

tures, therefore, could be described just like individuals as emphasizing one or another of Freud's stages of infant development.

Mead also argued that since "a large part of psychic life is inaccessible to the acculturated and functioning individual in any society," other means must be used to understand aspects of a culture. Therefore, she said, "rituals, myths, films, popular art, as unconscious expressions of a society," could be "analyzed in accordance with this assumption" and thereby enable any observer to understand that culture. From this detached vantage it was thought unnecessary for researchers to visit the culture or even to speak the language. It was, in terms used for the title of the book which emanated from the project, *The Study of Culture at a Distance* (1953). Another important influence was that of psychoanalyst Abraham Kardiner, with whom Benedict had jointly taught a course; according to Mead, Kardiner assumed "that human cultures are projections of the conflicts and sufferings of childhood and their symbolic solutions, in such a way that childhood experience is assigned a definitely causal role in history."

Mead had begun fusing cultural and psychoanalytic constructs as early as 1931 while working among the Arapesh when, according to her recollection, she began "to work seriously with the zones of the body" and sent home for abstracts of the work of Karl Abraham, the definitive psychoanalytic codifier of the oral stage of development. By the time Mead began work in Bali in 1936, she was convinced that early childhood experiences were essential determinants of human behavior. According to anthropologist Marvin Harris, "her Balinese field trip . . . was the first occasion for which Freudian principles provided the major research frame," and the book which came from this research, *Balinese Character*, "is indeed saturated with psychoanalytic terms, concepts and nuances."

There is also evidence from Mead's personal life that Freud's theory was extremely important to her in the years following the war. Her daughter recalled Mead directing a drama project in which friends performed "the scenes in Shakespeare that seemed to prefigure contemporary psychiatric insights. . . . The primary emphasis in the choice of the scenes from *Hamlet* was Oedipal." In 1946 Mead began a series of regular trips to teach in the department of psychiatry at the University of Cincinnati School of Medicine, a bastion of traditional Freudian thought. In 1963 she added the Menninger School of Psychiatry in Topeka to her annual itinerary, and on one occasion rebuked the Men-

ningers for not training psychoanalysts more thoroughly; poor training
in psychoanalysis, warned Mead, would inevitably lead to "a coarsening of the whole intellectual approach." Mead was treated regally by
the Menningers; as early as 1943 William Menninger had approvingly
cited her Samoan work in a paper he wrote on the anal phase of psychosexual development. According to biographer Jane Howard,
"Wherever Mead went in the world . . . she made it known that she
wanted to meet the psychoanalysts, many of whom became her good
friends."

Mead also recommended psychoanalysis for everyone. She placed
an easel next to her daughter's bed enabling her to rise and immediately paint what she had dreamed. In 1946 Mead persuaded Gregory
Bateson, her third husband, to go into psychoanalysis, which he did,
with a female analyst. The analyst, however, "did not maintain the cus-
tomary analyst's detachment from patients," and on one occasion
"went along on a vacation to Nova Scotia with Bateson" and two
other people. By the early 1950s, according to one of Mead's staff at
the American Museum of Natural History, " 'the [office] atmosphere
was so analytically oriented it permeated everything'. . . . Her colleagues would come back from their psychiatrists and spend 'hours
talking over what had happened.' " Everyone who worked for Mead
was in psychoanalysis except for Mead herself; "for the upper ten percent of the upper ten percent," she once contended, "there is no analyst."

Since the Research in Contemporary Cultures project was heavily
dependent on Freud's theory of infantile sexuality, the outcome of the
project was preordained to be merely culturally descriptive on the best
of days and patently ludicrous on the worst. In work that Ruth Benedict did for the Office of War Information, for example, Benedict had
characterized kite-flying, a national pastime in Thailand, as symboliz-
ing the relationship of men and women because the heavier "male"
kites try and dominate the lighter "female" kites. She also said that
Burmese people "psychologically show a quasi-spontaneous criminality," and explained that Romania had switched sides during the war
because of inconsistencies in child-rearing methods. On the other
hand, Benedict produced a widely praised description of Japanese culture in *The Chrysanthemum and the Sword* (1946), which reflected her
considerable lyrical skills while including almost no psychoanalytic formulations.

The member of the Research in Contemporary Cultures group
who became best known—and most ridiculed—for his cultural analy-

considerable renown among his colleagues by successfully swallowing "an unborn mouse" in a tribal rite, Gorer became close "and quite happily platonic" friends with Margaret Mead. Gorer was unabashedly Freudian in his thinking and, like Benedict and Mead, believed that psychoanalytic constructs could be used to explain cultures.

In Gorer's analysis of Japanese culture for the Office of War Information in 1943, he forthrightly acknowledged: "I have never been in Japan; I cannot read Japanese; I have no special qualifications for discussing Japanese culture." Nevertheless, based on "books written in European languages" and "interviews with some two score informants who were either wholly or partially Japanese or who have had prolonged and intimate knowledge of Japanese life," Gorer went on to analyze Japanese child-rearing methods ("the most consistent and most severe aspect is cleanliness training, training in control of the sphincter"), which he contended were related to obsessive and compulsive personality traits, repressed rage, and "a deeply hidden, unconscious and extremely strong desire to be aggressive." According to Gorer, "This theoretical construct gives the best explanation for the striking contrast between the all-pervasive gentleness of Japanese life in Japan, which has charmed nearly every visitor, and the overwhelming brutality and sadism of the Japanese at war." Gorer's fanciful use of toilet-training practices to explain the Japanese character elicited derision from many colleagues and almost certainly contributed to Benedict's omission of such theories in *The Chrysanthemum and the Sword*.

Gorer was not to be dissuaded, however, by the skepticism of others. In 1949 he coauthored a book in which he claimed to have identified the key to the Russian character—the institution of swaddling infants. His sources of information consisted of interviews with Russian émigrés in New York, Russian novels, and Russian films. The swaddling of Russian infants, Gorer contended, led to such adult personality traits as impassivity, controlled rage, guilt, suspiciousness, and despotism. According to Marvin Harris, "Gorer attempted to show that such phenomena as the Bolshevik revolution, the Stalin purge trials, the confessions of guilt at these trials, and many other events of recent Soviet history were 'related' to the generalized rage and guilt feelings associated with swaddling."

The swaddling hypothesis, the idea for which had come from Margaret Mead and which she vigorously defended, quickly and derisively became labeled by critics as "diaperology." The entire Research in Contemporary Cultures project was increasingly questioned, especially when postwar research in Japan found that toilet-training practices were not significantly different from those in America and when further information on Russia showed that swaddling varied both geographically and over time. How could one postulate overriding personality characteristics for a people as varied as Russians, it was asked, and claim that they are related to child-rearing practices which are equally varied. Other critics pointed out the inconsistencies in the studies; for example when it was asserted that "the cradleboard [among Indians] in the Southwest is said to have induced 'passivity,' but among the Plains Indians 'aggressiveness.' " By the late 1950s the idea of using Freudian theory to explain cultures had all but howlingly passed into oblivion.

Despite the failure of the Research in Contemporary Cultures project to provide anthropologists with a platform for a new world culture, the infusion of Freudian ideas into anthropology helped to spread them more broadly and lent them an aura of social science respectability. By the time that "diaperology" had waned, Ruth Benedict had died, but an impressive number of younger anthropologists had been influenced by Freudian theory. In a 1961 survey of Fellows of the American Anthropological Association, 11 percent of those responding to a questionnaire reported having been in psychoanalysis; the list of those identified as showing "varying knowledges of and dispositions to use psychoanalytic insight" included an impressive number of anthropologists who would provide leadership in that profession during the next two decades. And though Ruth Benedict and Margaret Mead did not succeed in marrying Freud with cultural anthropology, they did succeed in installing Freud as anthropology's mistress, appropriate for discussion with close friends but not to be brought to professional meetings.

The other party in the affair between psychoanalysis and anthropology was also profoundly influenced by the encounter. Karen Horney, a refugee psychoanalyst from Hitler's Berlin, had settled in New York City in 1934 and soon thereafter met both Ruth Benedict and Margaret Mead. Horney was already questioning Freud's dogma regarding penis envy and the castration complex, and she was further influenced by the cultural relativism being propounded by her friends. Erich Fromm, a lay psychoanalyst who also had come from Berlin, joined this threesome, and during 1935 they met regularly to discuss the relationship of psychoanalysis and anthropology.

In the autumn of 1935 Karen Horney taught a course entitled "Culture and Neurosis" at the New School for Social Research. Her writings at the time presented neuroses as culturally influenced, and she

person's psychosexual development; Horney's beliefs were thus heretical by the standards of traditional psychoanalytic dogma. In her writings Horney footnoted Benedict's *Patterns of Culture* as well as Mead's *Sex and Temperament*. When Horney published her widely read book, *The Neurotic Personality of Our Time* (1937), the assistance of both Benedict and Mead was acknowledged; significantly, the original title of the book was to have been *Culture and Neurosis*. Two years later Horney published *New Ways in Psychoanalysis*, in which she rejected additional aspects of Freud's teaching, including the libido theory, the Oedipus complex, and the death instinct.

The great schism in American psychoanalysis became inevitable and finally took place in 1941: Karen Horney and a group of dissident psychoanalysts resigned from the New York Psychoanalytic Institute and created what became the American Academy of Psychoanalysis. Erich Fromm was part of the dissident group as was Ruth Benedict's friend, Harry Stack Sullivan. The dissident group was commonly referred to as "the culturalists" or neo-Freudians. Although there have been splinter groups which have broken off from both the traditional Freudians and from the culturalists in the intervening years, the fundamental breach between these two groups has continued to the present to divide American psychoanalysts.

It can be seen, then, that anthropology fundamentally influenced psychoanalysis just as psychoanalysis shaped anthropology. Marvin Harris observed that "the two disciplines tended to reinforce the inherent tendencies toward uncontrolled, speculative and histrionic generalizations which each in its own sphere has cultivated as part of its professional license." Essentially they have reinforced each others' mythology; Freud's teachings about infantile sexuality were modified by the writings of Benedict and Mead, which in turn had been strongly determined by their own social and sexual agendas. The meeting of the two disciplines was an encounter between Freud's fables and anthropology's apologues, with both sides claiming that each other's mythology proved the truth of its own.

The Media Discovers the New Freud

Ever since Freud's theory arrived in America in the early years of the 20th century, newspapers and magazines had found him to be of curiosity. His name was a veritable phallic symbol, an opportunity to titillate readers' erotic interests beneath the guise of scientific discussion. Following World War II, however, the media discovered a new and more serious Freud, the same one being promoted by the New York intellectuals. The new Freud was a man who would not merely redeem man's sexual life but who also promised to save his soul.

The pacesetter for media presentations of the new Freud was the *Life* and *Time* publishing empire of Henry and Clare Boothe Luce. Ms. Luce had undergone psychoanalysis in 1929 in an attempt to "shake off her psychic angularities." As a playwright in the 1930s, she incorporated psychoanalysis into her productions; for example, in *The Women*, according to Sievers' *Freud on Broadway*, "psychoanalysis becomes a fashionable fad for these wealthy women, satisfying their need for attention, flattery and someone to listen to their malicious chatter." It is unclear whether Henry Luce also underwent psychoanalysis, although it is known that he hired Freud's nephew, Edward L. Bernays, as a New York public relations agent in the 1920s. In 1943, when Clare Boothe Luce was elected to Congress, she also hired Bernays as her public relations advisor.

In view of the Luces' interest in Freud, it was not surprising when in 1947 *Life* magazine ran a nine-page article simply titled "Psychoanalysis." "A boom has overtaken the once obscure and much maligned profession of psychoanalysis," the article began. "It merely reflects the increase in popular knowledge and acceptance of psychiatry, and especially psychoanalysis, as a cure." The article went on to explain in detail basic Freudian concepts—id, ego, superego, the Oedipus complex, transference, et cetera—and emphasized "repressed sexual desires" and "infantile experiences" as the cause of most unhappiness. Patients who were benefiting most by psychoanalysis were said to be "middle-aged men and women oppressed by a sense of futility and searching for some meaning in their lives; men and women still in the grip of an Oedipus complex, and homosexuals of both sexes." The cost of psychoanalysis was acknowledged to be very high, but "these rates are justified by the fact that a psychoanalyst's training begins where a doctor's ends." Case histories of successful "cures" were included along with pictues of Freud, Jung, Adler, Brill, Franz Alexander, Karl Menninger, Karen Horney, Gregory Zilboorg, and others. Freudian theory was said to "already have profoundly influenced human knowledge"; furthermore, "future generations of parents may rear generations of neurosis-free children by avoiding the mistakes that plant the origins of neurotic trends." This enthusiastic article was *Life*'s psalm of Freud, a psychoanalytic supplication for a new and better world.

cles inspired by psychoanalysis following the war. "The True Freudians" appeared on September 10, 1945, complete with a picture of Dr. Abraham A. Brill, smoking a prominent cigar, which had underneath the caption: "In the beginning was sex." "For the Psyche" followed in *Time* on September 2, 1946, "Freudians and Catholics" on August 4, 1947, and "Couch and the Confessional" on August 2, 1948. By October 25, 1948, psychoanalysis had become an extended cover story in *Time*, which featured claims that it had "had some extraordinary results," a sidebar of psychoanalytic "lingo" for the lay person, and psychoanalyst William Menninger proclaiming that brain damage "of the same kind can be done by a bullet, bacteria, or a mother-in-law." Psychic trauma had indeed arrived.

The editorial milieu at *Life* and *Time* magazines accurately reflected an increasing Freudian focus in other articles. According to Ezra Goodman, a writer who contributed frequently to *Time*, by the 1950s: "The emphasis on two-bit psychiatry was particularly pronounced because most of the personnel at the magazine had been or were in analysis and were preoccupied with the subject. Often the stories in the magazine read more like psychiatric than journalistic reports. At one time the magazine seriously considered adding a new department entitled 'Psychiatry.' " It was *Time*'s practice to submit their articles on famous personalities to psychoanalysts "for a snap analysis. . . . Sometimes the cover-story material was shown to a West Coast psychoanalyst, sometimes to an East Coast one. After a while *Time* was not only running out of Hollywood subjects but out of psychoanalysts to psychoanalyze them."

Although *Life* and *Time* led the media parade to psychoanalysis in the late 1940s and 1950s, most popular magazines included articles extolling Freud and his theory. Some of these contributions were written by psychoanalysts, such as Gregory Zilboorg's "Psychoanalysis and Religion" in the *Atlantic Monthly* or Erich Fromm's "Oedipus Myth" in *Scientific American*. Others were written as personal "Freud saved my life and career" accounts, such as entertainer Sid Caesar's "What Psychoanalysis Did for Me" in *Look* magazine. Caesar, pictured holding a cigar, described his problems as "caused in a large part by a profound feeling of inadequacy which had a basis in my childhood," but "once I made clear to myself that those childhood incidents were in the past, I found I could start a new life. . . . My work began to improve."

By the mid-1950s the flow of Freud's ideas through popular magazines approached fluvial proportions. One measure of this interest was the first volume of Ernest Jones's biography of Freud, which became a "freak bestseller" when it was published in 1953. As measured by the number of articles about Freud or psychoanalysis cited by the *Readers Guide to Periodical Literature*, interest peaked from 1956 to 1959, during which time there were more articles on Freud (calculated as a rate per total number of published articles) than at any time since the period 1915–1922. The number of Freudian citations in the late 1950s was in fact two and a half times the number found in the same magazines ten years previously. They ranged from "How I Got Caught In My Husband's Analysis" (*Good Housekeeping*) and "Psychoanalysis Broke Up My Marriage" (*Cosmopolitan*) to "Dialogue of Freud and Jung" (*Harper's*), "Moses in the Thought of Freud" (*Commentary*), and "Psychoanalysis and Confessions" (*Commonweal*).

Newspapers in the postwar period also reflected the new interest in Freud. This was especially true of those published in New York, the fountainhead of Freudian orthodoxy. The *New York Times Magazine* began publishing regular articles on Freud and psychoanalysis in 1946. In an article titled "Dreams—Fantasies or Revelations?" the author noted: "Because of the increasing influence of psychoanalysis many a man today believes that his dreams are important. . . . It all started with Sigmund Freud." In another 1947 article, "Analysis of Psychoanalysis," William C. Menninger noted that "never before has psychoanalysis been so much a matter of general discussion. . . . The principles of psychoanalysis more and more are being applied on a wider basis."

In a 1949 article by psychoanalyst Franz Alexander, "Wider Fields for Freud's Techniques," he noted that "the gist of Freud's theory is that the human personality is molded by experiences, most importantly those of childhood" and urged that this theory be used for "the betterment of the world." In a 1953 article, "Analysis of Sigmund Freud," Freud was likened to a prophet of the Old Testament; the article concluded that "few men have had a greater influence on their age." By 1956 the *New York Times Magazine* was consistently directing reverential awe toward Freud. In "The Freudian Revolution Analyzed," Alfred Kazin wrote: "In the same way that one associates the discovery of certain fundamentals with Copernicus, Newton, Darwin, Einstein, so one identifies many of one's deepest motivations with Freud. His name is no longer the name of a man; like 'Darwin,' it is now synonymous with a part of nature. This is the very greatest kind of influence that a man can have."

At the same time as popular magazines and newspapers were introducing large segments of the American public to Freud's ideas in the postwar period, the world of the New York theater continued to promote him. As W. David Sievers noted in *Freud on Broadway*: "In attempting an overview of drama of the last decade . . . one fact is most impressive: The drama now regularly turns to psychoanalytic psychology for source material."

Tennessee Williams was a case in point. Although he did not begin his own psychoanalysis until 1957, Williams used Freudian themes liberally in his earlier plays. Sievers deemed *A Streetcar Named Desire*, first produced in 1947, "the quintessence of Freudian sexual psychology. . . . Williams arranges in a compelling theatrical pattern the agonized sexual anxiety of a girl caught between id and ego-ideal." Esther Jackson, one of Williams's biographers, noted that "he uses the Freudian language as a system for designating reality" and said that the "psychological myth" underlying his plays was deeply Freudian in content. The overt sexual symbolism of *Cat on a Hot Tin Roof*, first produced in 1955, has been especially cited in this regard.

William Inge, who also underwent psychoanalysis, was another example of Freud's effect on playwrights. *Come Back, Little Sheba*, his first major success, produced in 1950, exhibited, according to Sievers, "considerable Freudian influence"; in fact, "several psychiatrists read and approved the manuscript before it reached production." Similarly, Inge's 1953 Pulitzer Prize-winning *Picnic* was said to "draw upon Freudian insights without succumbing to the obvious or the trite." Inge himself acknowledged Freud's profound influence on playwrights because of "the feelings and viewpoints that Freud has exposed in us."

Eugene O'Neill was yet another major playwright who had been psychoanalyzed and who used Freudian themes liberally in his work. The revival of *The Iceman Cometh* and the newly produced *Long Day's Journey Into Night* were two of the most successful plays of the 1950s. According to critic Lionel Abel, much of the popularity of these two plays was based on the fact that "the members of the audience were going to analysts, and both plays involved analysis and self-analysis of a spontaneous kind by all the characters. . . . one of the enjoyments people felt in these plays of O'Neill was the feeling that they were justified in sacrificing whatever sums they were spending on therapy. Art has always had as one of its functions the justification of life, and the way of life of Americans in the fifties was, by and large, a psychoanalytic one."

There are suggestions that people who attended plays in the postwar period were indeed well-versed in the ideas of psychiatry. Charles Kadushin, in a 1960 study of patients in long-term psychotherapy in New York City, found that 90 percent of them frequent plays several times a year. Such data lend credence to T. S. Eliot's assessment of the success of *The Cocktail Party*, a 1950 play that centered on a mysterious guest who later emerged as a psychoanalyst/Christ figure. Eliot attributed the play's success "to the fact that its theme touched on what people then were most interested in: drinking and psychoanalysis." Eliot himself apparently did not undertake a personal psychoanalysis although he had been hospitalized in Lausanne, Switzerland, for three months in 1921 for psychiatric symptoms which apparently included depression. The lead character in *The Cocktail Party* "appears to be modeled upon the actual Swiss physician who treated him in 1921," according to Jeffrey Berman. When in the play one of the characters asks, "What is hell?" and immediately responds, "Hell is oneself," the New York audience probably nodded and made a mental note to discuss this issue with their own analysts at the next session.

Lillian Hellman also was deeply influenced by psychoanalysis. Her first play, *The Children's Hour*, produced in 1934, had caused a sensation by depicting lesbianism "as a tragic flaw rather than a loathsome anomaly." In 1940 Hellman began a psychoanalysis with Dr. Gregory Zilboorg that continued for seven years; later in her life she underwent psychoanalysis a second time with another therapist. Hellman's 1946 play, *Another Part of the Forest*, which ran for 191 performances, was dedicated to "my good friend, Gregory Zilboorg" and was said by biographer William Wright to contain a "plethora of Freudian and mythic reverberations" and "Oedipal–Electra vibrato."

Hellman's case also illustrates the continuing connections between psychoanalysis and liberal politics in postwar America. Hellman acknowledged that she had joined the Communist Party and had visited Russia at the height of Stalin's show-trials in 1937 (although she claimed she knew nothing about them) and again in 1944, activities which brought her to the attention of the House Un-American Activities Committee in 1951. Among her lovers had been magazine publisher Ralph Ingersol, a Marxist, and another was Dashiell Hammett, mystery writer and active member of the Communist Party.

Both Hellman and Ingersol were in psychoanalysis with Dr. Zilboorg, who had been raised in Russia and had been a secretary to the

boorg was a flamboyant figure in the New York psychoanalytic community and had as another of his patients Marshall Field III, wealthy heir to a department store fortune. According to William Wright, "whatever Zilboorg did for Field's mental health, he was said to have played a major role in politicizing him." With Zilboorg playing a coordinating role, Marshall Field agreed to bankroll *PM*, a new left-wing newspaper, which began publishing in June 1940, with Ingersol as its editor and Hellman and Hammett involved in its planning. Zilboorg himself was a stockholder in the paper and wrote articles for it under a pseudonym. Zilboorg regularly became involved in the lives of his patients outside of the office—he was Hellman's guest at her farm on numerous occasions—and was censured by the American Psychoanalytic Association for such activities.

Another of Zilboorg's patients was George Gershwin, who maintained the tradition of linking psychoanalysis and Broadway musicals. Richard Rodgers was said to have used psychoanalytic ideas in his work as early as 1926 and Oscar Hammerstein II had also "shown an early affinity" for psychoanalytic themes. Sievers claimed that their collaborative efforts, such as *Carousel* and *Oklahoma*, continued to incorporate Freudian ideas. Similarly, Alan Jay Lerner and Frederick Loewe were said to have been "clearly indebted to psychoanalysis" for their moderately successful musical, *The Day Before Spring*, which was produced in 1945. An unsuccessful psychoanalysis may have preceeded the writing of another musical, *Inside U.S.A.*, which was staged in 1948; in one scene a dancer guides her analyst to the edge of a cliff and pushes him over it.

Prior to the founding of the Los Angeles Institute for Psychoanalysis in 1946, virtually all films which depicted Freudian themes seriously were taken directly from Broadway plays. In the 1930s these included *Reunion in Vienna* (1933), based on Robert E. Sherwood's play, in which traditional European values had "been replaced by rational values embodied most perfectly by psychoanalysis." *Blind Alley* (1939), based on a play by James Warwick, portrays an omniscient psychiatrist who lectures an escaped killer on unconscious motivation and dream interpretation. Other movies of the 1930s treated Freud less charitably. For example, in *The Front Page* (1931), based on the play by Ben Hecht and Charles MacArthur, an eminent psychiatrist is brought from Vienna to help the police solve a crime and in trying to do so clumsily gets himself shot. In *Carefree* (1938) Fred Astaire played a dancing psychiatrist who used psychotherapy, according to Gabbard and Gabbard's

Psychiatry and the Cinema, as "primarily a pastime for members of the idle rich who are a little short on common sense"; Astaire was regularly referred to in the movie as a "quack," even by the ducks he encountered in Central Park.

The influence of the New York theater on American culture increased rapidly following World War II, as New York writers and producers migrated to Hollywood to work in the burgeoning movie industry. Included in their luggage was *The Basic Writings of Sigmund Freud* (1938). The arrival of these writers and producers, steeped in Freudian theory, coincided with the founding of the Los Angeles Institute for Psychoanalysis by European psychoanalysts who had fled Hitler's Europe.

The Los Angeles Institute was from its inception a blend of Freud and Marx. Its most influential founders were Otto Fenichel and Ernst Simmel, both refugees from the Berlin Psychoanalytic Institute. Fenichel had been a member of the Communist Party, while Simmel was said to be "a dedicated Socialist" who was so thoroughly committed to egalitarianism that at his sanitarium near Berlin he had insisted that all his employees undergo psychoanalysis. Fenichel and Simmel were joined in Los Angeles by the psychologist Max Horkheimer and the socialist Theodor Adorno, both men exiles from the Frankfurt Psychoanalytic Institute who "cherished the idea of synthesizing Marx and Freud in their critique of modern society."

It was, then, a Fabian version of Freud that flowered in the Southern California sun in the years following the war. Many leaders in the movie industry, especially those with Jewish roots, identified Freud as one of their own. As post-Holocaust guilt and agonizing self-appraisal seeped into many of these leaders, they sought atonement in analysis. Freud had previously represented merely a euphemism for Hollywood's sex—Samuel Goldwyn had called Freud "the greatest love specialist in the world" and offered him $100,000 in 1922 "to cooperate in making a film depicting scenes from famous love stories of history"—but after the war Freud assumed personal, political, even spiritual dimensions as well.

"Hollywood found that psychoanalysis was fun," wrote Otto Freidrich in *City of Nets: A Portrait of Hollywood in the 1940's.* "Hollywood was full of neurotic people who wanted the meaning of their lives explained to them and who had lots of money to pay for the explanations." The men at the top were almost all in analysis: David Selznick rang his psychoanalyst's "doorbell at midnight and, standing outside, demanded to be heard"; and Darryl Zanuck's psychoanalyst was seated immediately to his left during some working lunches at Twentieth-

that "psychoanalysts were everywhere in evidence. There was one occasion when a redheaded starlet was caught speeding in a foreign sports car with a gentleman friend. She informed the police that the gentleman in question was her psychiatrist and that she was traveling fast because she was emotionally disturbed. The cops, somewhat baffled, let the matter drop." By the mid-1950s Bedford Drive in Beverly Hills had become known as "headshrinkers' row," and *Newsweek* reported that "in Los Angeles psychoanalysis had become a fad. . . . Everyone talks about his analysis or analyst. . . . Conversation is pervaded with psychoanalytic jargon."

American movies began reflecting Hollywood's new preoccupation with ids and egos even before the war ended. In 1944 Moss Hart's successful Broadway tribute to his psychoanalyst, *Lady in the Dark*, was made into a movie about a career girl (Ginger Rogers) who turns to her psychoanalyst (Ray Milland) for help in choosing between competing suitors. The psychoanalyst in the movie "was not only compassionate but intelligent, sophisticated and suave."

The real harbinger of Hollywood's psychoanalytic future was *Spellbound*, which was released in 1945. The idea for the movie emerged from producer David Selznick's personal psychoanalysis with Dr. May E. Romm, who was listed in the film's credits as a "psychiatric advisor." The script was written by New York playwright Ben Hecht, who had moved to Hollywood to become a screenwriter and was also in psychoanalysis at the time. Alfred Hitchcock was hired to direct it, Salvador Dali retained to paint the background for the dream sequences, and Gregory Peck and Ingrid Bergman cast in the lead roles.

Spellbound was in essence an advertisement for psychoanalysis. It opened with a prologue which explained:

> *Our story deals with psychoanalysis, the method by which modern science treats the emotional problems of the sane. The analyst seeks only to induce the patient to talk about his hidden problems, to open the locked doors of his mind. Once the complexes that have been disturbing the patient are uncovered and interepreted, the illness and confusion disappear. . . .*

The plot of the movie was a murder mystery with Hitchcockian twists set in a private psychiatric hospital. During the course of unraveling the mystery, various psychoanalysts expounded on the techniques of analysis; for example, one scene in which Ingrid Bergman plays a psychoanalyst is said to "contain the first detailed discussions of counter-transference in the American cinema." Psychoanalysis was shown to be clearly effective because the mystery was solved at the end of the movie by using clues from a patient's dream.

The Golden Age of psychiatry in the movies, as Gabbard and Gabbard referred to it in *Psychiatry and the Cinema*, really began with *Spellbound*. Although there were occasional films that were unflattering to psychiatrists, over the next 20 years the majority of them depicted psychiatrists, and especially psychoanalysts, as "authoritative voices of reason, adjustment and well-being." For example, Stanley Kramer's 1949 *Home of the Brave* featured a kindly and skillful psychiatrist who successfully treated hysterical paralysis in a war veteran by having him re-experience his psychic trauma. In Kramer's next film, *The Caine Mutiny*, released in 1954, a psychiatrist explained Captain Queeg's personality as the consequence of childhood problems, which included his parents' divorce. The symbolism of Queeg's compulsive rolling of two steel balls in his hand as he became more and more agitated was readily understood by Freudian aficionados who flocked to this popular film.

It is also of interest that the movie version of *The Caine Mutiny* treated psychiatrists more kindly than had Herman Wouk's original novel. The same can be said for *The Shrike*, a 1955 film based on Joseph Kramm's Pulitizer Prize—winning play of three years previously. In the play the psychiatrist was portrayed as a co-conspirator with a woman in helping her to involuntarily hospitalize her husband (José Ferrer), who wants to leave her. In the film, however, the psychiatrist persuades the woman to undertake psychotherapy and tells her husband to give their marriage another try; thus the Broadway psychiatrist, who was merely a jailer, became transformed by Hollywood script writers into "the oracular psychiatrist capable of identifying and healing psychologically wounded people."

Psychiatrists in general and psychoanalysts in particular continued to be portrayed as sympathetic, wise, and competent individuals in most movies in the late 1950s. Two of the most noteworthy examples were *Fear Strikes Out*, in which a benign psychiatrist apparently cured baseball player Jimmy Piersall (Anthony Perkins) of his mental illness, and *The Three Faces of Eve*, in which another competent psychiatrist helped a woman with multiple personalities (Joanne Woodward) resolve the conflicting claims on her psyche and live happily ever after. *The Three Faces of Eve* had overtones of *Spellbound* as an overt advertisement for psychiatry; it opened with Alistair Cooke solemnly intoning that what the viewer was about to see was the truth.

Life, *Time*, and the *New York Times Magazine*, but also had become a star of stage, screen, and radio. He had been featured as early as 1947 on "One Man's Family" in radio episodes in which the "omniscient psychiatrist clears up a bad case of juvenile delinquency, reorients the entire family, and converts Father Barbour (and you know Father Barbour!) apparently in about three interviews." The followers of Freud were ready for a broader assault on American culture. In accomplishing this goal they would be helped enormously by beachheads that had been established in the areas of child rearing and criminology.

6

Freud in the Nurseries

Most of the damage we have seen in child rearing
is the fault of the Freudian and neo-Freudians who have
dominated the field. They have frightened parents and
kept the truth from them. In child care I would say
that Freudianism has been the psychological crime
of the century.
—DR. LOUISE B. AMES, CO-DIRECTOR OF THE
GESELL INSTITUTE OF CHILD DEVELOPMENT

*B*enjamin Spock probably did more than any single individual to disseminate the theory of Sigmund Freud in America. Through the sale of 40 *million* copies of his book *Baby and Child Care* and his abundant writings in popular magazines, Spock persuaded two generations of American mothers that nursing, weaning, tickling, playing, toilet training, and other activities inherent in childhood are not the innocuous behaviors they appear to be on first glance. Such activities, according to Spock, are really psychic minefields that determine a child's lifelong personality traits, and maternal missteps on such terrain can result in disabling and irrevocable oral, anal, or Oedipal scars. Throughout his career Spock was deeply imbued with Freudian doctrine, and in a 1989 interview he acknowledged, "I'm still basically a Freudian."

Spock's child-rearing manual is said to follow only the *Bible* and the works of Shakespeare as the all-time best selling book in the

English language. Since its publication in 1946, it has also been translated into 39 languages, ranging from Croatian to Urdu. In 1990, when *Life* magazine published its list of the century's most influential Americans, Spock was prominently included. Spock had become a method, a belief system, and an institution; the man who was once merely Spock the pediatrician had undergone typographic transubstantiation and become Spock The Pediatrician.

It is important to realize that Homo sapiens had successfully raised children for several thousand years before child-rearing experts began proffering advice. In 1914 *Good Housekeeping* published an article, "Mothercraft: A New Profession for Women," and proclaimed a new era in which "the amateur mother of yesterday" would be replaced by "the professional mother of tomorrow." American mothers began looking to professionals for advice, and it was at this time that Sigmund Freud's theory arrived from Europe.

In historical retrospect it is surprising that it took so long for Freud's theory to become incorporated into American child-rearing practice. Freud himself had noted the implications of his theory for child rearing, and many of his early followers, especially his daughter Anna, urged that Freudian principles be used for the process by which "unrestrained, greedy and cruel little savages" are turned into "well-behaved, socially adapted, civilized beings." The key to child rearing, said Anna Freud, is the realization "that children, no less than adults themselves, are dominated by their sexual impulses and aggressive strivings."

Early supporters of Freud in America also pointed out the importance of his theory for child rearing. Max Eastman, in his 1915 paean to Freud in *Everybody's Magazine*, remarked on the "new wisdom [Freud] brings to the art of educating young children." That same year *Good Housekeeping* carried a Freud-inspired account of how excess fondling of young children or allowing them to view anything sexual was likely to be psychically damaging. Such Freudian warnings about child rearing were largely ignored, however, except among a small coterie of social workers and educators who were involved in the mental hygiene and child guidance movements.

Several studies have shown how little influence Freudian theory had on child rearing in America prior to World War II. One study examined 455 articles on child rearing published in popular magazines between 1919 and 1939 and specifically looked for references to the unconscious, infantile sexuality, or the Oedipal conflict. Remarkably few such references were found, leading the author to conclude that

Freud had had virtually no impact on the popular literature on child rearing during those two decades. Another study examined all articles on psychoanalysis published in popular periodicals between 1910 and 1935 and found that "only 17 percent of these articles said anything at all about the meaning of these psychological ideas for rearing young-sters." A third study analyzed all child-rearing manuals published in the 1920s; only 6 of 42 such books mentioned the oral or anal stage, two mentioned the Oedipal stage, and only one "appeared to be written avowedly and systematically according to Freudian theory."

John Watson and His Periscope

The major reason why Freud exerted so little influence on American child rearing prior to World War II was the popularity of behaviorism in general and the theory of John B. Watson in particular. Basing their ideas on animal models of behavior developed by Russian scientists Pavlov and Bechterev, behaviorists viewed children as stimulus-response machines who could be trained just as animals could be trained. Watson viewed children's brains as *tabulae rasae*, blank sheets upon which any story could be written depending on the environmental conditions. To illustrate his theory Watson claimed that he could take any "dozen healthy infants" and with proper training, "I'll guarantee to take any one at random and train him to become any type of specialist I might select—doctor, lawyer, artist, merchant-chief and yes, even beggarman and thief, regardless of his talents, penchants, tendencies, abilities, vocations, [or] race of his ancestors."

In its theory of child rearing, emphasizing nurture rather than nature, Watson's behaviorism was compatible with Freudian theory and laid the groundwork for the latter's acceptance by American mothers following World War II. In actual child-rearing practices, however, the two approaches differed widely. Watson's method of child rearing was characterized by strict schedules and unvarying routines toward the goal of producing independent and self-reliant adults. Watson was especially fond of comparing the rearing of children to the training of dogs: "If you expected a dog to grow up and be useful as a watch dog, a bird dog, a fox hound, useful for anything except a lap dog, you wouldn't dare treat it the way you treat your child." Watson believed that children should have to overcome obstacles as part of their training. Watson himself had had to overcome such obstacles; as a child his alcoholic father had abandoned the family and Watson had been arrested twice as a teenager. Later, as a professor of psychology at Johns

university, at which point he turned his ideas to commercial advantage and became an executive with an advertising agency.

Watson's 1928 child-rearing manual, *Psychological Care of Infant and Child*, was very influential among American mothers and sold over 100,000 copies. The *Atlantic Monthly* called the book "a godsend to parents" and *Parents' Magazine* said that it ought to be "on every intelligent mother's shelf." By current standards of child rearing it is a remarkable book. It recommended strict four-hour feeding schedules for infants, toilet training to begin at six months, and physical punishment from the earliest months. At night it recommended "a quiet goodnight; lights out and door closed. If he howls, let him howl."

Watson was never to be accused of spoiling children:

> There is a sensible way of treating children. Treat them as though they were young adults. Dress them, bathe them with care and circumspection. Let your behavior always be objective and kindly firm. Never hug and kiss them, never let them sit in your lap. If you must, kiss them once on the forehead when they say good night. Shake hands with them in the morning. Give them a pat on the head if they have made an extraordinarily good job of a difficult task.

He recommended preparing children for the vicissitudes of life from their earliest months, especially avoiding the "over-conditioning" which he said comes from a mother's love:

> If you haven't a nurse and cannot leave the child, put it out in the backyard a large part of the day. Build a fence around the yard so that you are sure no harm can come to it. Do this from the time it is born. When the child can crawl, give it a sandpile and be sure to dig some small holes in the yard so it has to crawl in and out of them. Let it learn to overcome difficulties almost from the moment of birth. The child should learn to conquer difficulties away from your watchful eye. No child should get commendation and notice and petting every time it does something it ought to be doing anyway. If your heart is too tender and you must watch the child, make yourself a peephole so that you can see it without being seen, or use a periscope.

To avoid too close an attachment Watson advocated that nursemaids be rotated regularly. "Somehow," he added, "I can't help wishing that it were possible to rotate the mothers occasionally too!" Such an arrangement might help counteract "mother love," which Watson labeled as "a dangerous instrument . . . which may inflict a never healing wound."

Given this strict behaviorist milieu, it is understandable why Dr. Spock's gentler advice was eagerly seized upon by mothers in later years. But despite the marked difference in approaches between Watson and Spock, they did share one important assumption that made acceptance of Freud's theory more likely. Watson, like Spock, assumed that early childhood experiences are the important determinants of personality, and he stated explicitly that "there is no such thing as an inheritance of capacity, talent, temperament, mental constitution and characteristics." Mothers should ask themselves: "Isn't it just possible that almost nothing is given by heredity and that practically the whole course of development of the child is due to the way I raise it?" It was a 20th-century version of John Locke and Jean Jacques Rousseau, who had viewed children's minds as blank slates at birth.

Benjamin Spock: The Perils of the Potty Chair

Benjamin Spock grew up in a strict New England household dominated by his mother. His father, whom Spock described as going "out of his way to be companionable with me," a "grave but just" man who was "fair and calm in any case of disapproval," was a lawyer for the New Haven Railroad. Spock's mother by contrast was "too controlling, too strict, too moralistic. . . . I was intimidated by her." On one public occasion Spock even said, "My mother was a tyrant." His education included "a private elementary school in an unheated tent where the children sat in felt bags to keep warm." As a child he was extremely timid, "worried about lions in the thicket in the large vacant lot next to our house" and afraid that he was going to be kidnapped by Italian women who came into his neighborhood to dig dandelions with which to make dandelion soup; when he saw them coming he would "rush into the vestibule with palpitating heart . . . [and] slam the door." Spock was, by his own admission, "a goody-goody . . . [who] never committed a mildly delinquent act, even on Halloween."

Spock appears to have been especially inhibited in sexual matters. "My mother frequently warned us against masturbation. . . . We must also not have 'naughty thoughts.' My mother led me to believe they might cause physical malformations in my children." By age 17 Spock "had never even kissed a girl—[he] had never had a date." He completed three years at Yale University before he had took his first drink.

Spock's success at Yale came academically and as a member of the university rowing team, which won a gold medal in the 1924 Olympics. Following Yale he went to medical school at Columbia University, where he graduated first in his class and then went on to a residency in pediatrics at Cornell Medical College. During these years he married Jane Cheney, a political activist from a wealthy family, and fathered the first of his two children.

Spock's introduction to the theory of Sigmund Freud took place through his wife, who began psychoanalysis shortly after their marriage. He recalled being especially interested in the psychological aspects of his patients' illnesses during his medical training, so he wrote to several pediatricians to inquire where he might get special training in this field. He decided to take a residency in psychiatry for a year at Cornell Medical Center but found the experience disappointing. "Taking care of schizophrenic and manic-depressive patients wasn't at all what I needed. . . . I learned that the staff members . . . who could make sense out of the psychotic patients' behavior and talk were those who had had Freudian psychoanalytic training. So I resolved to get that training for myself."

Spock began the first of his three personal psychoanalyses in 1933 with Dr. Bertram Lewin, an orthodox Freudian and one of the earliest members of the New York Psychoanalytic Society. He also enrolled in two seminars at the New York Psychoanalytic Institute, which he continued for five years, and undertook the supervised psychoanalyses of three patients, including one young woman whom he saw five times per week for three years. Spock's attempts at becoming a psychoanalyst were, by his own admission, singularly unsuccessful, with all three of his treatment cases being failures. Spock recalled that in his analysis of the young woman, "Over a three-year period we did not make significant progress. . . . the analysis was blocked by the patient's rivalry with me." Twenty years later Spock attempted under supervision to psychoanalyze two more patients, but confessed again, "I was not successful in either case." In 1941 he undertook his second personal analysis, this time with Dr. Sandor Rado, who had been one of Freud's favorite followers. Spock's third personal analysis lasted six years and took place in the 1980s, after Spock had divorced Jane, his wife of 48 years, and married Mary Morgan, who was 40 years his junior.

Spock's first psychoanalysis led him to become a devout believer in Freud's theory. He acknowledged that "all children are born with somewhat different temperaments" but argued that "personalities are formed from the impact of the environment—of which the parents

constitute a very signficant part—on children's inborn temperament."
In keeping with psychoanalytic theory Spock viewed children's early
experiences as crucial in shaping their later development.

Like most adherents to Freud's theory, Spock found "proof" of the
theory's correctness within his own psychoanalysis:

> *When I began psychoanalysis with Dr. Bertram Lewin at the age of*
> *thirty, I complained endlessly about my mother's criticalness and domi-*
> *nation for several months. Dr. Lewin urged me to remember my*
> *dreams, and I discovered that my mother hardly appeared in them at*
> *all—the ogres and lions and kidnappers led by association to my father*
> *and other father-figures. This is in accord with Freud's observation that*
> *the most basic rivalry—and most subtle fear—is what the son feels*
> *toward his father, mainly in his unconscious mind. So although my*
> *mother's criticalness and warnings contributed much, I think, to mak-*
> *ing me a timid child, my deeper instincts made me fear my father's*
> *anger more, even though I never saw it come out.*

Being scientifically oriented, Spock later attempted to prove the
validity of Freud's theory. In 1959, with a $30,000 foundation grant, he
recruited 21 families who were expecting their first child. Each family
was counseled twice a month for six years by one of eleven eminent
faculty members from the departments of psychiatry and pediatrics at
Case Western Reserve University in Cleveland, where Spock was
working at the time. The psychoanalytically oriented faculty members
included several who had been trained by Anna Freud. Spock's
hypothesis was that "with such skilled counselors consulting with each
family a full hour every other week, we could prevent difficulties"
especially in such areas as breast feeding, thumb-sucking, toilet training, and sibling rivalry. The children were then followed up for at least
thirteen years.

The results of the study provided no support whatsoever for
Freud's theory and, not surprisingly, little of the data was ever published. Spock acknowledged that despite the intense psychoanalytically
oriented counseling, "the children in the Study had just as many problems as any other children." (For example one child "was still wetting
the bed at twelve years in spite of all the psychiatric and pediatric inge-
nuity expended on his case for nine of those years.") Providing psychoanalytically oriented counseling for the mothers for one hour every
other week was supposed to make child-rearing problems easier, but
instead made the problems more difficult. Toilet training of the study

anticipated." The mothers in the study began toilet training of their child early in the second year, but "at the first signs of resistance from the child they stopped their efforts and postponed the business for a while." Spock described the mothers' efforts as "inconsistent and vacillating" and said that "most of them became discouraged and apprehensive as soon as their children showed any resistance."

Spock further noted that it turned out to be virtually impossible to predict which children would have problems later in childhood on the basis of early experiences with their parents, and specifically cited two very disturbed young children who "turned out quite well." The study, probably the most ambitious ever undertaken to prove that child rearing based on Freud's theory can ameliorate developmental problems, had completely negative results. Despite such conclusions, Spock continued to promote Freud's theory, suggesting that his belief was immune to refutation by objective data. In a recent interview Spock reiterated that "the whole Oedipal situation is proven again and again."

Baby and Child Care

In many respects, Spock's *Baby and Child Care* has been an invaluable book for American parents. Opening with the words: "Trust yourself—you know more than you think you do," the book provides useful and reassuring advice for parents on a wide variety of child-rearing problems. As the preeminent child-care manual for almost half a century, *Baby and Child Care* is both credible and comforting.

It is also fundamentally Freudian, which is precisely what Spock intended it to be. Spock acknowledged that "the theoretical underpinning of the whole book is Freudian" and Lynn Bloom, one of his biographers, noted that "Dr. Spock believes his psychoanalytic perspective is his most significant contribution to advice on child rearing." From the time he opened his private pediatric practice in 1933, Spock recalls "trying to take the psychoanalytic concepts I was studying and somehow fit them together with what mothers were telling me about their babies." Believing that "children and parents were entitled to the fruits of the newest medical and psychodynamic principles," Spock became a Freudian missionary on the pediatric frontier. Not surprisingly Spock's pediatric practice was comprised of parents who "were disproportionately psychoanalysts, psychologists, social workers and people who had been analyzed themselves"—that is, parents who reinforced his psychoanalytic belief system. Spock also admits to a personal reason for wishing to incorporate Freudian thinking into child rearing: Believing that his own Oedipal conflict had led to his unhappiness as a child, he wanted to "find ways to bring up children without quite as many kinds of uneasiness as I had experienced."

Spock first committed his Freudian vision of child rearing to print in 1938 in a chapter in a medical text which he co-wrote with Dr. Mabel Huschka. Entitled "The Psychological Aspects of Pediatric Practice," it included extensive discussions of breast feeding, toilet training, punishment, and the Oedipal situation from an orthodox Freudian point of view. Five years later an editor at Pocket Books suggested that he write a manual for parents, which, with the help of his wife, Jane, he did over the next three years. The book was an immediate success, reaching a sales plateau of approximately one million copies a year within three years of publication and maintaining that level up to the present.

Spock's advice to parents departed radically from the rigid scheduling of Watson's behaviorism, which had been in fashion during the 1920s and 1930s. Breast feeding was encouraged for its psychological benefits, but the baby should be allowed to set its own schedule. Weaning and toilet training should be permissive, when the baby indicated readiness, never before. In discussing toilet training, Spock supported his position with classical Freudian reasoning:

> When a baby gets into a real battle with his mother, it is not just the training which suffers, but also his personality. First of all he becomes too obstinate . . . too hostile and too 'fighty'. . . . Then there's over-guiltiness. . . . he may come to dread all kinds of dirtiness. . . . If this worrisomeness is deeply implanted at an early age, it's apt to turn him into a fussy, finicky person.

Persistent bed-wetting by children was said by Spock to be an unconscious regression by the child.

Spock also reflected Freudian theory in writing that little boys have a fear of castration while little girls have penis envy. Throughout the book Spock avoided psychoanalytic terminology but rather explained Freudian concepts in lay language. For example, on boys' purported fear of castration Spock wrote:

> He's apt to say, "Where is her wee wee?" If he doesn't receive a satisfactory answer right away, he may jump to the conclusion that some accident has happened to her. Next comes the anxious thought, "That might happen to me, too."

Freud's Oedipal conflict was discussed at length without using the term: "We realize now that there is an early stirring of sexual feeling at this period which is an essential part of normal development." Parents were urged to handle childhood masturbation very delicately because of its potential traumatic consequences: "He may develop such a morbid fear of anything sexual that he grows up maladjusted, afraid, or unable to marry or have children." Physical punishment of children was generally discouraged as unnecessary, especially spanking, which Freud had said would lead to sexual perversions ("libido might be forced into collateral roads"), since one's hindside was viewed as an erogenous zone.

By the early 1960s *Baby and Child Care* had become an indispensable handbook for American mothers. As Spock's popularity grew, he increasingly began to lecture to parents' groups and write for popular magazines. Foremost among these was *Redbook*, for which he began a monthly column in October 1963, two years after Margaret Mead had also begun a column in the magazine. Frequently the two columns appeared on adjoining pages; the advice from Spock and Mead were remarkably similar.

In his *Redbook* columns Spock invoked psychiatry in general and psychoanalysis in particular to justify his opinions and advice. The columns were punctuated with expressions such as "psychoanalysis may show" (July 1966), "psychoanalysis reveals" (December 1968), "child psychoanalysts have reported" (July 1975), and "there's lots of psychoanalytic evidence" (February 1979). At no point did he indicate that such "evidence" was purely speculation and based on no scientific support. For example, regarding masturbation, Spock warned, "Psychoanalysts believe that many worries about bodily inadequacy start from unnecessarily stern warnings about masturbation." On penis envy Spock wrote: "Freud taught—and I still believe—that girls acquire in early childhood an unconscious sense of bodily inferiority to one degree or other."

But it was the Oedipal conflict that appeared to preoccupy Dr. Spock the most, and he wrote about it regularly.

> *[It was] that crucial stage of early sexual development, roughly between three and five years of age. . . . A boy decides that his mother is the most attractive woman in the world and that he wants to marry her and have a baby with her. . . . The little boy gradually becomes aware that he has a powerful rival for his mother's love in his father. . . . Rivalry stirs up hostile feelings. A boy of four or five, though he*

> *admires and loves his father, nevertheless resents the father's advantages*
> *in size and knowledge and his many privileges, including the relation-*
> *ship with the boy's mother. The analysis of boys' drawings and dra-*
> *matic play and dreams shows that they feel particularly rivalrous and*
> *resentful about their fathers' much larger penises. . . . At times boys*
> *feel like doing something drastic to their fathers' penises. This hostility*
> *brings feelings of guilt.*

This period, Spock advised, is crucial to children's emotional development, and if it is not mangaged skillfully all kinds of untoward consequences may ensue, including "excitability, overactivity, sleeplessness,
frequent masturbation, school failure, difficulties in parental management and sexual behavior of which the parents disapprove." Furthermore, Spock maintained that "psychoanalytic experience has taught us
that adolescent rebelliousness doesn't consist of just an impatience to be
more independent of the parents. . . . In the deeper, unconscious layers
of the mind it has more of the quality of a resentful rivalry with the
parent of the same sex."

Spock offered practical advice to parents on how to navigate the
treacherous sexual shoals of the Oedipal period. For example he said,
"I think it's wise to make the rule 'Don't ever let a child sleep in your
bed.' " He also argued strongly against allowing children to see parents
nude; a small boy, said Spock, "is apt to be upset . . . the marked discrepancy in size of genitals arouses resentment." Spock also warned
about "parental tickling and roughhousing: They are fun for some parents, exciting for children, but not good for children" because they
stimulate children's sexual feelings. Spock said that it was permissible to
tickle a baby "momentarily with the hand once a day" but advised
against a parent's burying his face against a baby's abdomen—"that's too
much for any baby." Spock also recommended strongly against "tossing
the baby in the air" despite the fact that most babies "laugh noisily and
ask for more."

Other dangerous parental practices, according to Spock, are for
fathers to "pretend to be lions with their small children" or to engage
in "pillow fights with older children. Child psychiatrists have discovered that in some cases, at least, this is too exciting. It stirs up turbulent
emotions that don't settle down again completely and lead to nervous
symptoms." Spock said that for a father to "play lion" with his son is
wrong because "in the boy's unconscious his father is too threatening a
person and shouldn't reinforce his awesomeness with such play." Spock
claimed to know whereof he spoke on such matters; to a biographer he

with his son Mike and later Mike had had to undergo "nine years of psychoanalysis."

Spock never hesitated to recommend psychiatric care—and specifically psychoanalysis—for any child who seemed to be having problems. Both *Baby and Child Care* and his *Redbook* columns contain numerous references to situations in which professional help should be sought. For example, for school children "who run into study blocks despite high motivation and intelligence, the first step is to discuss the situation with the school personnel and get their recommendations. This often includes psychiatric consultation. The psychiatrist usually advises psychoanalysis or a less intensive but prolonged course of psychotherapy." When asked on one occasion to specify when a child needs psychiatric help, Spock responded, "Parents should ask for help when they are dissatisfied with some aspect of a child's behavior and haven't themselves been able to find the reason for it. This implies that it doesn't matter whether the problem seems great or small." Spock thus lay the groundwork for the widespread contemporary exploitation of child counseling and psychotherapy, especially as seen in unnecessary psychiatric hospitalizations for many children with problems.

Was Spock Permissive?

Just as Benjamin Spock was introduced to psychoanalytic theory by Jane Cheney, his first wife, so too were liberal politics. Spock had been "brought up in a Republican family and voted for Calvin Coolidge in 1924." Jane, however, "had been a Socialist in college" at Bryn Mawr and was active in the American Labor Party. Following their marriage Jane subscribed to the Marxist publication *New Masses*, and in later years she was active in efforts "to alleviate the plight of the children in Republican Spain" and worked for the American Civil Liberties Union, the Americans for Democratic Action (she was a member of the national executive board), and the National Committee for a Sane Nuclear Policy.

During Spock's 48-year marriage to Jane, he shifted progressively to the political left. In the 1930s he supported Roosevelt's New Deal, and in later years campaigned for Adlai Stevenson in 1956, John Kennedy in 1960, and Lyndon Johnson in 1964. He joined the National Committee for a Sane Nuclear Policy and became increasingly convinced that the war in Vietnam was both immoral and likely to lead to a nuclear holocaust. Because of such beliefs Spock became active in the anti-war movement and in October 1967, at age 65, was arrested at an anti-war demonstration for the first of many times. In a New York City jail he shared a holding pen with Allen Ginsberg, "who taught us all to meditate. We were all chanting, 'OMMMMM.' "

In January 1968 Spock was indicted with four other activists "for conspiracy to counsel, aid and abet resistance to the draft." He was tried, convicted, and sentenced to two years in prison plus a five thousand dollar fine; the conviction was later reversed by a court of appeals. In 1972 he ran for president on the Peoples Party ticket because he was convinced "that a new political movement must be built to the left of the Democratic Party."

The response of Republicans to Spock's political activism was predictable. They equated his liberalism with his child rearing theories, which they called "permissive," then blamed him for the student activism of the 1960s. Vice President Spiro Agnew said that such students "were raised on a book by Dr. Spock and a paralyzing permissive philosophy pervades every policy they espouse." Norman Vincent Peale, a friend of then–President Nixon, publicly characterized Spock's advice as follows: "Feed 'em whatever they want, don't let them cry, instant gratification of needs. And now Spock is out in the mobs leading the permissive babies raised on his undisciplined teaching." This was, said Peale, "the most undisciplined age in history."

Was Dr. Spock permissive and at least partly responsible for the student activists of the 1960s? At a personal level Spock was clearly not permissive. In his writings he has inveighed against dating too early, premarital sex, adultery, and pornography; in reading him one has the impression that in much of his social behavior Spock has never moved far beyond the strict New England admonitions of his mother. On numerous occasions Spock has publicly denied the charge of permissiveness saying, "I've never considered myself even remotely a permissivist."

At a theoretical level, however, Spock can be accused of being permissive insofar as he has promoted the theory of Sigmund Freud. Parents were told to be exceedingly careful not to damage puerile psyches by disciplining them, for undoing such damage might entail years of psychoanalysis if it could be accomplished at all. In Spock's view children are fragile reeds who are extremely sensitive to minor parental inconsistencies and emotional innuendo. Indeed, if one takes psychoanalytic theory seriously it is possible to imagine that bad days for a mother at a crucial stage of childhood development can irrevocably arrest the child's development at one stage or another. The inevitable outcome of Spock's—and Freud's—theory of emotional development

paralyzed by inaction, and too often unable to say "no."

Spock has acknowledged that attempts to follow his child-rearing prescriptions frequently lead to permissive parents at the same time as he has vehemently denied that he is personally permissive. The study of 21 mothers that he conducted at Case Western Reserve University illustrated the consequences of Freudian theory all too clearly.

According to Spock the mothers' inability to make demands on their children was due to their fear "that they would arouse antagonism in their children." "The parents were afraid to impose toilet training firmly for fear it would create hostility between themselves and their children." Spock described some of the mothers as being completely passive and nondirective because of fear of harming their child:

> *All the mothers ignored, in their apprehension, some of the evidences of their children's general readiness and several of them even ignored specific signals of bowel or bladder urgency. In two families in which the mothers believed that their children were still unready for training (at over two years), staff members saw the children tug at their mothers' skirts and say the family word for toilet. In one of these cases the counselor called the mother's attention to what he had just observed and suggested that she take the child to the bathroom; but she demurred, explaining that it was no use because he often balked when he got to the bathroom.*

Perhaps the most significant finding that emerged from Spock's study of mothers was that those with the most education were the most permissive. Among the 21 mothers in the study, 4 were not college-educated and these 4 were *more* successful in toilet training their children than the 17 who were college-educated. Furthermore, when Spock compared the results of his 21 mothers with mothers in another pediatric clinic in the same hospital "in which the families are predominantly working class" and "few of the families are college-educated or interested in psychology," he found that the other mothers had successfully toilet trained their children a full year before the mothers in his study, and this had been done "without severity, without struggle, and without evident harm to the personality." Spock summarized these findings in a 1964 *mea culpa* that he coauthored with Mary Bergen:

> *The fear of arousing painful tensions between mother and child can be traced, in part, to a vague awareness among college-educated parents of*

Freud's discovery, in the analysis of neurotic adults, that the child's possessiveness about his bowel movements in the second year may lead to conflicts with his mother, and that exaggerated hostility, the compulsion neurosis, and compulsive character traits sometimes develop from a prolonged training struggle. This concept has been included in parent-education materials such as those of one of the authors. It seems in retrospect that this particular knowledge has hampered many more parents than it has helped.

By 1989 Spock was stating the problem in more blunt and pessimistic terms: "It's professional people—like me—who have gotten the parents afraid of their children's hostility, and I don't know if we can undo it. Pandora's Box has been opened."

Despite his efforts, Spock has never been able to resolve the dilemma between adhering to Freudian theory for child rearing and producing guilty mothers with spoiled children. Within a few years following publication of the original edition of his book, Spock "was sure that it [the book] should be revised [in order] to counteract a growing tendency toward overpermissiveness among certain parents." Among the families in his own practice he saw increasing numbers of children with sleeping and feeding problems, such as those who were allowed to "throw food on the floor" or to hit their parents. Spock decided that whereas previously the most common problem in child rearing had been excessive parental rigidity, "there now were predominantly difficulties that came from much parental hesitancy."

Accordingly, Spock substantially revised many sections of his book, including those on discipline and sibling rivalry, for the second edition in 1957. For example, Spock says he "took care to tone down greatly the descriptions of the hostility and the personality disorders that are occasionally brought about as a result of training conflicts because I thought too many parents had become unnecessarily worried about them." In the 1968 edition, Spock admitted frankly that parental problems with child rearing "occur mainly in families with college background or with a definite interest in child psychology," the same parents who were reading both Freud and Spock. And yet just the previous year Spock had written in *Redbook*, in classical Freudian terms, that a child's bowel movements are "the first medium of exchange for the child, the first equivalent of money. And throughout the rest of life, at unconscious levels of the mind. . . BMs and money are, to a degree, interchangeable symbols of what is valuable." Spock cited stamp collec-

being told to be more relaxed about toilet training and at the same time that one false step—or one traumatic bowel movement—may make little Johnny or Sarah a miser for life.

Freud in the Classroom

Benjamin Spock was not, of course, the only child development specialist promoting Freudian theory in the postwar period. Because of the popularity of his book and his political activities, Spock was merely the most visible. Edith Buxbaum's *Your Child Makes Sense*, published in 1949 with a foreword by Anna Freud, included "a generous lacing of case histories which reveal the disastrous errors of parents who have thereby created pathologically sick children. . . . The effect of books of this type was to create anxiety in the parent—especially the mother— to a degree unparalleled in child-care history." Similarly, *The Magic Years: Understanding and Handling the Problems of Early Childhood* was published in 1959 and sold over a million copies; the author, psychoanalyst Selma Fraiberg, waxed hyperbolically against such practices as spanking and the psychic perils of toilet training: "This vitreous monster with its yawning jaws does not invite friendship or confidence at this age. The most superficial observation will reveal that it swallows up objects with a mighty roar, causes them to disappear in its secret depths, then rises again thirstily for its next victim which might be— just anyone." One article by a psychiatric social worker was titled "Spoil That Baby," and another by a psychologist advised parents not to discipline their children under any circumstances: "Instead of getting the meanness out [by using discipline] we stamp it in. It goes down deeper and deeper and becomes more fixed in our child's personality and character."

It is also important to realize that the Freudian child-rearing practices being promoted by Spock were consonant with public advice being given by Mead, Benedict, Erik Erikson, and leaders in progressive education circles. Mead and Spock became acquainted when she sought him out to be the pediatrician for her child because he had been psychoanalyzed and recommended to her "by my child development friends." Mead held the idyllic Samoans up as models for child rearing and claimed that they were almost completely permissive, especially in children's sex education and experiences. Practices such as spanking, Mead said, were "unusual to find [in] a primitive society. . . . Corporal punishment, administered self-righteously, tends to beget not a healthy avoidance of the next smack but a deep and often murderous hatred of all authority."

Spock was present when Mead's baby was born in 1939, one of a crowd so large that the event was described by a Mead biographer as having "the air of a Nativity pageant." The actual birth was delayed for ten minutes until the photographer arrived since Mead apparently wanted a full record of those first, psychoanalytically crucial, minutes. Mead was as permissive in raising her own child as she urged others to be, feeding her on demand "at the slightest whimper." According to her daughter, Mead set few limits; for example, "She never insisted that I brush my teeth because she had found the process unplesant and painful as a child." For almost 20 years Benjamin Spock and Margaret Mead went hand-in-hand through monthly *Redbook* columns promoting a Freudian theory of permissive child rearing.

Mead, Spock, and Ruth Benedict were also involved with the efforts of Dr. Carolyn Zachry to apply Freudian concepts to progressive education. Zachry, who obtained a doctorate in educational psychology at Columbia Teachers College and trained at the New York Psychoanalytic Institute, set up the New York Institute on Human Development in 1938 in affiliation with the Progressive Education Association. It was intended to be "a training institute for professionals concerned with child care" and was classically Freudian in its orientation. Both Mead and Benedict participated in Zachry's courses, and Spock was one of the institute's faculty members. Such efforts further disseminated Freud's theory of child rearing among New York's intellectual community and especially within the education field.

The incorporation of Freudian theory into education had been proposed from the earliest days of the theory's arrival in America. In 1910 Ernest Jones had published an article on psychoanalysis and education in the *Journal of Educational Psychology*; nine years later Wilfred Lay's *The Child's Unconscious Mind, The Relations of Psychoanalysis to Education, A Book for Teachers and Parents*, was published. This trend was encouraged by the mental hygiene and child guidance movements, which viewed schools as "a kind of mental hygiene laboratory. . . . From the mental hygiene viewpoint education implied, not a narrow focus upon intellectual attainment, but the child's evolution into a balanced, emotionally stable, and creative personality. The classroom teacher had to understand that the acquisition of intellectual skills was less important than emotional rapport." At Columbia Teachers College it was recommended "that academic credit be given for personal [psycho]analysis" and a group was "actively promoting the introduction of

The consequences of this infusion of Freudian theory into education were felt in the post–World War II period. Guidance counselors, who had once advised students on vocations and colleges, instead began concentrating on counseling students with personal problems. School psychologists, who had once tested students to identify cognitive problems and then recommended remedial courses, evolved instead into being psychotherapists. The assumptions of these school counselors and psychologists were predominantly Freudian in derivation. By the mid-1970s it was estimated that there were more than sixty thousand school guidance counselors and seven thousand school psychologists in the United States and, as the director of school mental health in Pittsburgh phrased it, "Schools are our community mental health centers."

Freud's influence in American education reached its pinnacle in the alternative or "free" schools that spread across intellectual communities and American communes in the 1960s. Such schools were based on principles developed by Zachry's institute in New York and also on the work of Alexander S. Neill of Summerhill School in England, as described in his book, *Summerhill: A Radical Approach to Child Rearing*. Neill, a Scottish educator, had undergone psychoanalysis three times, including an analysis with Wilhelm Stekel in Vienna. (Stekel, one of Freud's original followers, was described by Paul Roazen as "a dubious character with a dirty-minded interest in case material . . . his interest in sexuality remained quasi-pornographic.") In addition to his three personal psychoanalyses, Neill also studied for six weeks with Wilhelm Reich, who provided, according to Neill, a course of "Vegeto therapy," Reich's idiosyncratic brand of psychoanalysis. Neill attempted to psychoanalyze his pupils at Summerhill, a name which became synonymous with student permissiveness in all spheres. The traditional school, said Neill, "carries on the life murder of the young; it instills obedience, perhaps the worst of the seven deadly virtues. . . . it inhibits the child's natural sincerity by forcing it to 'respect' its teachers, to act a part, really to be a hypocrite."

Beyond free school and the transformation of school guidance counselors and psychologists into psychotherapists, however, Freudian theory did not influence American education nearly to the degree that it influenced child rearing practices. Education has standardized tests and a product that can be measured over time. These act as natural deterrants to educational innovation that is based on untested theory;

the product of child rearing, by contrast, is in the eyes of the beholder.

Perhaps the most important consequence of Freudian influence on child rearing has been the generation of parental guilt. Such guilt permeates the lamina of American consciousness to a degree probably unparalleled elsewhere in the world. It is as if the sins of the children have been visited on their parents, a reversal of the Biblical order of things. Such guilt creates in its turn the need for expiation and atonement, which many parents seek through psychotherapy. Thus Freudian theory has both effectively created a need and then furnished a product with which to fulfill that need, a merchandising stratagem as elegant in its simplicity as it is in its profitability.

Spock himself pointed out how much more guilty American parents feel than European parents in one of his *Redbook* columns:

> *A Frenchwoman who is a child psychiatrist in Paris wrote a revealing paper on how differently American and French parents act when they first come to her office. American parents typically begin: "We realize that we must have caused the problem and that it's probably we who need the psychiatric treatment." They are like dogs with their tails between their legs. French parents have no such guilt; they have indignation. They say to the psychiatrist, "We've given this child all the advantages—good home, good clothes and schooling, fine family. Yet he shows no appreciation. He is bad. We have scolded him, deprived him, spanked him, but nothing does any good. Now you make him behave."*

Isaac Rosenfeld, a prominent 1950s intellectual in Chicago, also captured the era of permissive child rearing in a vignette. His academic friends, he said, raised their children according to "Spock and Gesell with an assist from Bruno Bettleheim," and "one sure way of telling whether you are visiting an academic or nonacademic household is by the behavior of the children, and the extent to which you can make yourself heard above their clatter. If it is still possible to conduct a conversation, you are in a nonacademic household."

7

Freud in Jails and Prisons

I believe fully in the discarding of the concept of
responsibility. The idea of responsibility is, as I have said,
largely a metaphysical one.
—Dr. William A. White, *Forty Years of Psychiatry*

The theory of Sigmund Freud was incorporated into America's jails and prisons much more quickly than it would be welcomed into its nurseries. Whereas child rearing had to wait for Benjamin Spock's post–World War II writings to become Freudianized, criminology and corrections accepted Freud's ideas as they arrived from Europe even prior to World War I. Over the intervening years American thinking about criminal behavior has become so infused with psychoanalytic pigment that it is difficult to visualize what it once looked like in its natural state.

Until the arrival of Freudian theory in America it was widely believed that inheritance played a major role as a cause of criminal behavior. In the late 19th century, for example, the theories of Cesare Lombroso were widely discussed. Lombroso was a professor of psychiatry in Italy who contended that some individuals are born as criminals and could be recognized by certain physical signs (e.g., epilepsy) or stigmata (e.g., abnormal ears or long arms). Such theories of criminal behavior received support from influential publications such as Richard L. Dugdale's *The Jukes: A Study in Crime, Pauperism, Disease and Heredity* (1877). Based on a study of inmates in upstate New York county jails, the pseudonymous Jukes family was said to have produced 140 individuals who had been convicted of criminal offenses (including 60 chronic thieves and 50 prostitutes) among 709 descendants over a one-hundred-year period.

It should be noted that even the staunchest believers in the genetic roots of criminal behavior acknowledged that such theories accounted for only some cases. Lombroso, for example, contended that heredity accounted for 40 percent of criminals, the remainder being caused by poverty and passion. He campaigned for the differential treatment of born criminals, whom he believed could not be rehabilitated, from criminals of poverty and passion, whom he believed could be rehabilitated. In the 19th century it was also widely recognized that occasional criminal behavior was the product of insanity and recommended that such criminals be treated differently. This was codified in England in 1843 as the M'Naghten rule, when an obviously insane man of that name shot Sir Robert Peel's secretary and was judged as not having understood the nature of his act; this version of "not guilty by reason of insanity" was subsequently implemented in most of the American states.

Healy, Glueck, and White

The man most responsible for introducing Freudian theory into American thinking about criminology and corrections was Dr. William Healy, a British psychiatrist, who had discovered Freud's work while studying in Vienna and Berlin in 1906 and 1907. Healy emigrated to the United States and settled in Chicago, where in 1909 he began work with the Juvenile Court evaluating young offenders. Healy became increasingly convinced that most juvenile delinquency was caused not by heredity but rather by social forces, family relationships, and especially by early childhood experiences.

Healy gained recognition among his peers in 1915, when his book *The Individual Delinquent* was published. The most important cause of delinquent behavior, Healy wrote, was "mental conflicts and repressions," the source of which in "most cases [was] hidden sex thoughts or imageries, and inner or environmental sex experiences." Two years later his second book, *Mental Conflicts and Misconduct*, demonstrated Healy's continuing move toward a classical psychoanalytic explanation of juvenile delinquency. Illustrated with extensive descriptions of individual cases, Healy cited "the complex" as being "a great determiner of thoughts and action," according to Freud. The cause of most complexes was said to be sexual ideas:

ourselves have been utterly surprised at the development of so much delinquency of various sorts from beginnings in unfortunate sex knowledge which has come into the mental field as a psychic shock, producing emotional disturbance. To be sure, we have heard some widely experienced, intelligent observers of delinquents, knowing nothing of mental analysis, state that they consider sex affairs to be at the root of a tremendous number of criminalistic impulses.

Healy recommended that modified psychoanalysis be used to treat the offenders.

Healy's ideas regarding juvenile delinquency circulated widely among corrections officials, especially after he was recruited to Boston in 1917 to be director of the Judge Baker Guidance Center affiliated with Harvard University. Among his earliest supporters were William James and Adolf Meyer. In Chicago Healy had confined his work to juvenile delinquents, but in Boston he accepted referrals from schools and individual families and also incorporated social workers as an integral part of the "treatment" team. Healy's work with juvenile delinquents became the cornerstone of the mental hygiene and child guidance movements that were evolving at this time and were dominated by Freud's supporters (*see* Chapter 1). In 1920, when the National Committee for Mental Hygiene approached the Commonwealth Fund with a request to set up child guidance clinics in five cities, it was Healy's clinic in Boston that they cited as their model. Not only would such clinics prevent juvenile delinquency, it was claimed, but the clinics would also be useful in preventing other mental disorders by educating mothers about Freudian child-rearing principles.

Healy's incorporation of social workers as members of the treatment team permanently changed the social work profession. Social workers had previously concentrated on social and economic problems of poor people, but with the advent of child guidance clinics many of them became junior psychiatrists. By the late 1920s it had become common for social workers to undertake a personal psychoanalysis; one social worker recalled colleagues who "peered anxiously into the faces of their comrades with the unspoken question: have you been psychoanalyzed?" A social work textbook published at this time asserted "that all social case work, in so far as it is thorough and in so far as it is good case work, is mental hygiene." Henceforth the social work profession would provide a small but important Freudian fastness so that, in later years, one critic would claim, "American social workers confused

Freud with the Declaration of Independence." According to John C. Burnham's unpublished thesis "Psychoanalysis in American Civilization Before 1918," William Healy's "impact on criminology, sociology and social work was profound."

At the same time as William Healy was applying Freudian theory to juvenile offenders, Dr. Bernard Glueck was introducing the theory for adult offenders. Trained in psychiatry and psychoanalysis under Dr. William A. White, Glueck began work in 1916 at New York State's Sing Sing Prison on a research project sponsored by the National Committee for Mental Hygiene. Glueck examined 608 consecutive admissions to the prison and reported that "not less than two-thirds, or 66.8 per cent, had already served one or more terms in prisons or reformatories prior to their present confinement." Appalled at this high recidivism rate, Glueck demanded "a new type of effort . . . to locate the causes" which would "seek out the particular determinants in the make-up of the criminal himself." In studying the prisoners for such "determinants," Glueck found that "59 per cent of our 608 cases, in addition to evincing various conduct disorders—the direct cause of their imprisonment—also exhibited some form of nervous or mental abnormality which in one way or another had conditioned their behavior."

Glueck believed that he knew the source of the "nervous or mental abnormality" found so frequently among criminals; in his book, *Studies in Forensic Psychiatry* (1916), he made it explicit. The cause was "an offending pathogenic experience, or in other words what has been termed a psychic trauma. . . . The facts elicited by psychoanalysis point to a strict determinism of every psychic process. . . . Freud has thrown very convincing light on this subject." Glueck acknowledged that some criminals were incorrigible, although he called the idea of a born criminal a "wide-spread superstition" that was without foundation. He argued that the majority could be rehabilitated if their psychiatric problems were addressed. This could only be done by psychiatrists and others trained in mental health, especially Freudian, principles, which meant that "the penal problem is in a very large measure a psychiatric one."

Glueck's specific recommendations for decreasing recidivism and criminal behavior in general was the establishment of "a system of psychopathic clinics in connection with the criminal courts and the penal institutions of the state whose findings, based on more or less intensive study and observation of every convicted felon, shall guide judges and probation officers as well as prison and parole administrators in their decisions and detailed treatment of such offenders." Glueck especially favored indeterminate sentences under which a criminal would be released only if a mental health professional declared him to be cured of his underlying psychiatric problem. "Criminology," said Glueck, "is an integral part of psychopathology, crime is a type of abnormal conduct which expresses a failure of proper adjustment at the psychological level." Criminology and corrections would become a part of psychiatry. "Indeed, the whole idea of punishment is giving way to the idea of correction and reformation."

While William Healy was advocating Freud's theory for use with juvenile delinquents and Bernard Glueck was applying it to adult offenders, Dr. William A. White was promoting Freud's ideas more aggressively than either man and arguing that prisons should be completely replaced by psychiatric treatment centers. A psychiatrist like Healy and Glueck, White had been appointed Superintendent of St. Elizabeth's Hospital in Washington, D.C., in 1903. As early as 1906 he translated into English psychoanalytic papers of Alfred Adler, and the following year he met Carl Jung. By 1909 White had become one of the staunchest American promoters of Freud. In 1913 he cofounded the *Psychoanalytic Review*, the first English language journal on psychoanalysis, and two years later he urged the nomination of Freud for the Nobel Prize in medicine. Twice president of the American Psychoanalytic Association, and also president of the American Psychiatric Association, White was professionally influential both through his widely used textbook, *Outline of Psychiatry*, which appeared in fourteen revised editions over three decades, and through psychiatrists he trained, such as Bernard Glueck. White also wielded considerable political influence through prominent Washingtonians who were his private patients and through his wife, the widow of United States Senator Thurston.

White began publishing articles on the application of Freudian theory to criminology in 1913 and eventually published two books, *Insanity and the Criminal Law* (1923) and *Crimes and Criminals* (1933). "It has been demonstrated," White wrote, "that a given act of a given individual is an end product in this individual's life and can only be understood by knowing that individual's past," especially "the region which has been called the unconscious. . . . It is only by a study of these unconscious factors that conduct in general, criminal conduct in particular, can be understood, that any method of procedure can be devised that has any reasonable prospect of influencing it." White depicted criminals as psychological hostages being held by their unconscious and as such neither truly responsible nor blameworthy. "Some one has aptly said," White wrote, "that the murderer who sees his victim lying before him and a smoking revolver in his hand is probably, of all those who may be present, the most surprised."

White fully embraced the logical consequences of such unconscious determinism and called for "the discarding of the concept of responsibility" for criminal actions. White also urged "that prisons and punishment should both be abolished" as anachronistic; instead, there should be a "gradual transformation of prisons into laboratories for the study of human behavior and the conditioning of human conduct" with psychiatrists, of course, in charge. Fixed sentences would be replaced by indeterminate sentences, "making the return to freedom conditional upon some change in the individual that gives one a right to suppose that perhaps he will function more effectively as a social unit than he has in the past."

By the early 1920s, then, the theory of Sigmund Freud was being widely discussed by individuals in criminology and corrections. The question was whether criminals were born as such or whether they were the products of early childhood experiences and social forces. Those who favored a genetic theory were more often politically conservative, while proponents of psychoanalytic or social explanations were more often liberal; this dichotomy gradually increased as the arguments about criminal behavior became part of the broader nature-nurture debate as discussed in previous chapters. And in August 1924 the question of responsibility for criminal behavior became an issue that transfixed the nation.

The Leopold and Loeb Murder Case: The Treachery of a Teddy Bear

On May 22, 1924, the body of 13-year-old Bobby Franks was found stuffed into a drainpipe near Chicago; he had been killed by a blow to the head. Tracing a pair of glasses inadvertently dropped near the body, the police were led to 19-year-old Nathan F. Leopold and Richard Loeb, one year younger, who both rapidly confessed to the murder. Highly intelligent and well-educated sons of wealthy Jewish families, Leopold and Loeb were said to have committed the murder to prove that they could carry out a perfect crime. They were generally described as being spoiled and arrogant, had been influenced by Nietzsche's writings about a superman, admitted that they had carried out previous burglaries, and had premeditated the crime for several weeks.

Loeb's family immediately approached Clarence Darrow, with whom they were acquainted, to defend both Loeb and Leopold. "Save our boys," they pleaded with Darrow, "only you can save them from being hanged." Darrow was at the time well known for his defense of unpopular and politically leftist figures, including Eugene Debs in the American Railway Union case (1894), labor organizer "Big" Bill Haywood, who was accused of murdering the former governor of Idaho (1907), and the defendants in the *Los Angeles Times* bombing case (1911). No person whom Darrow had defended had ever received the death penalty.

The defense of Leopold and Loeb presented Darrow with a challenge such as he had not previously encountered. The defendants had confessed to the crime with premeditation and there were neither political nor economic circumstances which might be construed as mitigating factors. The case would become, according to a Darrow biographer, "a milestone in criminal justice. . . . Psychiatric evidence for the first time formed the major basis of a defense." As Darrow himself wrote shortly after the trial: "For the first time in a court of justice an opportunity was presented to determine the mental condition of persons accused of crime, according to the dictates of science and modern psychiatry, without arbitrary and unscientific limitations imposed by archaic rules of law."

Darrow was an excellent choice as a lawyer to mount a psychiatric defense. He had shown a strong interest in unconscious thought processes as early as 1899 in a short story he had published. Furthermore, at the turn of the century, he had astonished the inmates of the Cook County Jail by telling them that "they were in jail simply because of circumstances for which they were in no way responsible." Darrow had written frequently about the lack of responsibility of criminals for their actions, culminating in his book, *Crime: Its Causes and Treatment* (1922), in which Darrow argued:

> *It seems to me to be clear that there is really no such thing as crime, as the word is generally understood. Every activity of man should come under the head of "behavior." In studying crime we are merely investigating a certain kind of human behavior. Man acts in response to outside stimuli. How he acts depends on the nature, strength, and inherent character of the machine and the habits, customs, inhibitions and experiences that environment gives him. Man is in no sense the maker of himself and has no more power than any other machine to escape the law of cause and effect. He does as he must. Therefore, there is no such thing as moral responsibility in the sense in which this expression is ordinarily used.*

Darrow also advocated that "all prisons should be in the hands of experts, physicians, criminologists, biologists, and, above all, the humane. . . . Sentences should be indeterminate. . . . The indeterminate sentence can only be of value in a well-equipped prison where each man is under competent observation as if he were ill in a hospital." Darrow's recommendations were remarkably similar to those of William Healy, Bernard Glueck, and William A. White. The defense of Leopold and Loeb, therefore, would be based on Freudian theory, and the chief defense witnesses recruited by Darrow were Healy, Glueck, and White.

The Leopold and Loeb murder trial was a public spectacle unprecedented in American jurisprudence. The *Chicago Tribune* offered to broadcast it live over its own radio station, while the rival *Evening American* countered with a suggestion that the trial be held in the Chicago White Sox baseball stadium. Both newspapers offered Sigmund Freud "any sum he cared to name" to come to Chicago and psychoanalyze Leopold and Loeb for their readers; William Randolph Hearst, owner of the *Evening American*, even offered "to charter a special liner so that Freud could travel undisturbed by other passengers." The newspapers realized that not only Leopold and Loeb were going on trial, but Freud's theory was also. Freud, in poor health following surgery for cancer of the mouth, which had been diagnosed the previous year, declined the offers. The courtroom was crammed with the most renowned trial reporters in the country, who would report on Leopold and Loeb as well as on Freudian theory.

Darrow's defense strategy was to plead his clients guilty and so avoid a jury trial. The trial, held before only a judge, then became one about mitigating circumstances for the crime. Darrow argued that Leopold and Loeb were not responsible for their actions because particular events in their childhood led to emotional immaturity. Dr. William Healy described in general terms how childhood experiences had been damaging; as summarized by the Boston *Herald*, "The behavior of today is builded [sic] on yesterday's impressions." There were darkly sexual overtones to Healy's testimony when he hinted at a homosexual relationship between Leopold and Loeb. Leopold, Healy said, was "mentally diseased," and he quoted Leopold as having told him, "Making up my mind whether or not to commit murder was practically the same as making up my mind whether I should eat pie for supper." Loeb, said Healy, was also mentally ill: "To my mind the crime itself is the direct result of diseased motivation of Loeb's mental life. The planning and commission [were] only possible because he was abnormal mentally, with a pathological split personality."

Dr. Bernard Glueck emphasized the inevitability of the crime given the mental state of the defendants. Leopold, he said, was "governed to such a large extent by his delusional thinking" that he "could not have done otherwise than has been carried out in connection with this crime." No evidence of truly delusional thinking, however, was ever presented at the trial. Loeb, said Glueck, "is suffering from such a profound discord between his intellectual and emotional life as to be incapable of appreciating the meaning, the feeling of the situation." The murder represented "the inevitable outcome of this curious coming together of two pathologically disordered personalities." In an interview shortly after the trial, Glueck said he believed that Leopold's and Loeb's parents were ultimately responsible for the crime because "no one ever knew these boys as 'persons.' "

Dr. William A. White had just taken office as president of the American Psychiatric Association when he testified as the star witness at the trial. It would be his job to spell out in more detail the nature of the childhood events which were said to have warped the personalities of Leopold and Loeb. White "was permitted to testify for seven hours without interruption, detailing the personality and mental life of the two boys." He referred to both defendants by their childhood names in order to emphasize their immaturity, and said that they had been emotionally deprived by their parents despite having been given everything materially: "Dickie [Loeb] told me he thought his family had more or less neglected him but he thought that their intentions were perfectly good." Both Leopold and Loeb had had governesses as children, and White implied that these women had been a major cause of the boys' problems. Loeb's governess, for example, had been "prudish and austere," with "repressions" that had caused Loeb to be overprotected, and she had "pushed him tremendously in his school work." Indeed, according to one account of the trial, "Miss Struthers and Mathilda Wantz [the governesses] seemed as much on trial as [Leopold and Loeb]." As a consequence of such childhood experiences, White said, "Dickie with his inferiority complex developed definite antisocial tendencies."

Dr. White dwelt at great length on the emotional immaturity of Leopold and Loeb, always emphasizing the greater pathology in the latter. As a child Loeb had had a teddy bear, White said, to which he confided his fantasies. Because of the psychic traumas of his childhood Loeb had not been able to give up his fantasies. According to White,

Loeb was "still a little child emotionally, still talking to his teddy bear"; his emotional age was approximately five years despite his high IQ, which White acknowledged was approximately 160. White introduced pictures of Loeb as a young boy dressed up as a policeman and a cowboy and emphasized the seriousness of expression on young Loeb's face. "There is a tendency," said White, "for the fantasy life, an abnormal fantasy life, to realize itself in reality." According to White, Leopold had similar problems but they were not as severe, and his emotional development was estimated to be eight years. Leopold had flattered White on their first meeting remarking: "I am so glad to meet you, Dr. White. I know the exact number of lines you take up in *Who's Who*." White later confided to colleagues that both Leopold and Loeb "were perfectly charming" in his interviews with them.

White's "teddy bear revelation," as some newspapers called it, was widely ridiculed. One editorial noted: "If the fathers and mothers in general were not gifted with common sense, there would be a lot of worrying in Chicago over the teddy bear revelations in the Loeb–Leopold case. Thousands of youngsters with healthy imaginations whisper their childish fantasies to a toyland companion and grow up to a clean-minded law-abiding" adulthood. A New York newspaper headlined: "Loeb Only 5, Leopold 8, Says Alienist But Each is Owner of Giant Ego." The New York *Herald-Tribune* noted that "putting the boys in a class with doll-playing babies was one of the most unusual analyses ever pronounced in a criminal court." But Dr. White held firm to his opinion and could not be shaken on cross-examination. White claimed that his examination of Leopold's and Loeb's minds had provided "the clarity that attends an X-ray examination of the body" and denied that it was possible for "patients examined under psychiatric examination" to deceive the examiner. The only time White lost his composure during his long testimony was when the district attorney "finally forced an admission from White on the stand that he was being paid $250 a day" [by the families of the defendants; $2,500 a day in current dollars] despite being a full-time employee of the federal government, at which point White became "flushed and angered."

The ruling in the Leopold and Loeb case was in favor of the defendants and, implicitly, Freud. The judge called "the careful analysis made of the life history of the defendants [a] valuable contribution to criminology." Although he acknowledged "that similar analysis made of other persons accused of crime will probably reveal similar or different abnormalities," he found the evidence sufficiently compelling to be a mitigating factor. Accordingly, rather than being sentenced to death,

Leopold and Loeb were sentenced to "life plus ninety-nine years" with a recommendation never to be paroled. Loeb was subsequently killed in prison by another inmate, but Leopold was paroled in 1958 after serving 34 years.

The major effect of the Leopold–Loeb murder case was to legitimize Freudian theory in American courts. Clarence Darrow, highly praised by psychiatrists, was the principal guest speaker at the American Psychiatric Association convention in 1925. At the convention it was reported that "the real responsibility" for Leopold and Loeb's behavior rested with their parents, "according to the consensus of opinion" expressed by the psychiatrists.

William A. White was subsequently investigated by Congress for receiving private fees while functioning as a government employee, but was exonerated after he agreed to take vacation time for the period of the trial. (Darrow later asked White to testify at the Scopes trial on the teaching of evolution in 1925 but White declined.)

Darrow continued to be a strong supporter of Freudian theory for the remainder of his life and was one of the principal speakers (along with White, Abraham Brill, and Theodore Dreiser) at a 1931 banquet at New York's Ritz-Carlton Hotel to celebrate Freud's 75th birthday.

In addition to legitimizing Freudian theory, the trial widely publicized it. According to one historian, the case provided "a sort of crash course in psychoanalysis for the ordinary American . . . [and] for the first time brought to American readers extended coverage of Freudian thought." By linking Freudian theory to the liberal political persona of Clarence Darrow, the trial also helped solidify the emerging association of Freudian theory with leftist politics in America. The names of Leopold and Loeb became permanently linked to Freud, an ironic and euonymic outcome considering that Nathan F. Leopold's middle name was in fact Freudenthal.

Karl Menninger

The concept of criminal responsibility in American jursiprudence was fundamentally altered by the Leopold and Loeb murder case. In September 1924, one month after the trial, when Ralph Edwards was arraigned in Baltimore for writing bad checks he told the judge: "Maybe I did. I was tried in New York on a similar charge and an alienist [psychiatrist] discovered I didn't do it. It was my sub-conscious mind." By 1926 the Associated Press reported that psychoanalysis was replacing spanking as a "disciplinary measure for children." In light of

... century, reflected the new concept of ... after having undergone psychoanalysis twice: "I remember thinking how absurd had been my muckracker's description of bad men and good men and the assumptions that showing people facts and conditions would persuade them to alter them or their own conduct." The new morality taught that there were neither bad men nor good men, merely marionettes of unconscious minds.

William A. White emerged from the Leopold and Loeb murder case as the leading spokesperson for Freudian approaches to criminology, but he was approaching 60 years of age and retirement. Younger leaders were needed to implement these ideas, and so, shortly after the Leopold and Loeb case concluded, White, in his capacity as president of the American Psychiatric Association, established a Committee on Legal Aspects of Psychiatry. For its first chairman White selected Dr. Karl Menninger, a young psychiatrist who would make the implementation of Freud's theory to criminology his principal life's work. Dr. Menninger would become to corrections what Dr. Spock would be to child rearing—a popular evangelist of psychoanalytic scripture.

Menninger, like Spock, had been the eldest child in a prosperous Protestant family, except the Menningers lived in Kansas rather than Connecticut. Menninger was also pathetically dependent upon his controlling, puritanical mother. While in college Menninger had doubts about himself and wrote his mother: "Please mamma, tell me what you think. I am convinced that you know me better than I." Many years later, reflecting on the control which Menninger's mother held over him, his estranged wife wrote to him: "I guess it would have been better far if I had walked out the night you told me you'd written home to your mother and asked her if *we* could have a baby now." Menninger, like Spock, was also regarded somewhat as a social outcast by his peers in school and college; as a freshman at Washburn College he was blacklisted by a fraternity he longed to join because one of his friends told them that he was "feebleminded."

Such an assessment was hyperbole. Although never considered to be more than an average student, Menninger graduated from college, then went on to Harvard Medical School and psychiatric training at the Boston Psychopathic Hospital before returning to Topeka to establish the Menninger Clinic with his father and younger brother, Will.

And though he showed an initial interest in biological aspects of psychiatry, and in fact published two papers on the post–influenzal schizophrenia syndrome which are still considered to be important, Menninger was drawn to Freud's theory while he was a medical student. As a psychiatric trainee he undertook a brief trial of psychoanalysis with Dr. Smith E. Jelliffe, a close friend of William A. White and probably the person who brought Menninger to White's attention. Menninger considered going to Vienna to be completely psychoanalyzed but decided against it because, as he explained in a letter, "I don't think that I have the courage to stay away from my mama that long."

Through his work with patients at the Menninger Clinic, Karl Menninger's interest in Freud continued to grow, and in 1931 and 1932 he undertook a formal psychoanalysis with Dr. Franz Alexander in Chicago. Alexander was said to have been "among the very best" of Freud's pupils and is thought to have been entrusted with the psychoanalysis of Freud's son, Ernest. According to Menninger, Alexander "encouraged him to have a mistress" to compensate for his failing marriage. At the time, Menninger was having an affair with the wife of a Topeka judge, who in turn was having an affair with Menninger's wife. After the termination of this "foursome," as Menninger characterized it, he began an affair with his secretary, Jeannetta Lyle, which lasted almost ten years, until he finally divorced his wife and married her. Alexander's advice to Karl Menninger was apparently not unique; Will Menninger was also encouraged by Alexander, while in psychoanalysis with him, to have an affair, and when Will's wife met with Alexander "he had even urged her to take a lover."

In 1934 Karl Menninger met Sigmund Freud. Accompanied by Franz Alexander he visited Vienna while in Europe to attend the International Congress of Psychoanalysis. At Freud's home Alexander talked privately with Freud first, during which time, according to information Menninger received later, Alexander "told Freud that Karl was very narcissistic . . . [and] cautioned Freud not to praise his analysand very much." Freud was "utterly impersonal" to Menninger; when Menninger told him about the Menninger Clinic, at which he was using Freud's techniques to treat seriously mentally ill patients, Freud replied that he had "never had success with psychoanalysis on severely mentally ill patients."

Menninger's meeting with Freud was bitterly disappointing; he later recalled that "my narcissism received a terrific blow." He subsequently wrote to Freud and also cabled him, inviting him to come to Topeka, but he received no replies. Despite these rebuffs Menninger

of anything else that has been proposed, that I have nailed my banner on his mast, and I'll defend it against assault for the rest of my life." According to his biographer, Menninger's meeting with Freud "launched him on a personal crusade to prove that he could thoroughly master and explicate what he regarded as the most important collection of ideas in modern times." In the *New Republic* Menninger compared Freud to Plato and Galileo; when Freud died in 1939, Menninger wrote an obituary in the *Nation* praising his "genius . . . gentleness and sweetness of character" and saying that "he was so nearly unique an individual that it is difficult to find anyone with whom to compare him." Karl Menninger became, in his own estimation, "more Freudian than Freud."

By the time of Freud's death, Menninger had undertaken still another psychoanalysis, this time in New York City with Dr. Ruth Brunswick, who, as an orthodox Freudian, "stressed the role of the mother in personality development." An American, Brunswick had worked in Vienna from 1922 to 1938, during which time she had been intermittently in analysis with Freud. According to Paul Roazen's *Freud and His Followers*, Brunswick "was unquestionably Freud's favorite in Vienna," and for several years was "closer to Freud than his own daughter Anna." Considered to be "one of Freud's most brilliant pupils," Brunswick was given the responsibility of psychoanalyzing the Wolf-Man, one of Freud's most famous patients, after Freud had partially completed the treatment. While in analysis with Freud and with his encouragement, Brunswick had divorced her first husband and married Mark Brunswick; Freud then simultaneously psychoanalyzed both husband and wife. In the final years of her long psychoanalysis with Freud, Ruth Brunswick had become addicted to morphine and other drugs. (During Brunswick's psychoanalysis of Menninger she was severely drug-addicted and, according to Menninger, had a "tendency to fall asleep and even to receive telephone calls during his analytic hour," occasionally "ordering things for her household from department stores." Six years later Brunswick died of injuries received from a fall while under the influence of drugs.)

Despite three attempts at psychoanalysis, Karl Menninger continued throughout his life to have major problems in relations with others. He was said by biographer Lawrence J. Friedman to be "chronically insecure" and "moody and unpredictable." In 1951 he precipitated the only strike by psychiatrists in American history when European emigré psychiatrists at the Menninger Clinic declared a work stoppage until he agreed to pay them on a basis equal to their American counterparts. In 1965 Menninger was unceremoniously removed from his position as head of the clinic by a revolt of the staff, who had tired of his imperious governance, and he spent the remainder of his life in semi-exile in Chicago.

What Karl Menninger derived from his personal psychoanalyses and his study of Freud was a single overriding belief that dominated his thinking about all human behavior, including criminal behavior. Menninger believed that mothers are the cause of virtually all psychopathology, both at a personal level and as reflected in problems of civilization, and he developed this belief in two remarkably misogynistic articles published in the *Atlantic Monthly* in February and August 1939. Menninger's concept of original sin was "the childhood experience of having been frustrated too considerably and too rapidly" by mothers. This experience of frustration leads in turn to "the hate that burns in the child's heart," for the mother is "the one who first stirs up bitterness and revenge wishes in the child." Menninger continued: "It is the mother toward whom the deepest layer of resentment attaches itself. . . . If we penetrate the many layers of hatred we come eventually to the deepest hurt of all—'Mother failed me.' " The hatred engendered by mothers then became, in Menninger's view, the direct cause of neuroses, psychoses, alcoholism, sexual problems, marital problems, personality disorders, crime, and even war—indeed virtually everything wrong with both individuals and civilization had its roots in mothers' frustration of their children.

As William A. White's chosen heir to implement Freud's theory in criminology and corrections, Karl Menninger proved to be an adept pupil. In addition to White's writings, Menninger drew upon the ideas of William Healy, whom he had known in Boston, and also utilized ideas of Franz Alexander, Menninger's second analyst, who with Healy coauthored *The Roots of Crime: Psychoanalytic Studies* (1935).

Menninger's writings on criminal behavior date to 1925, the same year Dr. White appointed him chairman of the APA's Committee on the Legal Aspects of Psychiatry. In an article, "Psychiatry and the Prisoner," in which Menninger was identified as "Professor of Criminology, Washburn College, Topeka," he decried the "medieval stupidities" of the traditional legal corrections system and urged that a "scientific (psychiatric) attitude must sooner or later totally displace the existing legal method." The reason that this was necessary, Menninger said, is that crimes are

'just'? Or cancer, or gravity, or the expansion of steam? What criteria of justice can be applied to a broken arm or a weak mind?"

What was needed, according to Menninger, was treatment rather than punishment. "The physicians took surgery away from the barbers a century ago; now they are taking criminology and penology away from politicians, wardens and lawyers." Jails should be put "under expert medical management" in which "a 'sentence' will be as unthinkable for a murderer as it is now for a melancholic. . . . Release before complete recovery will be as irregular and improper for a thief or rapist as it now is for a paretic or a leper." Menninger concluded that "the future of the American prison system is in the hands of the psychiatrists and their allies, the social workers."

Karl Menninger continued to be the principal voice urging the psychiatrization of criminology for the following 65 years until his death in 1990. In an article in 1937 he praised "Alexander, Healy, White, Glueck and others . . . who have carried forward the scientific study of the psychology of crime with an eye to its more humane and effective control." The basis of such scientific study was said to be "the deductive genius of Freud that outlined for us in scientific terms the concept of a malignancy within the organism, an instinct in the direction of destructiveness," an instinct which was activated by mothers' unthinking behavior toward their young children. At that time, with Hitler threatening to draw Europe into an armed struggle, Menninger emphasized his thesis that mothers activate the death instinct and are thus the cause of wars, and noted that "it is on the basis of Freud's work that others have proposed applications of our psychological knowledge to the elimination of war." Similar themes were sounded in Menninger's *Man Against Himself* (1938) and *Love Against Hate* (1942).

Karl Menninger's influence as Freud's ambassador to the nation's jails and prisons increased over the years as the Menninger name became more widely known. In 1930, following publication of *The Human Mind*, Menninger was asked to write a monthly question-and-answer column on "Mental Hygiene in the Home" for *Ladies Home Journal*, the forerunner for Margaret Mead's and Benjamin Spock's columns in *Redbook* three decades later. The Menningers benefited strongly from the popularization of Freudian theory following World War II, with Will Menninger featured in a 1948 *Time* cover story; and two years later *Newsweek* labeled the Menninger Clinic "the world's best-known psychiatric center." In 1962 Walter Cronkite accompanied a CBS television crew to Topeka to film a story on the Menningers, and in 1979 their visages were institutionalized as the embodiment of healing in a stained-glass window in the Washington Cathedral. In 1981 Karl Menninger was awarded the Medal of Freedom, the nation's highest civilian honor. The Menningers came to represent Freud in America, and honor slowly evolved into veneration.

Given this emerging milieu it was not surprising that Karl Menninger's *The Crime of Punishment* was widely acclaimed when it was published in 1968. The *New York Times* called it "a thunderous, plain-speaking indictment of traditional law enforcement," while *Life*, deeming it "a model of rationalism," said: "Read this book and weep—and pray that Karl Menninger may be heard in our chambers of injustice and the pesthouses we call penitentiaries." *Time* also praised it, noting that "the beginning of public wisdom is to understand the criminal's mind." The *Saturday Review of Literature, Saturday Evening Post*, and *Catholic World* each carried articles by Menninger summarizing his ideas and recommendations.

Menninger's message was in fact little changed from what he had been saying since 1925. The preface to *The Crime of Punishment* opened with a brief account of the Leopold and Loeb murder case, setting the scene to view criminals not as legal problems but rather as medical problems. The roots of crime, Menninger asserted, could be traced to early childhood experiences and especially to mothers. As he explained in a 1959 article in *Harper's*: "The offenders who are chucked into our county and state and federal prisons are not anyone's beloved children; they are usually unloved children, grown-up physically but still hungry for human concern which they never got or never get in normal ways. So they pursue it in abnormal ways—abnormal, that is, from *our* standpoint." To punish such offenders is therefore not only illogical, but it is also unfair. As Menninger summarized it:

> I suspect that all the crimes committed by all the jailed criminals do not equal in total social damage that of the crimes committed against them. *[Emphasis in original.]*

Menninger argued that "the great majority of offenders, even 'criminals,' should never become prisoners if we want to 'cure' them." Furthermore, "before we can reduce the self-inflicted sufferings from ill-controlled aggressive assaults we must renounce the ancient, obsolete penal attitude in favor of a modern therapeutic one." This attitude would emphasize fines rather than incarceration, indeterminate rather

... . prisons to therapeutic institutions run by psychiatrists and social workers. "This would no doubt lead to a transformation of prisons, if not to their total disappearance in their present form and function."

From William Healy to Richard Herrin and Willie Horton

Sigmund Freud's theory, applied to criminal behavior and promoted by William Healy, Bernard Glueck, William A. White, Karl Menninger, and other psychoanalytically oriented psychiatrists, seeped slowly into America's courtrooms, cellblocks, and parole offices. As early as 1921, Massachusetts, already a bastion for psychoanalytic influence, implemented the Briggs Law, which mandated psychiatric examination for felons to ascertain "the existence of any mental disease or defect which would affect his criminal responsibility." California and other states, influenced by psychiatrists and social workers, added provisions for indeterminate sentencing to their penal codes. By 1929 it was noted that "the present commissioner of corrections in New York State is a psychiatrist, and under him at Sing Sing Prison is a full-time mental clinic headed by a psychiatrist for the purpose of introducing this approach into the prison system of the state."

The full flowering of Freudian theory in American criminology and corrections did not take place until after World War II. In Nazi Germany genetic theories of criminality had been prominent, and by 1936 there had been established "fifty examination stations throughout Germany to explore the genetics and racial specificity of crime. . . . everyone serving a three-month sentence or longer was required to be examined." In the aftermath of the war, genetic theories of criminality were completely discredited as were other pieces of the Nazi panoply of prejudice and hate. Theorists about crime in America would henceforth talk about its social, economic, and familial antecedents but not about its genetic roots.

Within this milieu, the concept of crime as a consequence of early childhood experiences and social conditions quickly became more widely accepted. Treatment facilities to "cure" prisoners of their criminal behavior opened in California in 1954 (Atascadero State Hospital) and Maryland in 1955 (Patuxent Institute); Massachusetts, New York, and Wisconsin also moved aggressively in this direction, and almost all state corrections systems were influenced at least in part by the "treatment" concept. In Topeka, Karl Menninger established a psychiatric consultantship from the clinic to the Topeka Police Department and the Federal Bureau of Prisons.

Dr. Seymour L. Halleck, a psychiatrist who trained under Menninger and one of the architects of the Wisconsin "treatment" system, exemplified the strategy behind such thinking in his book *Psychiatry and the Dilemmas of Crime* (1967). "Predisposition to criminality," Halleck wrote, "has to be understood largely in terms of stresses emanating from the parents. . . . The major stresses of childhood related to criminal predisposition can be classified as parental deprivation, parental abuse, parental inconsistency and parental overcontrol." Such ideas were further reinforced by popular purveyors of Freud's theory, such as Dr. Benjamin Spock, who described in one of his *Redbook* columns the origins of criminality: "They [criminals] were never loved in their earliest years. They never felt they really belonged to their parents." Similarly, Margaret Mead, a regular visiting lecturer at the Menninger Clinic and a member of its board of directors, spoke of the familial and social roots of criminality: "We can accept the fact that prisoners, convicted criminals, are hostages to our own human failures to develop and support a decent way of living."

Freudian theory has continued to influence American criminology and corrections in three important areas: the concept of responsibility, the idea of crime prevention, and the use of punishment. Regarding the first, William A. White had frankly advocated "the discarding of the concept of responsibility" and Karl Menninger echoed White's idea. It had been long accepted under English law that individuals whose criminal actions had been the result of insanity or other brain diseases should not be held fully responsible, but it was the influence of Freudian ideology that led to the broad extension of this concept. This occurred most dramatically in 1954, at the height of postwar psychoanalytic ascendence, when Judge David Bazelon handed down what was known as the Durham decision in the District of Columbia:

> Our traditions also require that where such [criminal] acts stem from and are the product of mental disease or defect as these terms are used herein moral blame shall not attach and hence there will not be criminal responsibility.

Karl Menninger, in applauding the Durham decision, called it "more revolutionary in its total effect than the Supreme Court decision regarding segregation."

Under psychoanalytic theory, "the product of mental disease or defect" can be extended to virtually every criminal act. It is not surprising, therefore, that the Durham decision and its subsequent variations extending the insanity defense have spawned an inbroglio in American courts that regularly passes through regions of the ridiculous and reaches heights of the absurd. For example, in the 1969 trial of Sirhan Sirhan, accused of killing Senator Robert Kennedy, defense psychiatrists introduced testimony regarding Sirhan's early childhood traumas and his hatred of his father, which the press ridiculed as "a psychiatric circus—maybe part of the clown act." The chief psychiatrist for Sirhan was Dr. Bernard L. Diamond (who had as a 12-year-old boy avidly followed the Leopold and Loeb trial, about which he "read everything he could find on the case and determined on the spot that he too would grow up to be a psychoanalyst").

It is important to note that the 1954 Durham decision was a completely logical extension of Freudian thinking about criminal behavior. The concept spread rapidly with one study showing that acquittals on grounds of insanity increased fivefold between 1965 and 1976, with the average length of confinement for murderers found not guilty by reason of insanity only eight months. Insanity, in the broad Freudian sense, became increasingly used as a defense for crimes, with defense lawyers and psychiatrists becoming ever more creative in their descriptions of seemingly innocuous childhood experiences which they claimed were actually malignant antecedents of crimes.

A well-publicized example of such a trial was that of Richard Herrin, a graduate of Yale University who in 1977 brutally hammered to death his girlfriend, Bonnie Garland, when she tried to break off their relationship. Herrin acknowledged having had a "plan" to murder her, searching the house for a proper weapon, and concealing the weapon beneath a towel in case Bonnie's family, sleeping in adjacent rooms, happened to see him in the hall. Herrin also admitted purchasing a book a few days prior to the crime about the use of the insanity defense in a murder trial, although this fact was not revealed at the trial.

Herrin hired a well-known criminal lawyer and defense psychiatrists with a fund raised by his friends and the Catholic community at Yale University. Herrin's lawyer claimed that his client was experiencing an "extreme emotional disturbance" at the time of the crime and that he was not guilty by reason of insanity. Herrin had been valedictorian of his high school class in a Mexican-American suburb of Los Angeles, then won a full scholarship and subsequently graduated from

Yale University, during which time he showed no overt manifestations of mental illness.

The two psychiatrists hired to testify in Herrin's defense, Drs. John Train and Marc Rubinstein, were both fully trained in psychoanalysis. Both claimed that Herrin was suffering from a "transient situational reaction" when he committed the crime. Dr. Train detailed early childhood experiences which he said had contributed to Herrin's mental state; these included having had a skin condition on his legs, bedwetting, and an alcoholic father who abandoned the family. Most important was said to be the fact that Herrin's mother had begun seeing another man during the time when her son was in "the Oedipal situation in life," which, according to Dr. Train, is "a very tender period in the development of a child." According to Herrin's lawyer, these childhood events led to "an intense fear of abandonment by his loved ones," so that when Herrin was confronted by a girlfriend who wanted to end their relationship he murdered her. The jury apparently believed such testimony and found Herrin not guilty of murder but only of manslaughter because of his "extreme emotional disturbance."

Herrin's defense was logical from the viewpoint of Freudian theory. As psychoanalyst Willard Gaylin noted in writing about the case: "Psychiatrically speaking, nothing is wrong—only sick. . . . If an act is not a choice but merely the inevitable product of a series of past experiences, a man can no more be guilty of a crime than he is guilty of an abscess." Mrs. Garland, who found her daughter's body spattered with blood and brain tissue, was unable to understand such Freudian logic and commented bitterly: "If you have a $30,000 defense fund, a Yale connection and a clergy connection, you are entitled to one free hammer murder. . . . Heaven help girlfriends and boyfriends that are breaking up." Mrs. Garland's assessment echoes the remark of Vladimir Nabokov who said, "The Freudian faith leads to dangerous ethical consequences, such as when a filthy murderer with the brain of a tapeworm is given a lighter sentence because his mother spanked him too much or too little—it works both ways." In his 1982 book about the Herrin case, Peter Meyer put it in its proper historical perspective: "If Menninger was a spokesman for a revolution already under way, by 1978 Richard Herrin was its beneficiary and the Garlands may have been one of its victims." The insanity defense, once narrowly defined for use by individuals who were truly insane, had in 50 years of Freudian influence become an exonerating umbrella covering virtually all crimes.

Another area in which Freudian theory has profoundly influenced

American criminology is the concept of crime prevention. Since the 1920s it has been argued by mental health professionals that juvenile delinquents can be diverted from an impending life of crime if they are provided with Freudian-inspired counseling and psychotherapy. Understanding one's intrapsychic roots, according to this theory, helps humans to grow up straight rather than crooked. As Dr. Thomas Salmon explained in 1921, child guidance clinics, "ought to be expected to have as great an influence in diminishing delinquency as is shown in the results of the National Tuberculosis Association which has remarkably lessened the prevalence of that disease."

From 1937 to 1945 an extensive research project was carried out to assess the effectiveness of counseling and psychotherapy in preventing crime. Referred to as the Cambridge-Sommerville Delinquency Prevention Project, it involved more than 600 Boston-area boys judged as likely to become delinquent. The boys were randomly assigned by the toss of a coin either to a "no-treatment" control group or to a "treatment" group that consisted of an ongoing relationship with a social worker. The social workers met with the boys on an average of twice a month for five and a half years. The social workers were said to use both traditional psychoanalytic techniques and also nondirective techniques based on the theory of psychotherapist Carl Rogers. In addition, over half the boys in the treatment group were tutored in academic subjects, half were sent to summer camps, and one-third were referred for "medical or psychiatric help." The boys averaged ten years of age at the start of the treatment program and almost sixteen years at the end.

In 1948, three years after the project ended, an initial evaluation of the results revealed that "the boys who had received treatment were not less likely to have been brought to criminal court; nor were they committing fewer crimes." It was predicted, however, that the effectiveness of the treatment would become clearer over longer time. "The evaluation of the program in the 1950s . . . again revealed no benefits from the program." In 1975 a thirty-year follow-up was undertaken during which 95 percent of the study group was located. It was found that "as adults, equal numbers [of the treatment and no-treatment groups] had been convicted for some crime. . . . Unexpectedly, however, a higher proportion of criminals from the treatment group than of criminals from the control group committed more than one crime" and the difference was statistically significant. Further analysis of the study revealed that longer treatment increased the chances of later criminal behavior and more intensive treatment, in which counselors had focused on personal or family problems, increased the chances of later criminal behavior. As Dr. Joan McCord, professor of criminal justice at Temple University, summarized the results: " 'More' was 'worse.' "

It appears, then, that counseling and psychotherapy given to young juvenile delinquents do not decrease later criminal behavior. On the contrary, insofar as it has any effect at all, it appears to *increase* later criminal behavior. In the terms proposed by Dr. Thomas Salmon in 1921, it is as if a tuberculosis prevention program was found to *increase* the prevalence of that disease. Why counseling and psychotherapy should increase rather than decrease criminal behavior is a matter for speculation. Dr. McCord suggested, among other reasons, that the counselors may have increased dependency among the boys and that they were therefore less able to cope with life's problems on their own after it was terminated. Alternatively, she suggested that "the supportive attitudes of the counselors may have filtered reality for the boys, leading them to expect more from life than they could receive." Equally plausible is that Freudian-oriented counseling taught the boys that they were not responsible for their behavior because it was a product of childhood experiences and society; so exonerated, boys in the treatment group were therefore less troubled by their consciences than those in the nontreatment group.

The third area of criminology in which Freudian theory has influenced American thinking is the concept of punishment. If an individual is truly not responsible for his or her criminal behavior, then it is illogical to punish that person. William A. White reflected this in his recommendation that "prisons and punishment should both be abolished," and Karl Menninger later expanded this theme into "the crime of punishment."

The Patuxent Institution in Maryland embodied the goal of rehabilitation, rather than punishment of criminals, as thoroughly as any penal institution in the United States. Following its creation in 1955, its 500 inmates, carefully selected from applicants residing in other Maryland state prisons, were "treated" with both individual and group psychotherapy to help them understand the roots of their criminal behavior. In keeping with the goal of rehabilitation rather than punishment, all inmates at Patuxent were eligible for home furloughs, work releases, and parole from the day they arrived; the determining factor regulating how quickly an inmate was released was not how much time he had served but rather how quickly he acquired insight into his behavior and was thereby "cured." Under these guidelines, individuals sentenced to life imprisonment were released from Patuxent after an average of 8 years (some in as few as 4 years), compared with 20 years in other Maryland state prisons.

Despite having been in operation for 35 years, until recently there was virtually no published research data on Patuxent's effectiveness. It was claimed that the recidivism rate of its released prisoners was 18 percent, compared with a 47 percent recidivism rate for other Maryland prisons; however, this had to be weighed against the fact that Patuxent had selected for admission only those prisoners deemed to be good rehabilitation risks, including a large number of individuals who had killed family members and who had no prior criminal history, a group known to have a very low recidivism rate. In 1991, however, the results of a more complete study were released, demonstrating that "former Patuxent inmates are just as likely to be rearrested as are people released from other state prisons included in the study," despite Patuxent's ability to select its admissions. The counseling programs were said to "have no discernible effect on prisoner recidivism."

It is known that among the inmates Patuxent had treated and released as "cured," there were some tragic mistakes. Charles Wantland, paroled after serving 6 years of a 30-year sentence for murder, killed a young boy 3 weeks later. Billy Ray Prevatte was sentenced to life in prison for killing a teacher and wounding 2 others; after being released by Patuxent, he was convicted of assault with a deadly weapon and attempted murder, sentenced to a 40-year term, and then released a second time by Patuxent 4 years later. William Snowden, sentenced to 28 years in prison for rape and assault in 1974, was paroled by Patuxent in 1981. Arrested 5 months later for another rape, he was sentenced to 20 years, but released again by Patuxent in 1986; in 1990, he was rearrested for murdering a woman. In 1988 Patuxent came under public scrutiny when it was revealed that James Stavarakas, a convicted rapist who was on a work release, had raped another woman, and Robert Angell, a triple murderer whose victims included one picked at random on the street, had been given unsupervised furloughs. One public official derided Patuxent as "nothing but a psychiatric sandbox" and said that "public safety is too important to be left to psychiatry."

An extensive review of all "treatment" programs (as opposed to punishment) for adult offenders was published in 1975. After looking at "the effect of individual psychotherapy upon recidivism" in 13 different studies, the authors concluded that "no clearly positive or negative general statement can be made." When the type of psychotherapy was taken into account, however, it was found that "pragmatically oriented therapy," which "focused on personal, vocational, and social issues," was "more successful than psychoanalytically oriented therapy in reducing recidivism." They concluded that treatment of offenders "is likely to be unsuccessful and perhaps even harmful, if it is administered to nonamenable, younger offenders by unenthusiastic therapists with a psychoanalytic orientation." Additional studies since 1975 have confirmed these conclusions.

The failure of treatment programs for adult offenders is also reflected by state and national recidivism rates. For the United States as a whole, according to 1986 data, "nearly 63 percent of the inmates released from state prisons are rearrested for a serious crime within three years"; this recidivism rate is virtually identical to the two-thirds rate reported by Dr. Bernard Glueck at Sing Sing Prison 70 years earlier, before Freudian theory became prevalent in the criminal justice system.

Probably the best known recent failure of prison program which has been strongly associated with the treatment of prisoners was the Willie Horton case. Convicted in Massachusetts in 1975 of fatally stabbing a gas station attendant 19 times and dumping his body in a trash barrel, Horton showed no remorse for the crime and also had a prior conviction for attempted murder. He was sentenced to life in prison without the possibility of parole. Despite a history of drug use while in prison, Horton was recommended for unsupervised furloughs in 1986 by his caseworker. Massachusetts, known to have an extremely liberal furlough policy, has for many years ranked behind only California in its enthusiasm for the psychotherapeutic treatment of prisoners. While on furlough Horton escaped, broke into a home, and repeatedly raped and terrorized the occupants for 12 hours.

When Massachusetts Governor Michael Dukakis ran as a Democrat for president in 1988, Willie Horton became a major issue in the campaign. A television commercial for Republican candidate George Bush informed America that Dukakis's "revolving-door prison policy gave weekend furloughs to first-degree murderers not eligible for parole." A Republican fund-raising letter pictured Dukakis next to Horton with the question: "Is this your pro-family team for 1988?" Dukakis was painted as a classic liberal interested in rehabilitating criminals, not punishing them, and therefore a man who was soft on crime. The editorial page of the *Washington Post* noted that "many Democrats seem unwilling or unable to think about violent crime as a matter of public policy"; those who promote punishment for its own sake, it continued, "are regarded as some kind of sadistic relic of the dark ages."

The Willie Horton episode was merely the continuation of a debate that had been ongoing since William Healy introduced Freud's theory into corrections three-quarters of a century earlier. The release of Willie Horton was partially a consequence of the Healy, Glueck, White, and Menninger tradition. Governor Dukakis and the Democratic Party were merely recent and more prominent victims of that tradition.

8

Philosopher Queen and Psychiatrist Kings: The Freudianization of America

---

*Above all, though many of the events of the decade seem
to belong to another world—to a raucous party that lasted
long but ended badly—the sixties remain a tangible myth,
a point of departure for every kind of social argument, as
well as the source of values widely diffused throughout
our culture.*
—Morris Dickstein, The Gates of Eden

*I*t is not coincidental that social activists, political liberals, and Freud's influence all achieved their apogees in 1960s America. They shared many genes, had been raised on the principles of Benjamin Spock, counted Margaret Mead as an aunt, and Herbert Marcuse, Paul Goodman, and Norman O. Brown among their godfathers. They read the writings of the same New York intellectuals and went to the same plays and movies. Freud, accompanied by many of America's most respected liberals, moved decisively from Greenwich Village to Main Street during the 1960s. And if one looked closely at pictures of Freud in middle America, the eidolon of Karl Marx was sometimes visible hovering in the background.

Politically it was the decade of the New Left. The Old Left, terminally discredited when Russian tanks rolled into Budapest and Khrushchev publicly acknowledged Stalin's crimes in 1956, suffered from a continuing effluvium which seemed to emanate from the corpse of Joseph McCarthy. The New Left, by contrast, rallied behind Fidel Castro's Cuban regime, organizing Fair Play for Cuba committees, student visits to this new socialist paradise, and protests against American intervention there. Later in the decade Che Guevara and other revolutionaries became important only in turn to be swept aside by mass opposition to the war in Vietnam.

The New Left was not a product of Vietnam, although Vietnam became its symbol. As historian Richard Pells pointed out, "The radicalism of the 1960s was primarily cultural rather than political or economic." The civil rights of blacks in America dominated the agenda—from the lunch counters in North Carolina, Freedom Rides in Alabama, and voter registration drives in Mississippi to riots in Harlem, Watts, and Detroit. It was a decade of sit-ins, teach-ins, and marches organized by the Student Nonviolent Coordinating Committee and the Congress for Racial Equality and supported by the New Left. Hispanics and poor whites also achieved occasional celebrity, as in the farmworkers' boycott in Delano, California, and the Miners' Movement in Hazard, Kentucky.

The civil rights of students were another integral part of the New Left. The Free Speech Movement at Berkeley and student takeovers at Columbia University and on hundreds of campuses were part of a broad movement in which students demanded control over their courses and their lives. Liberalized sex and drug use were elements of this new social contract, and in support of their position the students invoked mentors such as Timothy Leary, Ken Kesey, Allen Ginsberg, Lenny Bruce, and Bob Dylan.

In this milieu of liberalism and social activism, the theory of Sigmund Freud became even more widespread. Psychoanalysis and insight-oriented psychotherapy, once exotic playthings of the rich, became much more readily available in wealthy suburban and university communities as psychiatrists, psychologists, and psychiatric social workers proliferated. As early as 1961 *Esquire* assessed the scene and reported that "everybody . . . is constantly [psycho]analyzing himself, and then, for good measure, his fiancée, his mother, his favorite poet, and his chief opposition." That same year the *Atlantic Monthly* published a special supplement, "Psychiatry in American Life," saying that the psychoanalytic "revolution has been incalculably great in the United States. . . . psychoanalysis and psychiatry in general have influenced medicine, the arts and criticism, popular entertainment, advertising, the rearing of children, sociology, anthropology, legal thought and practice, humor, manners and mores, even organized religion." And this was only the beginning.

Margaret Mead, Ex Cathedra

It was also in 1961 that Margaret Mead began writing a monthly column for *Redbook*, a popular women's magazine with a circulation of 4.5 million. Except for Benjamin Spock, Mead probably did more than any individual during the 1960s to both popularize Freud's theory and link it to mainstream liberal thought. She was a formidable force upon American culture, and named "Woman of the Year" by the American Association of University Women in 1961 and "Mother of the World" by *Time* magazine in 1969. On her 75th birthday, the American Museum of Natural History and her publishers took out a full-page congratulatory ad in the *New York Times*. By then she had adopted a forked walking stick and wardrobe of capes, the visage of a prophetess with the mantle of omniscience.

Mead went almost everywhere offering opinions on almost everything to almost everyone. During one representative nine-month period in 1963, she "gave more than 100 speeches in this country and abroad, participated in almost 50 radio or television programs, and wrote dozens of pamphlets and articles" in addition to her duties as associate curator at the American Museum of Natural History and teaching responsibilities at Columbia University. Her friends characterized her as "a woman of enormous self-confidence" and "liberally endowed [with] the courage of [her] convictions." Those less impressed said that "her egotism was ungovernable," called her "Mother-Goddess Mead," and passed along the joke about what Mead had said when introduced to the Delphic Oracle: "Hello, isn't there something that *you'd* like to know?"

Redbook was one of Margaret Mead's most effective podiums, for it reached a broader audience than her anthropological writings ever could. From 1961 until her death in 1979, Mead addressed an extraordinary variety of issues, including the fluoridation of water, prayer in public schools, legalization of off-track betting, why people overeat, the national debt, should Americans be permitted to travel to Cuba, the two-party political system, Jean Paul Sartre's refusal of the Nobel Prize, and the authenticity of Shakespeare's authorship.

The greatest number of Mead's contributions to *Redbook*, as indeed everywhere she wrote, concerned her theories about human behavior and the implications of these for marriage, child rearing, and other personal and social problems. Her ideas were classically psychoanalytic in derivation, yet she avoided the jargon connected with Freud's theory. For example, in "A New Understanding of Childhood," Mead described how the investigations carried out by Freud had "drastically" changed our understanding of childhood. In some columns she was vague about the precise nature of childhood problems, which she claimed were at the root of mankind's problems, as for example when she wrote: "The source of the child's difficulties may be a series of events that deeply affected his relations to people," or "There are also difficulties that result from terrible mishaps in upbringing." On other occasions she explicitly expounded Freudian doctrine:

> *Comparative studies suggest that the way people teach children to control their sphincters is closely related to the way they teach them many other things—about cleanliness in general, about whether the human body should be accepted joyfully or should be treated as something shameful, about the importance of property, about punctuality and routine.*

The treatment for problems originating in childhood development was, according to Mead, psychoanalysis, which she described as being another form of "modern scientific treatment" for the mentally ill or disturbed. "Freud's belief," Mead said, "was that if troubled individuals could recover what was lost and understand what went amiss, they could free themselves to become happier and more productive." "A willingness to look at oneself within some framework of analysis . . . can add immeasurably to an individual's sense of herself as a person." Although Mead acknowledged that she herself had not undergone psychoanalysis, she strongly advocated it for her friends, employees at the Museum of Natural History, and readers of *Redbook*.

As might be expected in a popular women's magazine, Margaret Mead wrote frequently about marriage, divorce, and child rearing. Her contributions included "What Does the American Man Expect of a Wife?" (May 1962), "A Continuing Dialogue on Marriage" (April 1968), "Double Talk About Divorce" (May 1968), "What Makes a Lousy Marriage" (February 1969), "Too Many Divorces Too Soon" (February 1974), "What Makes Marriage Such a Special Relationship" (May 1975), "Every Home Needs Two Adults" (May 1976), and "Can the American Family Survive?" (February 1977). She never explained where she obtained her credentials as an expert on marriage, however, nor how this expertise was related to her own three failed marriages, the longest of which lasted only six years. Mead's prescriptions for child rearing had a similarly hollow ring when juxtaposed with the fact that when her own daughter was between two and four years of age, Mead, by her own admission, never "spent three uninterrupted days with her daughter." Furthermore, when her daughter was thirteen, Mead went to the South Pacific "for almost a year . . . communicating primarily through common letters to be shared by her family and friends, letters that arrived duplicated by her office in single-spaced purple copies," and which her daughter said she read "barely or not at all."

And yet Margaret Mead was a ready-made prophetess for the 1960s, a bridge between Freudian dogma and the social agenda of the New Left. As a lifelong liberal she served on an advisory committee on women's issues for President Johnson, supported the social programs of the Great Society, and backed Hubert Humphrey's bid for the presidency in 1968. She glorified the youth culture, claiming, "There are no elders who know what those who have been reared within the last twenty years know about the world into which they were born. . . . The young, free to act on their own initiative, can lead their elders in the direction of the unknown." Such pronouncements, combined with her advocacy before a Senate subcommittee for the legalization of marijuana and her praise of mescaline and LSD as "aids to therapy," made her enormously popular on college campuses, at which she gave as many as 80 lectures a year.

Margaret Mead was also perfectly situated to respond to the issue of civil rights as it penetrated into the American consciousness. Her professional career had been spent attempting to demonstrate the dominance of nurture over nature, culture over genes. As the issue of race became a national one, Mead insisted again and again that human infants are all born "with the same range of potentialities" and that "the most powerful influence on man's mental achievement appears to be his culture." Groups called races, Mead said, have no "measurable differences in their capacity as groups of individuals to take on any civilization." Specifically regarding the question of intelligence, Mead said that there was no evidence to suggest that racial differences were biological rather than cultural in origin.

As she wrote her columns for *Redbook* and traveled from campus to campus, Margaret Mead's lifelong sexual agenda was never far from view. She condoned premarital sex, trial marriages, and praised nudism as a way to bring about "a reduction in puritanism and prudery that would ultimately lead to a decrease in neuroses and certain kinds of crime." She severely criticized laws against homosexuality, claiming that such laws, like those prohibiting drug use, merely lead to contempt for all laws and therefore encouraged criminal behavior.

Mead wrote and spoke about homosexuality on numerous occasions, insisting that "bisexual potentialities are normal." When an individual is exclusively homosexual, Mead said, it is "probably because of a trauma in childhood that made them fear or hate the opposite sex or, alternatively, deeply prefer the role of the opposite sex." In January 1975 she used her entire column in *Redbook* for a discourse on "Bisexuality: What's It All About?," in which she claimed that "a majority [of humans] are bisexual in their potential capacity for love" and called bisexuality "a normal form of human behavior." In an article on "Cultural Determinants of Sexual Behavior" she claimed explicitly that bisexuality was a superior condition to either heterosexuality or homosexuality: "If the term *natural* be taken to mean behavior of which all human beings are potentially capable, then we may also argue that the individual who is wholly incapable of a homosexual response has failed to develop one human potentiality." Despite such defenses of bisexuality and homosexuality, Mead never publicly acknowledged her own bisexuality, and in one *Redbook* interview she inanely disparaged the emerging public gay and lesbian movement by saying, "Running around wearing a button that reads 'I am a lesbian' has nothing to do with being a human being."

Marcuse, Goodman, and Brown

Margaret Mead's message was frequently interpreted by college students as a hodgepodge of Freud, sex, drugs, and student power. Freud was being widely read on campuses in the 1960s; indeed, even in the 1950s David Riesman noted that at the University of Chicago students "seemed to me to be too inclined to swallow [Freud] whole as part of their progressivism." In the 1960s the students also read the putative gurus of the New Left—Herbert Marcuse, Paul Goodman, and Norman O. Brown. Their message had many similarities to that of Mead and strongly reinforced the association between Freudian theory and liberal politics.

According to Morris Dickstein's *Gates of Eden: American Culture in the Sixties*, Marcuse, Goodman, and Brown were "the theorists whose work had the greatest impact on the new culture of the sixties." The three professed politically liberal beliefs—Marcuse as a Marxist, Goodman as an anarchist, and Brown as "a typical thirties radical"—and all three saw students as the vanguard of a new proletariat. All were also orthodox Freudians (Marcuse and Goodman publicly attacked Horney, Fromm, and the Neo-Freudians) and all believed that sexual repression underpinned most of society's ills. They depicted a Freudian utopia in which there would be no alienation and no unnecessary sexual repression, an erotic Erewhon in which Priapus would be king.

Herbert Marcuse was born to wealthy parents in Berlin in 1898, completed a doctorate in philosophy, then migrated to the United States when Hitler ascended to power. During World War II, he worked as a civilian intelligence officer for the Office of Strategic Services (OSS)—the predecessor of the Central Intelligence Agency—and then moved to the State Department for six years following the war. He next taught philosophy at Brandeis University for twelve years and, when his contract was not renewed, moved to the University of California at San Diego. *Eros and Civilization* (1955) and *One-Dimensional Man* (1964) became required reading for the New Left. The books were characterized by Greg Calvert, a leader in the Students for a Democratic Society (SDS) as "the most exciting works available" and in 1967 the SDS New York regional office held a conference on Marcuse's works.

In 1968 Marcuse became identified with the student uprisings at Columbia University. Student leader Mark Rudd had been inspired, it was said, by Marcuse's writings, which had been given to Rudd by Marcuse's stepson, a fellow student. One slogan of the marchers, "The riot is the social extension of the orgasm," was said to have been Marcusean in origin. The *New York Times* called Marcuse "the ideological leader of the New Left." Members of Congress and the American Legion publicly attacked Marcuse; following a threat on his life, he went into hiding.

Marcuse attempted to fuse the theories of Marx and Freud. He had written on Marx before emigrating from Germany, but as he grew older he appeared to become less classically Marxist and more, in the words of Leszek Kolakowski, "a prophet of semi-romantic anarchism in its most irrational form." When Angela Davis, who had been one of his students, was accused of smuggling arms into San Quentin prison (to Jonathan Jackson), Marcuse wrote her an open letter in *Ramparts*, justifying her actions as obeying "a moral imperative" and concluding that "your cause is our cause." According to historian Paul Robinson, Marcuse identified student protesters, civil rights workers, and antiwar demonstrators "as the true descendants of the classical Marxian proletariat."

Marcuse's Marxism was amalgamated with Freud's theory regarding the repression of sexual instincts as the price for civilization. In *Civilization and Its Discontents*, Freud had written: "Our civilization is largely responsible for our misery. . . . the price we pay for our advance in civilization is a loss of happiness through the heightening of the sense of guilt." In a similar vein Marcuse wrote: "The sickness of the individual is ultimately caused and sustained by the sickness of his civilization." Whereas Freud had been pessimistic that civilization could be changed, Marcuse not only believed that it could be but provided a blueprint for what he identified as "a new stage of civilization." Sexual repression, Marcuse argued, had originally been necessary so that workers could concentrate on production. With modern production techniques such repression was no longer necessary, so Marcuse designated it as "surplus repression." If such repression was abolished, mankind could return to an original state of infantile sexuality (Marcuse's "polymorphous perversity"), in which the entire body would once again become a source of sexual pleasure.

Marcuse advocated the return of civilization to an idyllic state in which there would be no alienation and in which "being is essentially the striving for pleasure." According to Richard King's *The Party of Eros*, Marcuse was a utopian who wished to bring about "a new erotic pastoral" in which "play, sensuous activity, and gratification were the desired human activities."

Because it was widely read, *Eros and Civilization* contributed significantly to making Freud's ideas respectable among the New Left generation of students. Many students were introduced to Freud for the first time by Marcuse; at Columbia, for example, future Freud biographer Peter Gay recalled being "smitten with Herbert Marcuse's *Eros and Civilization*," and later underwent psychoanalysis. The subtitle of Marcuse's book is in fact "A Philosophical Inquiry Into Freud," and the *New York Times* praised it as "the most significant general treatment of psychoanalytic theory since Freud himself ceased publication," with the exception of the work of Ernest Jones. The students were drawn to Marcuse because he elevated them to leadership in his utopian world, because he labeled society as "psychotoxic," and because he preached sexual liberation. The "polymorphously perverse," including homosexuals, were especially praised; as Paul Robinson noted, "The sexual deviant had been the hero of *Eros and Civilization*."

Sexual deviants play an even more important role in the work of

Paul Goodman. Born in New York, Goodman attended City College and later earned a Ph.D. in humanities at the University of Chicago. He taught at several colleges and wrote regularly for magazines such as *Partisan Review*. He also contributed to anarchist publications; one historian called him "the chief spokesman for the non-Marxist tradition of western radicalism." Like many anarchists he refused to support American efforts in World War II, arguing that the working class had more to fear from Roosevelt than from Hitler, and because of this position he fell out of favor with the New York intelligentsia.

Goodman was a Freudian with a strong admixture of ideas from Wilhelm Reich. Goodman linked sexual and political repression, arguing that "the repression of infantile and adolescent sexuality is the direct cause of submissiveness of the people to present rule of whatever kind"; this was derided as the "gonad theory of revolution" by two critics. Goodman himself was militantly bisexual; he had a wife and son but, like Margaret Mead, became involved in homosexual relationships. Goodman was also one of the first to participate in the postwar encounter group and group therapy movement, and in 1951 coauthored *Gestalt Therapy* with Fritz Perls and Ralph F. Hefferline, in which the theories of Freud and Reich are both prominent.

Goodman's major contribution to the New Left was his book *Growing Up Absurd: Problems of Youth in the Organized Society*. Nineteen publishers had rejected the manuscript before Norman Podhoretz, the editor of *Commentary*, serialized it and helped persuade Random House to publish it in 1960. The pretext of the book was laid out by Goodman on the first page: "I assume that the young really need a more worth-while world in order to grow up at all, and I confront this real need with the world that they have been getting." Goodman then proceeded to describe contemporary society in which individuals despite their affluence were alienated from schools, jobs, and especially from each other. Much of the problem, Goodman wrote, was that society constantly thwarted man's instinctual urges and ego development, thereby forcing him into conformity and repressed modes for the convenience of society.

A large portion of *Growing Up Absurd* is taken up with proposals for organizing society along lines Goodman had outlined in previous writings. Decentralization of authority to small units, worker say in management, small business development, town planning, civil liberties, for example, would produce an "organic integration of work, living and play." The vanguard destined to lead society into the future was the students, who were "in Goodman's terms America's most exploited social group and hence would have to serve as the cutting edge of social regeneration."

Within a few years of publication, *Growing Up Absurd* had sold over 100,000 copies and was considered as one of the campus "bibles" of the 1960s. In a survey of SDS leaders taken in 1965, it was found that half had read Goodman and Marcuse, far more than had read Marx, Lenin, or Trotsky. Goodman was an active participant in campus uprisings, including those at Berkeley, and in anti-Vietnam protests later in the decade. He also continued to be an outspoken proponent for homosexual rights, writing: "My homosexual acts have made me a nigger, subject to arbitrary brutality and debased when any out-going impulse is not taken for granted as a right." Goodman's autobiographical notes, *Five Years* (1966), reveal a man preoccupied with casual homosexual encounters and the attendant stigma imposed by society.

Norman O. Brown was the third, and least likely, of the trinity of 1960s campus gurus. Educated in England, where he earned a Ph.D. in classics, Brown emigrated to Chicago in 1936. He was "deeply stirred by the Spanish Civil War and the Gary [Indiana] steel strike" and "identified himself with the social movements on the Left." During World War II, he worked for the OSS, after which he went to Wesleyan University, where he became chairman of the classics department. Married and with four children, he was teaching courses on Latin, Greek mythology, and Western civilization in 1953 when Marcuse suggested to him that he read Freud, which he had not done previously.

According to Brown, "I have never had and will never have again the experience I had on reading Freud. . . . Everything etched itself on my mind. I read his writings six and ten times. . . . Freud had the effect of destroying my universe. . . . Freud had a rapid, explosive effect on me." In less than a year Brown began *Life Against Death: The Psychoanalytical Meaning of History*, which would make him widely known.

Brown's opening sentences summarized the book: "There is one word which, if we only understand it, is the key to Freud's thought. That word is 'repression.' . . . The essence of society is repression of the individual, and the essence of the individual is repression of himself." Brown attempted to apply classical Freudian theory to history, especially focusing on the life instinct, death instinct, and stages of infant sexuality. An entire section of Brown's book is devoted to studies in "anality," with chapters such as "The Excremental Vision." These views were summarized by Brown as follows:

Assuming that our capitalist civilization exhibits anal-neurotic traits on a mass scale and not just an individual scale, the orthodox psychoanalytical dogma can yield no program of social therapy except a change in toilet-training patterns. . .

Brown's solution to the repression of modern man was the emancipation of libidinous energy: "What the great world needs, of course, is a little more Eros and less strife." Like Marcuse and Goodman, he depicted a nonrepressive utopia to be achieved through psychoanalysis, although its details were vaguely sketched:

We, however, are concerned with reshaping psychoanalysis into a wider general theory of human nature, culture, and history, to be appropriated by the consciousness of mankind as a whole as a new stage in the historical process of man's coming to know himself.

Brown adopted Marcuse's goal of returning the human body to its original "polymorphously perverse" state, "delighting in that full life of all the body which it now fears . . . a purely sensuous life."

Although Brown's thought followed Marcuse's closely in most respects, he departed significantly in assigning to Christian theology an important role in achieving his utopia. In *Life Against Death* Brown wrote: "Here again Christian theology and psychoanalysis agree—the resurrected body is the transfigured body. The abolition of repression would abolish the unnatural concentrations of libido in certain particular bodily organs . . ." This idea was developed at length in Brown's *Love's Body* (1966), in which he attempted to fuse Freud, Marx, and Christianity: "Freud and Marx and Pope John: the thing is to bring them together."

Life Against Death was published by the Wesleyan University Press in 1959 and reviewed by only the student newspaper. A year later the book was by chance discovered by editor Jason Epstein and given to Norman Podhoretz at *Commentary*, who was "overwhelmed [and] convinced that we had stumbled on a great book by a major thinker." Podhoretz persuaded Lional Trilling to review it and Trilling wrote: "One of the most interesting and valuable works of our time. Dr. Brown's contribution to moral thought—and most especially where he touches on sexual behavior—cannot be overestimated. . . . It gives us the best interpretation of Freud I know." By 1966 the book had sold over 50,000 copies and was said to be "one of the underground books that undergraduates feel they must read to be with it." Theodore Roszak, in *The Making of a Counter Culture*, called the work of Brown (and Marcuse) "one of the defining features of the counter culture" and labeled Brown as "a very professorial prophet indeed: A Dionysus with footnotes."

Marcuse, Goodman, and Brown, then, contributed significantly to introducing 1960s students to Freudian theory and associating this theory with liberal ideology. Repression was the explanation for everything that was wrong, a Viennese Rosetta stone that could be used to decipher the ills of an individual or a whole society. According to Frederick J. Hoffman, "For the young men and women of the period [repression] served as a convenient label for all their grievances against society. . . . Repression became the American illness."

Two logical antidotes to repression included political activism and sex, which was the message of Marcuse, Goodman, and Brown that was probably heard most clearly on college campuses. Morris Dickstein, who was a student activist at Columbia at the time, reflected this assessment in *Gates of Eden*, his retrospective of the 1960s. In remembering his own reaction to Brown's *Life Against Death*, Dickstein noted:

> *I can recall how much excitement I once felt, and feel some of it again, but it's distressingly hard to know just what the excitement was all about . . . Brown and Marcuse's attack on the 'tyranny of genital organization' and their paean to the 'polymorphous perverse' sexuality of the infant. . . . I find it difficult to imagine what these things meant to me, who had scarcely attained to the tyrannized state that they were attacking. To me they meant not some ontological breakthrough for human nature, but probably just plain fucking, lots of it. . . . these men seemed to promise that good times were just around the corner.*

The American Intellectual Elite

Herbert Marcuse, Paul Goodman, and Norman O. Brown, in addition to being the most important intellectual influences on university students in the 1960s, were also part of a wider circle frequently referred to as the American intellectual elite. Charles Kadushin, in his study *The American Intellectual Elite* (1974), demonstrated that New York City was the group's capital because a third of the intellectuals had grown up there and over half of them lived in New York at the time of the study. Starting with lists of contributors to the leading intellectual journals, Kadushin identified approximately 200 leading

American intellectuals and subsequently interviewed 110 of them. Kadushin asked the intellectuals "to name intellectuals who influenced them on cultural or socio-political issues, or who they believed had high prestige in the intellectual community"—in brief: to rate each other. The result was a list of 70 "elders." The most highly rated of these 70 consisted of 21 individuals, who included the following: Hannah Arendt, Daniel Bell, Saul Bellow, Noam Chomsky, John K. Galbraith, Paul Goodman, Richard Hofstadter, Irving Howe, Irving Kristol, Dwight Macdonald, Norman Mailer, Herbert Marcuse, Mary McCarthy, Daniel P. Moynihan, Norman Podhoretz, David Riesman, Arthur Schlesinger, Jr., Robert Silvers, Susan Sontag, Lionel Trilling, and Edmund Wilson.

Some members of the intellectual elite such as Macdonald, McCarthy, Trilling, and Wilson, had been part of the prewar New York intellectual circle, which had its headquarters in the *Partisan Review*. The majority of them, however, were second- and third-generation members of the intellectual community but appeared to share two important characteristics with the original group: they had been strongly influenced by Freudian theory and their political inclinations were distinctly liberal. In order to test this impression, selected writings of each of the 21 intellectuals were examined and a questionnaire was sent to each of the 13 who were still alive in 1990. (The findings of this analysis are discussed in detail in Appendix A.)

Of the 21 leading intellectuals identified by Kadushin in 1974, there was insufficient information available on 2 (Irving Howe and Daniel P. Moynihan) with which to assess the impact of Freudian theory on their thought. Of the remaining 19, there were 6 for whom Freudian theory does not appear to have been a significant professional influence (Hannah Arendt, Noam Chomsky, John K. Galbraith, Richard Hofstadter, Irving Kristol, and Arthur Schlesinger, Jr.). The remaining 13 of these 19 intellectuals (68 percent) appear to have been significantly influenced by Freudian theory at some point in their careers, although in 3 cases (Saul Bellow, Paul Goodman, and Norman Mailer) the psycyhoanalytic derivations of Wilhelm Reich were more important than those of Freud himself. As Alfred Kazin summarized the situation in 1959: "Freud has been marvelous for intellectuals." It is also noteworthy that 12 of the 13 intellectuals who were significantly influenced by Freudian or Reichian theory (Saul Bellow excepted) have been identified at some point in their careers with liberal political views. The American intellectual elite of the 1960s and 1970s contributed to increasing Freud's respectability, disseminating his theory, and linking Freud's theory to liberal political thought.

In assessing the attraction of American intellectuals in the postwar period to Freudian theory, the effect of the Holocaust must be kept in mind. As discussed in chapter 5, the murder of millions of Jews was not an event that merely happened between 1942 and 1945, but was rather an event which continued to take place among intellectuals in the 1950s, 1960s, and indeed continues to take place even now. It was an event that has especially transformed Jewish history and consciousness in a way for which there is no Gentile equivalent. Since nearly half of the American intellectual elite were Jews, according to Kadushin's study, the Holocaust almost certainly exerted a profound effect on their personal needs and on their perception of the world.

Susan Sontag is a good example of how the Holocaust affected second- and third-generation American intellectuals. When she was 12 she chanced upon a collection of photographs of the concentration camps at Bergen-Belsen and Dachau. She recalled:

> Nothing I have seen—in photography or in real life—ever cut me as sharply, deeply, instantaneously. Indeed it seems plausible for me to divide my life into two parts. . . . Some limit had been reached, and not only that of horror; I felt irrevocably grieved, wounded, but a part of my feeling started to tighten; something went dead; something is still crying.

The continuing effect of the Holocaust on intellectuals was also visible in 1963 when Hannah Arendt published *Eichmann in Jerusalem*, an account of the trial of Adolf Eichmann that discussed responsibility for the evil perpetrated on the Jews. Among intellectuals, a historian noted, "Arendt's book provoked a hailstorm of criticism and enormous debate, well beyond mere political discussion," including a public forum that produced a bitterly argued debate. Even non-Jewish intellectuals like Arthur Schlesinger, Jr. facetiously observed: "next to Himmler, even Babbitt began to look good."

Just as there were important similarities between the first generation of American intellectuals and those who came later, there were also important differences that affected the dissemination of Freudian theory. Foremost was the fact that the intellectuals themselves became progressively more scattered across America, as they moved from a prewar concentration in the cafés of Greenwich Village to a postwar dispersal within the halls university campuses. Some intellectuals without

Ph.D.'s, such as Irving Howe and Alfred Kazin, were given professorships, while others such as Daniel Bell had conferred upon them Ph.D.'s by the university after they had been hired to teach, on the basis of published research. Philip Rahv was given a professorship at Brandeis University despite the fact that he did not even have an undergraduate degree. Kadushin noted in his study that 40 percent of the intellectuals were professors in colleges and universities.

Another important difference between the postwar generation of American intellectuals and those who had preceded them was that the postwar group was much more respectable. Part of this respectability was political; as a group, they moved toward the political center. Whereas those of the 1930s, such as Edmund Wilson, Lionel Trilling, Dwight Macdonald, and Mary McCarthy had clustered in leftist groups like the American Committee for the Defense of Leon Trotsky, those in the 1960s, such as Richard Hofstadter, John K. Galbraith, and Arthur Schlesinger, Jr., were mainline liberals. Even those who became most active in opposing the war in Vietnam—Norman Mailer, Susan Sontag, and Noam Chomsky, for example—were participants in what was increasingly regarded as the will of the American people. By the late 1960s, a few of the intelligentsia, such as Irving Kristol and Norman Podhoretz, had moved further to the political right and were being identified as neoconservatives. In the 1980s Kristol and Podhoretz actively supported President Ronald Reagan, an event which would have been inconceivable to their radical intellectual brethren 50 years earlier.

There were additional indices of respectability other than political centrism. As Alexander Bloom summarized the new status in his *Prodigal Sons: The New York Intellectuals and Their World*, "Being an intellectual became a profession rather than a way of life." Diana Trilling recalled that the intellectuals of the 1930s "had had to defend even having a comfortable chair to sit in," whereas by the 1960s "everybody had a well-upholstered sofa." And while the earlier group had written exclusively for limited circulation journals such as *Partisan Review*, the postwar intellectuals increasingly wrote for larger circulation journals such as the *New Yorker*. Indeed, according to one observer, *"Partisan Review* was getting to be like a farm team for the *New Yorker."*

The new respectability of postwar intellectuals was also symbolized by a *Time* magazine cover story on them in 1956, which included pictures of Lionel Trilling, Sidney Hook, and Walter Lippmann as well as quotes by Irving Kristol, Edmund Wilson, David Riesman, Daniel Boorstin, and Leslie Fiedler, among others. "What does it mean to be an intellectual in the United States?" *Time* asked, and then answered that many of the intellectuals "have come at last to realize that they are true and proud participants in the American dream." Even greater respectability awaited the intellectuals when John F. Kennedy moved into the White House in 1960, for "he began to flatter the intellectuals, which is to say he invited them to his house for supper." Some, like John K. Galbraith, Arthur Schlesinger, Jr., and Daniel P. Moynihan stayed on to work in the Kennedy administration. As Norman Mailer recalled the era: "We became just a touch like minor royalty."

The newfound esteem of the American intelligentsia inevitably affected everything they represented, including Freudian theory. It was no longer just the politically radical intellectuals at City College and Columbia who were offering up the doctor from Vienna as the savior of the world, but also mainstream liberal professors at the Universities of Buffalo and Brandeis. Such men and women were venerated— didn't *Time* itself approve?—and so their prescriptions for mankind deserved a careful hearing. And if the authors of articles in the *New Yorker* were in psychoanalysis, then surely there must be something to it. In this fashion Sigmund Freud rode the coattails of America's intellectuals out of New York and into the heart of academic America.

Freud and the Democratic Party

As America's intellectuals in the postwar period shifted their politics from Marxism toward the Democratic left, so too were Freud's other followers increasingly identified politically with the Democratic party. This trend was most clearly evident among psychoanalysts themselves. A 1966 study of psychoanalysts by Arnold A. Rogow showed that their preference in elections between 1948 and 1964 had been as follows:

1948	Republican (Dewey)	12%
	Democrat (Truman)	62%
	Progressive (Wallace)	22%
	not eligible—did not vote	4%
1952	Republican (Eisenhower)	21%
	Democrat (Stevenson)	79%
1956	Republican (Eisenhower)	15%
	Democrat (Stevenson)	85%

1960	Republican (Nixon)	6%
	Democrat (Kennedy)	91%
	not eligible—did not vote	3%

1964	Republican (Goldwater)	3%
	Democrat (Johnson)	95%
	not eligible— did not vote	2%

Freud's followers had voted for Henry Wallace's leftist Progressive party almost twice as frequently as they had voted for Thomas Dewey, and had given Richard Nixon only 6 percent and Barry Goldwater only 3 percent of their votes. (Furthermore in the 1964 election 62 percent of the psychoanalysts had contributed money to the presidential campaign, a record of monetary support probably unmatched by any professional group.) Even social scientists teaching in universities, a group well known for liberal political views, had given Goldwater 10 percent of their votes in the 1964 election. (There is no comparable record of voting preference for individuals who underwent psychoanalysis, but there are suggestions that it is in the same range as the voting preference of the psychoanalysts themselves.)

The attraction of Freud's followers to the Democratic party was not surprising. The party of Jefferson and Jackson—which for over a century had based its platform on low tariffs, a balanced federal budget, and a small federal bureaucracy—was transforming itself into the party of liberal social programs. In the 1924 elections, Democratic candidate John W. Davis had received only 28 percent of the popular vote, and by the late 1920s the Democrats were being called "a permanent minority" party; only two Democratic presidents had been elected since 1861. The political landscape changed abruptly in 1929 when Wall Street fell. Suddenly the Republican party—the party of business and corporate America—lost its appeal and the Democratic party rose from its knees. The Democrats were the party of the newly arrived immigrants, including the European psychoanalysts. The Democrats were also the party of the rapidly growing cities, where most followers of Freud lived. Most importantly, the Democrats were the party of social programs designed to improve living conditions and thereby ameliorate, according to Freudian theory, civilization and its discontents.

The association of Democrats with liberal social programs did not begin *de novo* with Franklin D. Roosevelt's New Deal. Woodrow Wilson's administration had spearheaded women's suffrage, passed a Child

Labor Act, and appointed the first Jewish justice (Louis Brandeis) to the Supreme Court. And in 1924 the Democrats had included in their platform a claim to being the party "concerned chiefly with human rights," as opposed to the Republicans, who were said to be "concerned chiefly with material things." The New Deal was an opportunity to put such rhetoric into legislative action through the Social Security Act, which provided federal grants for the aged, the blind, dependent children, and maternity and infant care as well as through unemployment insurance which provided assistance for the jobless. The federal government had accepted responsibility for the personal welfare of its citizens to a degree heretofore unknown; on taking office, Roosevelt had pronounced "social values more noble than mere monetary profit." One of the chief architects of Roosevelt's New Deal was Harry Hopkins, who at one time had been a social worker in New York City.

Following World War II, the linkage between personal welfare and Democratic politics continued under President Harry Truman's Fair Deal. The 1948 Democratic convention included the most comprehensive civil rights program ever written into a Democratic platform. Subsequent legislation supported by the Truman administration comprised increased social security payments, a comprehensive housing bill, an increase in the minimum wage, and nationwide health insurance. The emerging Democratic alliance included labor unions, immigrant and minority groups, civil libertarians, journalists, and intellectuals.

Mental health professionals had also become an integral part of the Democratic liberal coalition by the time of Truman's electoral victory. This was symbolized by the inclusion in the 1948 Democratic party platform of a proposal expressing explicit support for the Mental Health Act, a reference to the legislation that had created the National Institute of Mental Health two years earlier. The ostensible reason for setting up this institute was prompted by emerging evidence that state mental hospitals were veritable "snake pits" and that mental illnesses were much more common than had been previously realized, as reflected by the high rate of mental disorders found among draftees during the war; the National Institute of Mental Health was expected to promote research on mental diseases and support the training of additional professionals.

However, the architects of the National Institute of Mental Health believed that they already knew the causes of mental disorders—early childhood experiences and social conditions which had traumatized the psyches of Americans. The three principal architects who designed the legislation creating the institute—William Menninger, Francis J. Braceland, and Robert H. Felix—were all devoted to Freudian theory and were among the leaders of American psychiatry (each eventually would serve a term as president of the American Psychiatric Association). Menninger claimed that psychiatrists "have some knowledge of the unconscious dynamics" and should therefore "participate in community affairs in order to apply our psychiatric knowledge to civic problems." Braceland, equally immodest in his vision, said, "Modern psychiatry . . . no longer focuses entirely upon mental disease, nor [upon] the individual as a 'mental patient,' but rather it envisages man in the totality of his being and in the totality of his relationships." Felix added that community mental health programs "would require the active cooperation of other community agencies in carrying out, where indicated, plans for modification of the patient's environment." The spirit of Sigmund Freud had not only arrived in Washington but was ready to take on the nation's mental health as its special project. All that was needed was a president who would appreciate it and give it a chance.

The Joseph McCarthy–Dwight Eisenhower years were not propitious times for promoting a mental health agenda, especially one that had become associated with liberal political ideologies. Freudians consolidated their base in the nation's medical schools and psychoanalytic institutes and awaited the return of politicians who would recognize their special skills. They also contributed heavily to the congressionally appointed Joint Commission on Mental Illness and Health, a hydra-headed group which between 1955 and 1960 drew up a blueprint for future mental health services, including a recommendation for substantially increased federal involvement in financing such services.

Concurrently there emerged during the Eisenhower years the beginning of an organized, almost exclusively Republican opposition to the mental health lobby. Based primarily in right-wing political groups, anti-mental health sentiment issued strongly in 1956 in response to federal proposals to create a psychiatric hospital in Alaska; the anti-mental health groups charged that the bill was merely "a plot to establish a concentration camp" to be used for political prisoners. One witness testifying against the bill said that "psychiatry is a foreign ideology; it is alien to any kind of American thinking."

During the late 1950s, anti-mental health forces increasingly linked psychiatry to leftist political thought. *D.A.R.*, the official magazine of the Daughters of the American Revolution, claimed that "mental health is a Marxist weapon," while the newsletter of a rightist group in New York, the National Economic Council, stated, "Mental health is an inaccurate label for what is really a weapon being skillfully used by Communist propagandists to bring about conformity to the Marxist ideology." Demonstrations and literature opposing mental health eventually spread to more than half the states, although southern California and Texas remained the ideological centers. The anti-mental health forces increasingly became intermingled with those opposed to the fluoridation of water, desegregation of schools, and the use of polio vaccines as well as with explicitly bigoted groups opposing Jews and blacks. The effect of this was to vitiate and eventually discredit the anti-mental health movement, consigning it to the intellectual slag heap of the John Birchers. For many Americans, however, the anti-mental health movement made explicit suspicions that had previously been implicit: that psychiatry in general and Freud in particular were connected on the best of days to liberal Democratic causes and on the worst of days to Marxism and communism.

Mental Health in the Great Society

The election of John F. Kennedy to the White House in 1960 was a momentous political event for American psychiatry. Here was a president who could understand and appreciate them, an intellectual president who had published a Pulitzer Prize–winning book. Most importantly, here was a president whose own sister, Rosemary, was acknowledged to have been mentally retarded and, as a few insiders knew, mentally ill as well.

President Kennedy did not disappoint the followers of Freud. He assembled an interagency committee to make specific recommendations regarding mental health legislation—Daniel P. Moynihan was one of the key members—and in early 1963, Kennedy delivered a historic Special Message to Congress on "Mental Illness and Mental Retardation." Specifically, Kennedy proposed a program of federally funded community mental health centers (CMHCs) that would both allow individuals with mental illness to be treated closer to home and also prevent new cases of mental illness from developing by treating them early in the course of their disorder. With this projected improvement in the country's mental health, it was said, state psychiatric hospitals could gradually be emptied and ultimately closed. The CMHC legislation passed Congress easily and was signed into law by President Kennedy three weeks prior to his assassination.

Although community mental health centers were represented to Congress as a new treatment system for the mentally ill, the program was subverted from its inception by followers of Freud, who saw it as an opportunity to finally put their beliefs into action. Dr. William Menninger, one of the architects of the National Institute of Mental Health, had stated clearly that Freudian theory "serves as the only logical basis for preventive psychiatry—a valid mental hygiene." Given an opportunity, Freudian psychiatrists had never been shy about claiming that their special insight into human behavior qualified them to prescribe for the world's social problems. In 1941, for instance, Dr. George S. Stevenson, president of the American Psychiatric Association, had asserted that wars were "mental health problems" because their roots lay in "psychological and psychopathological factors." Similarly, in 1946, Dr. G. Brock Chisholm, president of the World Federation for Mental Health, had claimed that "if the race is to be freed from its crippling burden of good and evil, it must be psychiatrists who take the original responsibility."

From its beginning the federally funded community mental health centers program was used to try and implement the Freudian theory of human behavior on a mass scale. Dr. Robert Felix, director of the National Institute of Mental Health when the CMHC legislation was passed, urged his colleagues to combine psychoanalysis and public health, as he himself had done, and become involved with "education, social work, industry, the churches, recreation, the courts" so that mental health services would be "fully integrated into, and a regular and continuing part of, the total social environment."

Felix's vision of a government-sponsored mental health utopia was shared by other NIMH psychiatrists. For example, Dr. Stanley Yolles, who succeeded Felix as NIMH's director in 1964, claimed that the responsibility of mental health professionals was "to improve the lives of the people by bettering their physical environment, their educational and cultural opportunities, and other social and environmental conditions." Yolles pointed with pride to CMHC staff members who "in addition to treating the classic range of mental illness, are helping clients with problems about housing, bill collection, reading difficulties. . ." Yolles was especially emphatic about the responsibility of psychiatrists to combat poverty: "The conditions of poverty, since they constitute a breeding ground for mental disease, require the professional involvement of the modern psychiatrist."

However, there never has been evidence that poverty *causes* mental diseases, other than in rare cases in which severe malnutrition may lead to pellagra. What Yolles was reflecting was a modern version of Freud's *Civilization and its Discontents*, in which societal conditions are assumed to be the cause of mental disorders. This view was shared by many other prominent NIMH psychiatrists, including Dr. Leonard Duhl. Psychoanalytically trained, Duhl argued that mental illness is "a socially defined condition and mental health must be conceived of as a social problem." Psychiatrists, said Duhl, should help "construct a social system that produces mentally healthy individuals" and must look especially closely at conditions in the cities. "The totality of urban life is the only rational focus for concern with mental illness . . . our problem now embraces all of society and we must examine every aspect of it to determine what is conducive to mental health."

Freudian theory thus implicitly became official government policy; the "prevention of mental illness" and the "promotion of mental health" were to be carried out by federally funded psychiatrists in government offices with their utopian plans typed on official stationery. Never before had federal funds been used to pay a community activist to tell a CMHC how to promote mental health, as Saul Alinsky did in Chicago. Never before had federal funds been used to help a CMHC organize a community to obtain a red light installed at a busy intersection, as took place at a CMHC in Los Angeles. Never before had federal funds been used to help a CMHC community board define as its mission "to resolve the underlying causes of mental health problems such as unequal distribution of opportunity, income, and benefits of technical progress," as occurred in Philadelphia.

It is hardly surprising that the grandiose expectations for community mental health centers came to naught. The CMHCs which attempted to become social change agents quickly confronted political realities and found that other community leaders were not willing to accept their Freudian wisdom on faith alone. The CMHCs that preached social activism, such as Lincoln Hospital Mental Health Services in New York, were taken over by the social activists they had unleashed. By the early 1970s, most CMHCs had ceased trying to change society to conform to Freudian specifications and simply reverted to being traditional mental health outpatient clinics, providing counseling and psychotherapy to individual patients.

In retrospect it is interesting to note the similarities between the Freudian-inspired social engineering envisioned for CMHCs in the 1960s and the Marxian-inspired social engineering which was being promoted in the 1930s. Both were based on theories about human behavior for which there was little scientific basis; both stressed humanistic and liberal social values; and both were promoted with a zeal that often became religious fervor. Dr. Harold G. Wittington, one of the earliest CMHC directors, noted that as early as 1965, "The Soviet Union has had, for quite a few years now, psychiatric facilities which sound very much like our comprehensive community mental health centers . . . [which] have emphasized the social-rootedness of man and the social-relatedness of all human behavior." Three years later Dr. Lawrence Kubie criticized his CMHC colleagues for harboring "the Russian fantasy that all psychiatric illnesses are due to social inequities and can be both cured and prevented by curing the ills of our social order." The Soviet experiment was based on Marx, while the American CMHC experiment was an outgrowth of Freud, but their similarities further reinforced the association that had evolved between Freud and the American political left.

Although the influence of Freudian theory in the Kennedy administration focused primarily on the CMHC movement, it indirectly affected other aspects of social programming. Poverty, for example, was said by officials at the National Institute of Mental Health to be a major cause of mental illnesses, and in such a view the entire war on poverty, proposed under Kennedy and implemented under President Lyndon Johnson, could be viewed as a mental health program. It is also of interest that the three individuals who were the primary architects of President Johnson's 1964 Office of Economic Opportunity's campaign against poverty were all deeply imbued with Freudian theory. The three included Michael Harrington, whose 1962 book on poverty, *The Other America*, had been read by President Kennedy. Harrington believed that "the poor tend to [have] much higher rates of psychosis than any other class or stratum, because the life of poverty is a constant assault upon the psyche as well as the body." Harrington's own four-year psychoanalysis began in 1965, one year after he became involved in Johnson's Great Society, but already in 1964 he acknowledged that "I had read my Freud and lived in a New York world where the majority of my friends had been, or were being, psychoanalyzed."

The other two architects of Johnson's program to help the poor were Frank Mankiewicz and Paul Jacobs, who, together with Harrington, worked for "two frantic weeks of sixteen-and-eighteen-hour work days" in January 1964 to deliver a plan to Sargent Shriver, President Johnson's director of the Office of Economic Opportunity. Although Mankiewicz had not been personally psychoanalyzed, his father had been in analysis and so Mankiewicz "grew up in a home in which Freudian concepts were prominent." According to Mankiewicz, Paul

Jacobs was also "imbued with Freudian thinking," and it seems likely that both Mankiewicz and Jacobs shared Harrington's belief that mental illnesses were partly caused by poverty. The fact that Mankiewicz was a prominent liberal Democrat (he later managed George McGovern's 1972 presidential campaign), Harrington a leading Socialist, and Jacobs "an ex-Trotskyist and union organizer," placed this program squarely on the liberal side of the political spectrum.

The war on poverty was, in retrospect, one of the most humane and well-intended programs ever instituted by the federal government. There is evidence that it did reduce poverty, primarily among the elderly, through such programs as Medicare, Supplemental Security Income, and food stamps. And the Headstart program for children is generally acknowledged to have been a success by providing better nutrition, cultural enrichment, and parental involvement in the learning process of their children. Such successes, however, have been in spite of and not because of Freudian theory, which helped shape the programs either consciously or unconsciously. There is no evidence that the war on poverty reduced the rate of mental illnesses nor any evidence that it has promoted the ephemeral but oft-cited goal of "mental health."

Freud in the Universities

As Freudian-influenced New York intellectuals gravitated toward America's university campuses in the 1950s, Freud's theory became consequently incorporated into the academic curriculum. Students read Marcuse, Goodman, and Brown and listened to speakers like Margaret Mead extol Freudian theory as the course of emancipation toward a nonrepressed society. Increasing numbers of faculty undertook personal psychoanalyses and increasing numbers of students signed up for courses that subsumed Freudian concepts. The university became a nidus for Freudian spores, which were then carried back to every town in America by its educated young citizens.

The social sciences and humanities became especially strongly imbued with psychoanalytic teaching, and virtually every introductory course in anthropology included Mead's *Coming of Age in Samoa* and Benedict's *Patterns of Culture*. Courses on culture and personality were heavily subscribed and usually stressed anthropological writings by psychoanalysts such as Erik Erikson and George Devereux. Most departments of psychology, by contrast, resisted the rush to Freudian theory (which was regarded by many psychologists as lacking in scientific merit) merely including it as one theory of abnormal personality.

English departments in many universities also incorporated Freudian theory into courses on literary criticism. The prototypes for such efforts were *Hamlet and Oedipus* (1949) by Ernest Jones and the analysis of Edgar Allan Poe's work, *Life and Works of Edgar Allan Poe* (1949), by Marie Bonaparte, one of Freud's patients and supporters. By the early 1970s, journals such as *Imago* had become specialized in psychoanalytic studies of literature and textbooks were being published with such titles as *The Practice of Psychoanalytic Criticism*, *The Unspoken Motive: A Guide to Psychoanalytic Literary Criticism*, and *Out of My System: Psychoanalysis, Ideology and Critical Method*. This latter book, written by Professor Frederick Crews, chairman of the graduate program in the English department at the University of California at Berkeley, praised psychoanalysis as "the only thoroughgoing theory of motives that mankind has devised." Such books and courses used Freudian theory to explicate literary works or to analyze authors.

History was another university department into which Freudian theory made surprisingly deep inroads. The application of psychoanalytic principles to history had been demonstrated by Freud himself in his exegesis on history in general (*Civilization and Its Discontents*) and in his psychohistorical analyses of Leonardo da Vinci (*Leonardo Da Vinci and a Memory of His Childhood*) and President Woodrow Wilson (*Thomas Woodrow Wilson: A Psychological Study*). Freud's work on da Vinci, which was based on a single memory of a "vulture" attacking him as a young child, was later shown to be fundamentally flawed because the Italian word for kite (*nibbio*), a smaller bird which did not fit Freud's psychoanalytic formulations, had been inadvertently mistranslated as "vulture" (*Avvoltoio*). Freud's work on Wilson, written in collaboration with William C. Bullitt, who had been Freud's patient, was severely criticized by most historians; Barbara Tuchman, for example, said, "The authors have allowed emotional bias to direct their inquiry, which has led to undisciplined reasoning, wild overstatement, and false conclusions." Freud's own psychohistorical efforts, then, were not an auspicious beginning for the field.

The psychohistorical work that has been generally cited as a good example for how Freud's theory can enlighten history is Erik Erikson's *Young Man Luther: A Study in Psychoanalysis and History* (1958). Erikson, who had had no education beyond secondary school, became a well-known lay analyst and later was appointed professor of human development at Harvard University. In trying to understand why Luther rebelled against the Church, Erikson utilized characteristics in

Luther's personality which reveal the "active remnants of his childhood repressions." Specifically, Erikson wrote:

> We must conclude that Luther's use of repudiative and anal patterns was an attempt to find a safety-valve when unrelenting inner pressure threatened to make devotion unbearable and sublimity hateful—that is, when he was again about to repudiate God in supreme rebellion, and himself in malignant melancholy. The regressive aspects of this pressure, and the resulting obsessive and paranoid focus on single figures such as the Pope and the Devil, leave little doubt that a transference had taken place from a parent figure to universal personages, and that a central theme in this transference was anal defiance.

This is classical Freudian theory: making use of events in the person's early psychosexual development to explain that person's later behavior. Erikson elsewhere partially repudiated such simplistic thinking, calling it "originology . . . a habit of thinking which reduces every human situation to an analogy with an earlier one, and most of all to that earliest, simplest and most infantile precursor which is assumed to be its 'origin.' "

Following the publication of *Young Man Luther*, psychoanalytic theorizing spread rapidly in history departments in many colleges and universities. By 1977 a survey showed that more than 200 colleges, universities, and psychoanalytic institutes were offering courses on psychohistory; *The Journal of Psychohistory* was being published regularly; and annual meetings were held for historians interested in utilizing this approach. Articles and books psychoanalyzed historical figures such as John Quincy Adams, Napoleon, Stalin, Hitler, and Richard Nixon.

The common denominator in psychohistory is the utilization of a Freudian (or neo-Freudian) frame of reference to explain historical behavior and events. As historian Gertrude Himmelfarb wrote in a telling critique of the field: "Psycho-history derives its 'facts' not from history but from psychoanalysis." Yet the validity of the approach is only as good as the validity of Freudian theory. As Martin Gross noted in *The Psychological Society*, "in this psychoanalytically based rewriting of history, not battles and issues or men and their mistresses, but *mother*, *father*, and *siblings* emerge as the shapers of world history." Richard Nixon behaved the way he did as president because, according to James W. Hamilton, "Richard felt considerable rivalry and resentment towards [his younger brother] Donald, in particular, who arrived as the anal phase was beginning to dominate his psychosexual development and harbored death wishes toward him . . ."

Freud in the Publishing Industry

Since the 1930s, the publishing industry has had a relationship to the New York intelligentsia that has varied from being close to frankly incestuous. Intellectuals and publishers, often related or familiar with each other through marriage or shared consorts, recommend and review each others' work and pass promising manuscripts around for publication. As cited earlier, examples of such arrangements include Norman Podhoretz's serialization of Paul Goodman's *Growing Up Absurd* in *Commentary* and then persuading Random House to publish it, and publisher Jason Epstein's discovery of Norman O. Brown's *Life Against Death*, which he passed on to Podhoretz at *Commentary*, who in turn persuaded Lionel Trilling to review it. The New York intellectual community and the publishing industry are essentially two parts of a single whole.

It would be surprising, therefore, to find that the publishing industry has been less profoundly influenced by Freudian theory than have the intellectuals. Though a systematic survey of this question has never been conducted, inquiries were made in 1990 to a random group of New York psychiatrists who had been in practice for at least 30 years. The psychiatrists who responded estimated that 50 to 80 percent of book or literary journal editors had gone into psychodynamic psychotherapy at some point in their careers. Regarding the question of whether such psychotherapy is more or less common or about the same as 30 years ago, respondents were equivocal. One of the psychoanalysts commented upon the situation by noting that "thirty years ago the literary field was to a significant degree an extension of psychoanalysis."

There are also strong indications that Freudian motifs became prominent in literary fiction in the 1960s. Two reasons for this were the fact that presumably many writers had undertaken psychoanalysis following World War II and that editors and publishers, some of whom were also in analysis, were more interested in manuscripts that included Freudian themes. The new literary era was signaled in 1961 by an article in the *Atlantic Monthly* in which Alfred Kazin noted: "Just as so many psychoanalysts want to be writers, so many writers now want to be analysts." As if to illustrate Kazin's point, a flood of psychoanalytically inspired fiction followed.

Doris Lessing's 1962 *The Golden Notebook* is a case in point, in which a Jungian psychoanalyst, Mother Sugar, helps heroine Anna Wulf overcome a case of writer's block. Lessing later acknowledged that she was in psychoanalysis while writing the book and that Mother Sugar was based on her own therapist. In Sylvia Plath's *The Bell Jar*, published in 1963, an idealized psychoanalyst, Dr. Nolan, skillfully treats heroine Esther Greenwood. Dr. Nolan is in fact a portrayal of Plath's own psychoanalyst, who had treated her for depression in 1953 and again in 1958.

In 1964 *I Never Promised You a Rose Garden* was published. Written by Joanne Greenberg (who used the pseudonym Hannah Green), it depicts the successful psychoanalytic treatment of a young woman (Deborah Blau) by an all-knowing and kindly psychoanalyst, Dr. Fried. The book is based on Joanne Greenberg's actual treatment by Dr. Freida Fromm-Reichmann. The book has sold more than five million copies and was made into a successful movie.

Perhaps the best example of how heavily indebted fiction became to psychoanalysis in the 1960s was the work of Philip Roth. According to Jeffrey Berman's book on psychoanalysis in literature, *The Talking Cure*, Roth's "characters are the most thoroughly psychoanalyzed in literature." Dr. Otto Spielvogel, who listens patiently to Alex Portnoy's malcontent monologue in *Portnoy's Complaint* (1969), is modeled on Roth's own psychoanalyst, with whom Roth was in therapy "for many years." The psychoanalyst in *My Life as a Man* (1974) is also based on Roth's analyst; according to Berman, "no novelist has given us a more authentic account of psychoanalysis in its intellectual and emotional vagaries than Roth does" in this book.

There are exceptions, of course, to this literary hagiography of Freud. The most conspicuous exception—indeed, "the most virulently anti-Freudian artist of the century"—was Vladimir Nabokov. According to Berman, "no novelist has waged a more relentless campaign" against psychoanalysis than Nabokov, who called Freud a "Viennese quack" and deemed psychoanalysis "one of the vilest deceits practiced by people on themselves and on others." The Oedipus complex struck Nabokov as especially absurd: "Let the credulous and the vulgar continue to believe that all mental woes can be cured by a daily application of old Greek myths to their private parts."

Nabokov's opposition to Freudian theory apparently emanated from the period when he lived in Berlin in the 1920s. The Berlin Psycyhoanalytic Institute had been founded in 1920 by Max Eitingon, a wealthy Russian physician who had been psychoanalyzed by Freud "in the course of evening walks in Vienna"; and through the leadership of Karl Abraham the institute became very active. The Berlin Institute was known to be especially sympathetic to Marx and communism, to which Nabokov was bitterly opposed. In 1926 Nabokov attended a lecture on "the witchdoctor Freud," as he called him, and thereafter began his assault on "the vulgar, shabby, fundamentally medieval world of Freud with . . . its bitter little embryos spying, from their natural nooks, upon the love life of their parents."

Nabokov was especially critical of the Freudian movement "as a kind of internal Marxism" and called psychoanalysis "nothing but a kind of microcosmos of communism" because of its similarity to Marxist mind control. He argued that "the difference between the rapist and therapist is but a matter of spacing," gave psychoanalysts in his novels such names as Dr. Sig Heiler and Dr. Rosetta Stone, and challenged Freudians "to analyze or anal-ize his works." Nabokov advised Freud and his followers to "jog on, in their third-class carriage of thought, through the police state of sexual myth." Nabokov contributed devastating critiques of Freud and his theory, but in the literary milieu of the 1960s such critiques were rare.

Freud in the Film Industry

Given the influx of Freudian ideology into Hollywood following World War II, movies in the 1960s also reflected the Freudomania that was sweeping America. According to a psychiatrist who has practiced in Beverly Hills for over 30 years, "in the fifties and sixties almost everybody in 'the Industry' seemed to be in treatment." Although there has been no formal study conducted, a recent survey of some Los Angeles psychiatrists who have practiced there since at least 1960 estimated that 50 to 60 percent of writers, producers, and directors in the film industry go into psychodynamic psychotherapy at some point in their careers. One psychiatrist said that the practice of his Beverly Hills colleagues had "been supported for years by these film and TV folks." A 1989 listing of psychiatrists who belong to the American Psychiatric Association revealed that 131 of them practice in Beverly Hills, which has a population of 33,690; there is therefore one psychiatrist for every 150 adults who live there, which is as high a concentration of psychiatrists as can be found anywhere in the world.

According to Dr. Irving Schneider, characters as psychiatrists appeared in more than one hundred movies in the 1960s. The most notable of these was John Huston's 1962 film on Freud himself, which

was originally simply titled *Freud*. Huston had contemplated making
such a film since 1946 and, when given the opportunity, he recruited
Jean-Paul Sartre to write the screenplay. "Sartre," wrote Huston in his
biography, "knew Freud's work intimately and would have an objective
and logical approach. . . . He regarded Freud's studies as valuable for
what they discovered about the human mind." Sartre's initial lengthy
screenplay would have made a movie lasting more than five hours, but
it was rewritten to be under three hours, with Montgomery Clift cast
as Freud. Both Huston and Sartre wanted Marilyn Monroe to play
Cecily, the most important patient in the film, but "Monroe's own
analyst objected on the grounds that Anna Freud had not approved the
project," and Susannah York instead was casted.

If *Spellbound* in 1945 had lacked subtlety as a testimonial for psychoanalysis, *Freud* had the appearance of being a commercial advertisement made by the American Psychoanalytic Association. It began with
Huston's voiceover comparing Freud explicitly to Copernicus and
Darwin and implicitly to Christ by telling the audience that the movie
would depict "Freud's descent into a region almost as black as hell
itself—man's unconscious—and how he let in the light." Sartre's script
focused on Freud's theory of the stages of infantile sexuality and especially the Oedipal complex.

In the movie, when Freud has cured patient Cecily of her symptoms, which were said to be caused by her sexual attachment to her
father, Freud sits with Cecily's mother, with the camera "revealing
beween them a huge painting of Cecily's father." Such heavy-handed-
ness persisted to the end of the film, at which point Huston's voiceover
returned with a final message:

> *"Know thyself." Two thousand years ago these words were carved on
> the Temple at Delphi. "Know thyself." They're the beginning of wis-
> dom. In them lies the single hope of victory over man's oldest enemy,
> his family. This knowledge is now within our grasp. Will we use it?
> Let us hope.*

Freud was not a box office success when it was released in 1962, nor
the following year when it was re-released as *Freud: The Secret Passion*.
As enamored as the public was with Freud and his theory, Americans
were apparently not yet prepared to substitute Freud for Christ.

No other film ever deified Freud as did Huston's effort, although
many other films in the 1960s presented Freud's theory in a very favor-
able light. A psychiatrist was brought in at the end of Alfred Hitch-

cock's *Psycho* (1960) to explain the workings of Norman Bates's mind (he was clearly having Oedipal problems with his mother) and instruct the police how to find the bodies of his other victims. In Elia Kazan's *Splendor in the Grass* (1961), a wise psychiatrist helped Natalie Wood reconstruct her life. The film version of *David and Lisa* (1962) featured an all-knowing psychoanalyst who successfully did battle with David's insensitive and mean-spirited mother, the cause of his problems in a classical Freudian mode. According to Gabbard and Gabbard's *Psychiatry and the Cinema*, "More than two decades after it was made, many psychiatrists still refer to *David and Lisa* as one of the most 'realistic' depictions of psychiatric treatment."

In recent years, however, the presentation of psychiatrists and psychoanalysts in movies has been more mixed. On one hand, there have been very positive depictions in movies such as *I Never Promised You a Rose Garden* (1977), *Ordinary People* (1980), and *Lovesick* (1983), in which Alec Guiness plays the ghost of Freud and admonishes the psychoanalyst (Dudley Moore) to control his countertransference feelings. On the other hand, there have been movies like *One Flew Over the Cuckoo's Nest* (1975), *Dressed to Kill* (1980) and *Frances* (1982), in which psychiatrists were portrayed as sadistic, incompetent, and even homicidal. Perhaps Hollywood's recent ambivalence toward Freud is best embodied in the films of Woody Allen, who is clearly intrigued with psychoanalysis at the same time as he ridicules it. In *Annie Hall* (1979), Allen says that he has been in analysis for 15 years, and if he has not made progress in one more year he will go to Lourdes. Later he adds that he was once suicidal and would have killed himself, "but I was in analysis with a strict Freudian, and if you kill yourself they make you pay for the sessions you miss."

Freud Goes to Esalen

One of the most far-reaching social changes that came out of 1960s America was the personal growth movement, a movement fathered by Freud and mothered by affluence and leisure. Over three decades this movement has evolved into an abecedarian melange of psychoanalysis, Eastern philosophies, and exercise, which ranges from aikido, body energy therapy, creative aggression therapy, direct psychoanalysis, eidetic therapy, feeling therapy, and holistic therapy to Senoi dream group therapy, transactional analysis, vita-erg therapy, and zaraleya psychoenergetic therapy. The movement is sometimes referred to as the human potential movement, although the human potential that appears to have been demonstrated most clearly has been the potential for some humans to make large amounts of money from the loneliness, disappointment, and unhappiness of others.

The personal growth movement evolved from Freudian theory during a period in which the number of psychiatrists, psychologists, and psychiatric social workers in the United States was increasing rapidly. In 1945 there was a total of approximately 9,000 such professionals. In 1948 the federal government began providing grants to training programs; these grants peaked at $118.7 million per year in 1969. At present it is estimated that there are approximately 200,000 such professionals—40,000 psychiatrists, 70,000 psychologists, and 80,000 psychiatric social workers—a 22-fold increse during years in which the population of the nation did not quite double.

Virtually all of the new professionals were exposed to Freudian theory during the course of their training, and in 1970 it was estimated that one-third of the psychiatrists had undertaken additional formal psychoanalytic training. Among psychiatric social workers, a personal psychoanalysis was even more widespread; a survey of 30 social workers in New York City as early as 1951 found that 27 of them had undergone psychoanalysis. The emerging fields of psychotherapy and counseling became permeated with Freudian concepts of infantile sexual development, repression, transference, and the theoretical necessity of re-experiencing early childhood traumas to achieve personal growth.

As the professionals grew in numbers, affluent individuals increasingly sought help. The problems which brought such individuals to the psychiatrists, psychologists, and psychiatric social workers were not those of serious mental illnesses, such as schizophrenia and manic-depressive psychosis, as had been true before World War II, but rather what Charles Kadushin called the "psychoanalytic syndrome [of] sexual difficulties, feelings of lack of self-worth, and problems of interpersonal relations." Many such problems are part of the human condition and are therefore infinite in number, meaning that virtually everyone is a potential candidate for psychoanalysis, other forms of psychotherapy, or personal growth experiences of one brand or another. The personal growth movement was further fueled by the economic prosperity that followed the war; leisure and affluence are American prerequisites for striving to understand one's atman.

No single individual symbolized the evolution of the personal growth movement more clearly than Fritz Perls, a psychoanalyst who helped develop Gestalt therapy. A German by birth, Perls obtained an M.D. degree in Berlin in 1921 and then undertook traditional psychoanalytic training for seven years with three different psychoanalysts. Among Perls's psychoanalytic supervisors were Helene Deutsch, Karen Horney, Otto Fenichel, Paul Federn, Wilhelm Reich, and Edward Hitschmann; when asked how Freudian schools of thought differed from one another, Perls recalls Hitschmann obliquely answering: "They all make money."

With Hitler's coming to power, Perls fled to Holland, then migrated to South Africa to accept a job arranged for him by Ernest Jones. Following the war, Perls went to New York, where, with his friends Paul Goodman and Ralph F. Hefferline, he coauthored *Gestalt Therapy* and opened the Gestalt Therapy Institute on Central Park West. A decade later Perls left his wife and the institute and moved to the Esalen Institute in Big Sur, California; Perls and the institute were to make each other famous as the mecca of the personal growth movement.

As Perls's own writings make clear, he was first and foremost a follower of Freud. In his biography, written the year before he died of cancer, Perls called Freud "the Edison of Psychiatry . . . and also Prometheus and Lucifer, the bearers of light." He described his single meeting with Freud, which occurred in 1936 while Perls was presenting a paper at the International Psychoanalytic Congress. Perls made an appointment and arrived at Freud's door at the scheduled time. Freud opened the door.

> *Perls: "I came from South Africa to give a paper and to see you."*
> *Freud: "Well, and when are you going back?"*

Perls described himself as "shocked and disappointed" by their abrupt, four-minute conversation. Freud had not only failed to appreciate the brilliance of Perls's paper on "oral resistances" but had snubbed him as a nobody. Perls recalled his reaction as being "I'll show you—you can't do this to me." More than 40 years later, Perls still considered proving to Freud "the mistakes he made [as] one of the four main unfinished situations of my life." Perls wrote:

> *I am really beginning to enjoy myself. Especially writing this vignette hitting back at psychoanalysis. After all, Freud, I gave you seven of the best years of my life.*

Perls claimed that Gestalt therapy was an extension of, and improvement on, Freud's treatment method. "Freud took the first step.

. . . I accomplished the next step after Freud in the history of psychiatry. . ." Gestalt, which literally means configuration in German, focused on feeling rather than thinking, on awareness rather than insight, on the present rather than the past. As in psychoanalysis, eliminating resistances was said to be essential to the success of the therapy, and Perls claimed to be able to accomplish this faster than Freud's techniques could do.

Gestalt therapy was among the best-known variations of the 1960s human potential movement. Esalen was the movement's spiritual center, described by Art Harris as "a playground for promiscuity, a Cape Canaveral of inner space, a fat farm for pudgy egos." From Esalen vibrations of sensitivity training and encounter groups tripped out over middle America into women's groups, couples' groups, church groups, even groups in industry. Personal growth was a composite of the psyche with elements that varied from area to area; some very strange substances were added, but what usually lay at the core was Freudian theory.

Literally hundreds of variations of the personal growth movement have evolved since the early 1960s, and most show Freudian influences in one way or another. Primal scream therapy is another example. Developed—and patented—by Dr. Arthur Janov as the *only* effective means to self-knowledge, primal scream therapy consists of having an individual re-experience painful childhood memories. This process is in fact classic abreaction; in Janov's version it is often accompanied by piercing screams, writhing, vomiting, and general discomfiture. The ultimate "primal" is purported to be re-experiencing one's own birth.

Janov's theory is remarkably reminiscent of early Freudian writings. The painful childhood memories produce damned up impulses and cerebral pressures that affect various brain structures; indeed, after reading Janov one is uncertain whether to call a plumber or a brain surgeon to alleviate the problem. Primal scream therapy is said to be effective for "all addictions . . . sexual dysfunctions, some hormonal disorders . . . asthma, colitis, hypertension and migraine . . . [and] nonorganic psychosis." A study of primal therapy by a psychoanalyst reported that the two approaches share much in common; in terms of results, the study also found that individuals undergoing primal therapy have approximately the same rate of improvement as traditional Freudian psychotherapy. Primal therapy is, like psychoanalysis, expensive; for many years the fee was $6,000, to be paid in advance, and it was said that the person's first primal scream usually came when being presented with the bill.

One of the reasons for the success of therapies like Gestalt and primal scream and of the personal growth movement in general is that they promise insight, self-awareness, and an understanding of one's childhood experiences more rapidly than traditional Freudian therapy. When Freud began analyzing people, he did so in a matter of weeks or, at most, months. In the postwar period, the length of psychoanalyses became increasingly protracted, so that periods of five years or more were common. Gestalt, by contrast, promised *satori* (a state of illuminative insight) in a matter of days, and primal therapy's original period of treatment was to be three weeks. Other varieties of personal growth promised to bring about the "new you" in a single weekend. The personal growth movement is thus the ultimate Freudianization of America. In a nation that relies on Big Macs for sustenance, the idea of instant insight has a strong appeal, a kind of McDonald's for the mind.

McFreud in America

It is virtually impossible today for an educated person in America to avoid exposure to Freud's theory of human behavior. The introduction usually takes place in college through courses in psychology or anthropology, although Freud may also turn up in English, history, humanities, human relations, women's studies, or in a variety of other courses. In women's studies, for example, Margaret Mead's *Male and Female* is sometimes assigned reading, in which Freud's stages of sexual development are presented as established fact and specifically outlined. For example:

> As little boys and little girls reach the age at which they are experimenting with their own budding sexuality, they also reach a crisis in their relations with adults that in psychoanalytical theory has been technically called the Oedipus situation. . . . So as far back as before the baby is born the form of the Oedipus situation is foreshadowed and there is already an indication of how the potential rivalry between father and son, or mother and daughter, will be handled.

Freudian theory is usually presented sympathetically by college professors because many have themselves been in psychotherapy. Students who are having personal problems are frequently encouraged to utilize college counseling services which, like most counseling, is influenced by Freudian theory.

Individuals who get through college without being exposed to

Freudian theory in classrooms or in counseling will almost certainly be confronted with it shortly thereafter. Educated Americans are rare who do not have at least one friend among the estimated ten million people who consult psychiatrists, psychologists, and psychiatric social workers each year. One study of Harvard University graduates revealed that 31 percent had seen a psychiatrist within 25 years of graduation; another unknown number had seen psychologists and psychiatric social workers. Several studies have shown that mental health professionals are heavily concentrated in university towns and in affluent communities where the college-educated settle. The problems for which most of these people seek help are not serious mental illness, but rather intrapersonal (e.g., low self-esteem) and interpersonal (e.g., marital) difficulties. (Indeed, "It was recently reported that in one part of the United States [an academic setting] the prinicipal presenting symptom for patients who offer themselves to be analyzed by analysts-in-training is the difficulty of completing their Ph.D. dissertions.") Freud himself stated, "The optimum conditions for . . . [psychoanalytic psychotherapy] exist where it is not needed—i.e., among the healthy." Thus professional mental health resources are strongly oriented toward the worried well rather than the suffering sick.

The ten million Americans who annually seek help from psychiatrists, psychologists, and psychiatric social workers are merely the visible portion of the counseling and psychotherapy iceberg. Many million more Americans—nobody knows precisely how many—seek help from a variety of marriage counselors, family therapists, pastoral counselors, and self-appointed gurus, who range from being caring and competent individuals to being, in Shakespeare's terms, "rascally, yea—forsooth knaves." The common denominator of virtually all counseling and psychotherapy is the Freudian assumption that intrapersonal and interpersonal problems originate in childhood experiences, especially in one's relations with mother and father. Shyness, difficulty in making a commitment, depression, anxiety, obsessiveness, slovenliness, substance abuse, eating disorders, trouble-making friends, inability to find meaning in life—virtually all problems are said to have the same origin.

A case in point is codependence counseling, the latest in a long line of therapy fads in America. Pia Mellody, one of the leaders of the codependence movement, has explained that codependence is caused by parental "abuse" of children, also called "dysfunctional parenting" and further defined as "any experience that was less-than-nurturing or shaming." Another writer on codependence defined its cause as "child abuse, defined broadly to include any emotional or physical abandonment, disrespect or inadequate nurturance."

According to Mellody, "because of dysfunctional childhood experiences a codependent adult lacks the ability to be a mature person capable of living a full and meaningful life," which leads to problems in the relationship with the self and relationship with others. The treatment for problems of codependence is quintessentially Freudian: "Recovery involves reviewing your past to identify formative experiences in your early life that were less-than-nurturing or abusive." Freudian defense mechanisms and the unconscious are heavily emphasized and are mixed with recovery principles derived from Alcoholics Anonymous.

Most other types of counseling and psychotherapy are also based on Freudian theory; the labels and shape of the box vary, but the base substance in all the boxes is shipped from the same factory in Vienna. This is true for longer-lasting varieties, such as family therapy, which extends the blame for problems from mother and father to grandmother and grandfather. (Murray Bowen, one of the popularizers of family therapy, trained in psychoanalysis under Karl Menninger.) It is also true for "here today, gone tomorrow" therapy fads such as those put forward by John Bradshaw, whose lectures in 1990 were selling out the 3,000-seat Javits Convention Center in New York and who in 1991 was profiled in *Time* magazine ("Father of the Child Within"). Bradshaw's message focused on the assertion that all individuals are "abused children": "A lot of what we consider to be normal parenting is actually abusive," said Bradshaw. His advice to people who want to solve their problems was to find the "neglected, wounded child" in each person and then give it "the unconditional love withheld by parents." There appears to be no end to these therapy gimmicks, each claiming to be new and better. One observer called the therapy scene a "dogma eat dogma world." Purveyors of these new therapies would have us believe, to paraphrase one critic, that the nuclear in nuclear family is the same nuclear as in a nuclear explosion. Originally, Freudian theory was only sold at expensive Central Park West psychoanalytic boutiques to wealthy customers like Mabel Dodge and Marshall Field III. Like most luxuries in America, however, it has been repackaged and is now available to everyone at "McFreud" outlets.

Freudian theory will also be encountered by anyone who decides to become a parent in America. Even before the baby is born, mothers-to-be must make a decision about breast-feeding, which, according to Freud, is one of the crucial events of life. Americans have become so obsessed and anxious about this decision, which other mammals appear to carry out quite easily, that one book, *Preparation for Breastfeeding*, has 172 pages and advertises itself as having "over 175,000 copies in print."

Once the child has been born, another decision must be made regarding which child-rearing manual to use. Since the events of early childhood are widely believed to be so critical to later development, most parents use several manuals from among the plethora that are available. One observer noted in 1976 that "the average bookstore has at least 30 different child-care titles on its shelves," while libraries carry listings under this subject that often run to several hundred. Many other child-rearing manuals incorporate Freudian theory in a manner that suggests that it is established fact. For example, a discussion of the Oedipal relationship is found in Ellen Galinsky's *Between Generations: The Six Stages of Parenthood* (1981):

> *Children do, in the preschool years, become romantic and flirtatious with their parents. In the Freudian and psychoanalytic schema, this begins in infancy, when both boys and girls make a primary identification with their mothers. The boy in the preschool years competes with the father for the love of the mother, but eventually realizes that he is no match for this bigger, stronger man, and drops this courtship, instead deciding to take on his father's qualities, identifying with him. The girl child wants her father, but realizing that her mother has already won him, remains identified with the mother.*

Disappointingly few child-rearing manuals state explicitly that the events surrounding breast-feeding, toilet training, and sexual identification, which Freud described as crucial for a child, are in fact merely phases of normal childhood development. William Sears's *Creative Parenting* is an example of a manual that does this very well:

> *A word about the infamous Oedipus complex seems in order here. This is the love triangle in which the boy wishes to replace the father, and the girl the mother. In the majority of homes, this idea is an insignificant and passing phase. Children seem more intent on becoming like the parent than replacing the parent.*

When parents look for advice beyond traditional child-rearing manuals, they continue to encounter Freudian theory. In some cases it is explicit, as in books by psychoanalyst Alice Miller with titles such as *Prisoners of Childhood* and *Thou Shalt Not Be Aware: Society's Betrayal of the Child*. Miller emphasized "the decisive significance of early childhood" in personality development. In other cases, Freudian theory is more implicit, packaged between the lines, as when *Parents* magazine featured a story, "When Young Children Need Therapy," or even by the credentials of authors chosen to write stories that have nothing to do with therapy (for example, "Rebecca Shahmoon Shanok, M.S.W., Ph.D., is director of the Early Childhood Group Therapy Program of the Child Development Center and a therapist in private practice"). The message to parents is clear: Childhood is a psychic minefield through which only the most skilled parents, with much professional help and equal amounts of luck, can ever hope to successfully navigate.

One of the more recent and most commercially successful packagings of Freudian theory in America has been the "all children need therapy" idea. For-profit psychiatric hospitals catering to children and adolescents have proliferated across the American landscape, with billboards and radio advertisements implying that parents are negligent for failing to hospitalize children who have any problems, which is to say all children. This idea achieved terminal absurdity by extending such therapy to infants; according to *Newsweek*, "In clinics and private practitioners' offices around the country, troubled and troublesome children as young as 3 months old are receiving the benefits of cribside therapy." The *New York Times Magazine* carried a laudatory account of such infant therapy, acknowledging that "the legacy of Freud . . . is a cornerstone of the current work."

Freudian theory has also crossed new frontiers in risibility in court cases in which children have brought lawsuits against their parents for "malparenting." The most highly publicized such case was 24-year–old Tom Hansen of Boulder, Colorado, who sued his parents for $350,000. Hansen's lawyer explained that his client's mental health had been been impaired by "inadequate parenting . . . a pattern of psychological deprivation." Among specific allegations brought by Hansen was that his father had made him dig weeds in the yard when he had caught his son smoking marijuana. In another case, in Washington State, two teenage sisters sued their parents for psychological damage caused by parental fighting. Columnist Ellen Goodman astutely noted that cases such as Tom Hansen's fascinate the public because he "has done in public what so many others do in private: He has proclaimed his parents guilty for his own life."

Beyond college courses, counseling, and child rearing, one can find Freudian theory peeking its head through many cultural curtains. In 1990, a new musical, *Freudiana*, opened in Vienna with plans to bring it to New York. "Freud's theories of the Oedipus complex, the ego and the id" as well as dreams were said to be the basis for the song-book, including the refrain of one song: "The answers await in the darkness of childhood." Also in 1990, a concert of P.D.Q. Bach's music at Carnegie Hall included the oratorio Oedipus Tex, and the Paul Taylor Dance Company was opening "The Sorcerer's Sofa" at the New York City Center, using a psychoanalyst's couch labeled "A Chaste Lounge." Drama and movies continue to utilize Freudian theory prominently, with one film series at New York's Anthology Film Archives focusing on the portrayal of psychoanalysis in the movies.

In print Freudian theory appears to be ubiquitous. Psychology sections in bookstores include dozens of titles by, about, and derived from Freud. At the professional level, International Universities Press alone listed 150 books on Freud and his theory in its 1991 catalogue, including *Oedipus in the Stone Age, A Psychoanalytic Study of the Myth of Dionysus and Apollo,* and *Sigmund Freud's Dreams.* Readers who prefer fiction can find Freud prominently represented among the works of such novelists as Philip Roth and Doris Lessing, while those who enjoy magazines are confronted by articles on "Re-examining Freud" (*Psychology Today*), "Encountering Freud" (*Society*), "Confessions of a Head Case" (*Esquire*), "Van Gogh and Gauguin on a Couch" (*Art in America*), and "Dreams on a Couch" (*Newsweek*). Among magazines, the *New Yorker,* the *New York Times Magazine,* and the *New York Review of Books* remain as the most prominent purveyors of Freudian theory; between 1980 and 1988 these three magazines, constituting just 2 percent of all magazines listed in the *Readers' Guide to Periodic Literature,* accounted for 24 percent of all magazine articles on Freud and psychoanalysis.

Freudian theory is also prominent in newspapers and on television. For example, in 1990, Kitty Dukakis published an autobiography, *Now You Know,* and made the rounds of talk shows, explaining how her addiction to alcohol and drugs had been caused by her rejecting mother. In addition to the Dukakis case, within a single month in 1990, the following stories were found in newspapers: (1) a book about Joel Steinberg and Hedda Nussbaum (accused of having beaten to death their adopted daughter Lisa) that attempted to explain their behavior in terms of their childhoods; (2) a report on the trial of Edward Diggs, a prominent Annapolis lawyer accused of defrauding a corporation of millions of dollars, at which a psychiatrist testified that Mr. Diggs had been "an only child in a family where table manners were given more importance than more personal issues"; and (3) testimony in the trial of John E. List, accused of murdering his wife, mother, and three children, which attempted "to show that Mr. List, the product of a rigid family and strict religious training, killed his family out of love for them and concern for their souls."

Freudian theory has indeed pervaded our thinking and the air through which we view events to such a degree that we are barely aware of its existence. As Peter Gay noted: "It is commonplace that we all speak Freud today whether we recognize it or not." It has become an integral part of our culture with the freedom to blame parents and childhood experiences as American as the freedom of speech and freedom of the press. Freud's theory has become incorporated into the very soul of the nation. As W. H. Auden wrote in a poem about **Freud:**

> *To us he is no more a person*
> *Now but a whole climate of opinion.*

9

The Scientific Basis of
Freudian Theory

---

*For no other system of thought in modern times, except the
great religions, has been adopted by so many people as a
systematic interpretation of human behavior.*
—ALFRED KAZIN, "THE FREUDIAN REVOLUTION ANALYZED"

*F*or the ideas of Sigmund Freud in America, the closing years of the
20th century are both the best of times and the worst of times. His
concepts are "now found firmly implanted in the unexamined beliefs
of the average man and woman . . . . a basic part of our cultural substance." Harold Bloom even suggested in the *New York Times* that
Freud's ideas "have begun to merge with our culture, and indeed now
form the only Western mythology that contemporary intellectuals have
in common." One enthusiast compared Freud's theory to the quantum
theory and the theory of relativity and suggested that his work might
"in the end prove to be more decisive and far-reaching than the discoveries of Planck and Einstein. . . . Will the Twentieth Century go
down in history as the *Freudian Century?*"

While Freud's ideas have been inculcated upon American thought
and culture, these ideas have also been scientifically discredited. In his
recent book *Psychoanalysis: A Theory in Crisis*, Marshall Edelson
observed that "psychoanalysis, as a body of knowledge about human
beings or the human mind, has become the object of a dismissive, disil-

214

lusioned, and frequently derogatory polemic," with some people claiming that "an apocalyptic end of a failed and exhausted paradigm is occurring now." The current situation of Freudian theory having broad cultural acceptance but simultaneously being in scientific disrepute is reminiscent of Thomas Kuhn's description of changing ideas in his brilliant essay, "The Structure of Scientific Revolutions." Kuhn argued that scientific revolutions come about when existing paradigms, which he defined as "universally recognized scientific achievements," have "ceased to function adequately in the exploration of an aspect of nature to which that paradigm itself had previously led the way." The sense of malfunction then leads to crisis and precipitates a scientific revolution, which leads to the emergence of a new paradigm. Freudian theory has provided a paradigm with which most Americans have viewed human behavior for the last four decades, but the paradigm no longer functions adequately. It is as if Americans now unconsciously eat Freud's theory for breakfast and lunch and then dismiss it disparagingly over dinner. In Kuhn's formulation this is leading to indigestion.

Direct Tests of the Freudian Theory

During the same years in which Freud's theory was becoming integrated into American thought and culture, sporadic attempts were made to provide the theory with a scientific framework. It was important to do so because Freud's followers claimed that his theory was indeed scientific. Dr. Abraham Brill, for example, frequently spoke of the "laws" discovered by Freud and said that "psychoanalysis is the microscope for the study of the mind." Popular descriptions of Freudian theory and therapy commonly included the notion that they were scientific. In a 1915 *Good Housekeeping* article, for instance, the author assured his readers that "this new therapy is definitely scientific. . . . Once more it should be iterated that it is not a fad but a science." This line of "Freud as scientist" has been repeated like a litany over the years by supporters of Freud, including Peter Gay's recent biography, in which he called Freud "a hardheaded man of science."

Freud himself was much less certain that his theory rested on a scientific base. Although he had been trained as a scientist and on occasion invoked scientific metaphors to support his position, after the turn of the century Freud's interest in scientifically validating his theory appears to have waned. In 1900 he wrote to a friend: "I am not really a man of science, not an observer, not an experimenter, and not a thinker. I am nothing but by temperament a *conquistador*—an adventurer, if you want to translate the word—with the curiosity, the boldness, and the tenacity that belongs to that type of being." Freud criticized those who were concerned about validating his theory, saying that "those critics who limit their studies to methodological investigations remind me of people who are always polishing their glasses instead of putting them on and seeing with them." In later years, when a psychologist wrote Freud to tell him that he had found scientific support for Freud's theory, Freud responded testily that his theory needed no validation.

There is, then, an aspect of Freud's thinking that is more closely allied to revealed truth than it is to science. According to Dr. Heinz Hartmann, Freud said that "the basic concepts of science form rather the root than the foundation of science and ought to be changed when they no longer seem able to account for experience." It should also be recalled that Freud was actively conducting experiments on occult phenomena throughout these years, and it may have been within such a context that Freud spoke when he once proclaimed: "We possess the truth, I am sure of it."

Freud's critics noted the scientific shortcomings of his theory from its earliest days. Following one of Freud's first public presentations of his theory to his Viennese colleagues in 1896, one prominent physician dismissed it as simply "a scientific fairy tale." In 1900 the Vienna Medical Society satirized Freud in a skit in which the doctor on stage gravely declared, "If the patient loved his mother, it is the reason for this neurosis of his; and if he hated her, it is the reason for the same neurosis. Whatever the disease, the cause is always the same. And whatever the cause, the disease is always the same. So is the cure: twenty one-hour sessions at 50 Kronen each." A 1913 review of *The Interpretation of Dreams* noted that "this book is indicative of a total lack of the characteristics which lead to scientific advance." In 1916 an article in the *Nation* claimed that Freud's theory was "well founded neither theoretically nor empirically. . . . it conveys the impression of unscientific method." An article in *Current Opinion* the same year alleged that Freud's "sex theory" stood "upon the same ground as the green cheese hypothesis of the composition of the moon," and in 1917 psychiatrist Boris Sidis likened psychoanalysis to "Astrology, Alchemy, Chiromancy [palmistry], Oneiromancy [foretelling the future from dreams], and generally to Medieval symbolism. . . ."

What, if anything, is the scientific basis for Freud's theory? The question needs to be addressed because science continues to be invoked by Freud's followers to justify their methods of therapy as well as their prescriptions for social change. If there is no scientific basis for the theory—if it is, as British researchers Hans Eysenck and Glenn D. Wilson claim, merely "the premature crystallization of spurious orthodoxy"—then its promotion as science constitutes a kind of fraudulent claim.

The scientific validity of the oral, anal, and Oedipal stages of development has been reviewed by several researchers over the years. In 1937 Gardner Murphy and his colleagues noted:

> *Although we have now been exposed for some time to psychoanalytic and other psychiatric hypotheses regarding the effects of birth trauma, weaning trauma, extreme emphasis on early control of urination and* defacation, excessive attention from adults, dethronement by a second child, we have almost no objective records of the development of children going through these experiences, or of experiments controlling certain aspects of the problem.

Ten years later Harold Orlansky reviewed the pertinent "empirical data bearing on the theory that various features of infant care determine adult personality" and reported that "our conclusion has been largely negative." In 1952 Ernest Hilgard et al., although sympathetic to the Freudian point of view, acknowledged in a review that "anyone who tries to give an honest appraisal of psychoanalysis as a science must be ready to admit that as it is stated it is mostly very bad science, that the bulk of the articles in its journals cannot be defended as research publications at all."

In more recent years there have been three extensive reviews of studies bearing on the question of whether experiences in early childhood are crucial determinants of adult personality. Paul Kline, who himself did research trying to relate toilet training practices to later personality, concluded in *Fact and Fantasy in Freudian Theory* (1972) that there is evidence for clusters of personality characteristics generally referred to as oral (preoccupied with eating, overdependent, passive) and anal (orderly, parsimonious, obstinate), but that "only two studies give even slight support to the Freudian theory" relating these personality characteristics to "infant-rearing procedures" (one of the studies being his own). Kline added: "Freudian theory, so far as it is dependent on data at all, rests on data which by the criteria of scientific methodology are totally inadequate. These data are, for the most part, the free associations of patients undergoing therapy and their dream reports, and both of these sources are unquantifiable and riddled with subjective interpretation."

Drs. Seymour Fisher and Roger P. Greenberg, psychologists who were also sympathetic to Freudian theory, undertook an ambitious compilation of Freudian studies in *The Scientific Credibility of Freud's Theories and Therapies* (1977). Fisher and Greenberg, like Kline, found evidence to support a clustering of adult personality characteristics around themes called oral and anal personalities. They concluded that "it is true that little was found in the scientific literature to buttress Freud's view that the oral and anal patterns originate in crucial early oral and anal stages of development." The fact that clusters of personality traits exist in adults is widely accepted, but the existence of such clusters, by itself, constitutes no evidence in support of Freud's theory. If Freud was correct, then a relationship must be shown to exist between such clusters of personality traits and experiences in early childhood.

The third major review of scientific evidence for Freud's theory was published by Hans Eysenck and Glenn D. Wilson in *The Experimental Study of Freudian Theories* (1973). In this volume the authors found "not one study which one could point to with confidence and say: 'Here is definitive support of this or that Freudian notion; a support which is not susceptible to alternative interpretation, which has been replicated, which is based on a proper experimental design, which has been submitted to proper statistical treatment, and which can be confidently generalized, being based on an appropriate sample of the population.' After three-quarters of a century this is a serious indictment of psychoanalysis." In a later volume in 1985 Eysenck was less polite in his assessment of the scientific merits of Freud's theory. He likened Freud's ideas to "a medieval morality play" populated "by such mythological figures as the ego, the id and the superego . . . too absurd to deserve scientific status." Freud was, Eysenck wrote, "without doubt a genius not of science but of propaganda, not of rigorous proof but of persuasion, not of the design of experiments but of literary art. His place is not, as he claimed, with Copernicus and Darwin but with Hans Christian Anderson and the Brothers Grimm, tellers of fairy tales."

In addition to the studies reviewed above, there have also been longitudinal studies that have measured specific events in childhood and then followed the children into adulthood to ascertain the relationship between the childhood events and adult personality characteristics. The most thorough such study was carried out in Yellow Springs, Ohio, where, beginning in 1929, a total of 650 children were closely followed from birth until age 18. Among data collected in the study was information on breast-feeding and toilet training. Despite such data and despite the fact that Dr. Lester Sontag, the director of the study for over 40 years, had been trained psychoanalytically, no results were ever reported from this study corroborating Freud's theory. The reason for this, according to an interview with Dr. Sontag, is that very early in the study it became clear that genetic antecedents of behavior were much more important than experiences such as breast-feeding and toilet training. This conclusion was also reached by Dr. George Vaillant, who followed a group of Harvard University graduates for 30 years and noted that "people's childhoods affected adult development much less than one would expect."

Is there really no scientific evidence at all to support Freud's theory that childhood experiences are important determinants of adult personality characteristics? Of the three childhood stages, the anal stage, said to peak at 18 to 24 months of age, is the most convenient to study both because the timing and severity of toilet training are quantifiable and because the adult personality characteristics associated with it were clearly delineated by Freud himself. Freud's hypothesis was that toilet training that was "too early, too late, too strict or too libidinous" may lead to "fixation" of the child's personality at that stage of development. Adult personality characteristics associated with this fixation, as described in Freud's 1908 essay on the subject, "Character and Anal Eroticism," were a triad of orderliness (often associated with conscientiousness and trustworthiness), parsimony (which may become avarice), and obstinancy (which "can go over into defiance, to which rage and revengefulness are easily joined"). In addition to these personality characteristics, Freud noted that fixation at the anal stage may produce homosexuality, paranoia, constipation, and a preoccupation with money (which he believed was a substitute for feces). Ernest Jones later added that "all collectors are anal erotic." The hypothesis, then, as stated by Fisher and Greenberg, is that "persons who fit the anal character classification should have had special experiences during the anal phase."

A total of 26 published studies bearing on Freud's theory that toilet training practices influence adult personality traits were cited in one or more of the reviews mentioned above. (These studies are summarized in Appendix B.) It can be seen that the majority of them either did not test the theory (e.g., simply verified the existence of adult personality characteristics without relating them to previous toilet training), utilized adult personality traits not thought to be specifically related to the anal stage (e.g., aggression), or had major methodological flaws (e.g., used no control group).

Only four of the studies met minimal, acceptable scientific criteria for testing the theory. The results in all four showed no relationship between toilet training experiences and later personality characteristics (Sewell, 1952; Bernstein, 1955; Beloff, 1957; Hetherington and Brackbill, 1963). The results of the other 22 studies also suggest no support for the belief that childhood experiences determine adult personality traits. The entire group of studies are uniformly disappointing for Freud's followers, as can be seen from the fact that no additional studies have been carried out since 1970. The single most interesting finding in the 26 studies is in fact an unexpected association in three studies between childhood personality traits such as orderliness and similar personality traits in the child's parents (Beloff, 1957; Finney, 1963; Hetherington and Brackbill, 1963). This finding lends support to the data suggesting that adult personality characteristics have a large genetic component, as will be discussed below.

There is not a single study verifying Freud's theory that events in the anal stage of development determine adult personality characteristics. The same conclusion is reached when studies relating to the oral and the Oedipal stages are examined. There *are* studies showing that some individuals have clusters of personality traits called "oral" traits, but no study which relates such traits to breast-feeding or other events of the oral stage of development. Studies of the oral stage are particularly difficult to do because Freud himself did not describe oral personality traits (these were described later by Karl Abraham); in addition, there is evidence that the length of breast-feeding by a mother, often used as a measure of maternal acceptance of the child, does not necessarily correlate with how accepting or rejecting the mother is of the child.

Research on the Oedipal stage and its possible relationship to adult personality is similarly plagued with methodological problems despite the importance attached to this stage by Freud himself. In contrast with breast-feeding or toilet training, the events of the Oedipal period are very difficult to quantify. As has been pointed out by one researcher, a boy who wants his father rather than his mother to put him to bed could be said to be hostile toward the father (separating him from the mother), loving toward the father, fearful of the mother, or loving toward the mother (allowing her to rest). There *are* studies showing that small boys as well as small girls relate more closely to their mothers than their fathers, but this is hardly surprising given the mother's primary care-giving role in most families. Similarly, it has been observed that small children will sometimes relate sexually to the parent of the opposite sex, but there is no evidence that this is any more than a normal stage of development and part of their emerging sexual identification.

Some aspects of Freud's Oedipal theory have been convincingly disproven. The assertion that women are envious of men's penises has been tested by Fisher and found to be false: "It can be immediately declared that Freud was wrong in his assumption that the average woman perceives her body in more negative and depreciated terms than the average man." Similarly, young boys' identification with their fathers has been shown to have nothing to do with a fear of castration, as postulated by Freud, but rather to be "facilitated by a positive, nurturant attitude on the part of father." The core of the Oedipal hypothesis—that events in the Oedipal stage will crucially determine adult personality characteristics—has not been disproven, but neither are there any studies that support such a hypothesis. In this sense Freud's Oedipal theory is on precisely the same scientific plane as the theory regarding the Loch Ness monster—it has not been conclusively disproven, and one may turn up at any moment to prove it.

Indirect Tests of Freudian Theory

In addition to the direct tests of Freud's theory regarding early childhood experiences as crucial determinants of adult personality traits, indirect tests also suggest that this theory is incorrect. Childhood experiences vary profoundly from culture to culture; if such experiences are truly important in determining adult personality, then there should be cultural correlations of childhood experience with adult behavior. For example, in some cultures small children are indulged and allowed to roam freely; in traditional Albanian culture, however, during their first year children were formerly "bound securely to a wooden cradle customarily placed in the darkest corner of the room often with a cloth thrown over their heads so that no light is visible." Toilet training practices vary from those of the Tanala people in Madagascar, where training is begun before three months and "the child is expected to be continent at the age of six months," after which time accidents are "severely punished," to the permissive Siriono people in South America, where "not until a child has reached the age of six [years] does he take care of his defecation needs alone." According to Freud's theory, then, the Tanala should include an exceptional number of bookkeepers, cleaning ladies, and stamp collectors and should have villages which rival Swiss villages in immaculateness. Attempts by anthropologists to relate such child-rearing practices to adult personality traits in different cultures have been singularly unsuccessful and have now been largely abandoned.

Many cultures have other childhood experiences that are certainly traumatic and which, given Freud's theory, might be expected to have an effect on adult personality. For example, in some cultures young children have their feet or skulls systematically deformed by binding. In other cultures the child's ears, nose, and mouth are pierced and their bodies are scarified or tattooed by painful procedures. Male circumcision and female clitoridectomy rites, often done at puberty with considerable pain, are common in some cultures and should have provided dramatic illustrations of Freud's theory regarding castration anxiety. The widespread American custom of circumcising males in infancy, a painful and notable event from the vantage point of the young child, should also affect adult personality traits if Freud was correct; comparison of circumcised with uncircumcised males should show personality differences. Despite several decades of anthropological work in cultures around the world, much of it done by anthropologists who had themselves been psychoanalyzed and were sympathetic to Freudian theory, there is virtually no cross-cultural evidence that supports the theory regarding the importance of childhood experience in shaping adult personality.

Still another method for indirectly testing the validity of Freud's theory is to look at the results of Freudian-inspired preventive treatment. According to the principles of mental hygiene, as mothers become educated about Freudian theory they should become more skilled at avoiding the psychic pitfalls of the oral, anal, and Oedipal stages of development. Dr. Spock's study of 21 mothers in Cleveland was the most complete study done along such lines and it found that mothers with more knowledge of Freudian theory had *more* difficulties, not fewer difficulties, in raising their children. One might logically also expect that psychiatrists, psychologists, and psychiatric social workers, since they are the most knowledgeable about Freudian theory, would themselves raise children with fewer problems than professionals in other fields. Although this thesis has never been tested formally, discussion of it with mental health professionals elicits embarrassed smiles of the "you've got to be kidding" variety. Anecdotal accounts, such as the recently published *Children of Psychiatrists and Other Psychotherapists*, suggest that such children do not have fewer problems.

Nevertheless, followers of Freud have claimed that if enough professionals were available to treat millions of individuals along Freudian principles, then indicators of "poor mental health" such as suicide, divorce, and crime would decrease. Yet as the number of mental health professionals has increased, the suicide rate has remained virtually unchanged, while the divorce and crime rates have *increased* dramatically during the same years. Furthermore, as discussed in chapter 7, the one program which attempted to decrease crime by providing counseling to troubled youths failed to do so. Such outcomes do not engender confidence in the validity of either mental hygiene or Freudian theory.

The failure of both direct and indirect studies to support the Freudian theory of development has been an ongoing embarrassment to Freud's followers. It has elicited two principle responses. One is to invoke the Freudian concept of reaction formation to explain the failures. For example, in a study of breast-size preference among men, it was theorized along Freudian lines that men who were more dependent (because they were said to be fixated in the oral stage of development) should also prefer large-breasted women. When the results of the study showed exactly the opposite—the dependent men preferred small-breasted women—it was said to be due to a reaction formation among the men. Such an explanation is not science but merely psychoanalytic relativity in which all results can be interpreted as supportive of Freudian theory. Psychoanalysts have been regularly criticized for such mental gymnastics; for example, philosopher of science Ernest Nagel, in a 1959 essay, "Methodological Issues in Psychoanalytic Theory," commented: "A theory must not be formulated in such a manner that it can always be construed and manipulated so as to explain whatever the actual facts are, no matter whether controlled observation shows one state of affairs to obtain or its opposite."

The other reason given by Freud's followers to explain the failure of studies linking events of childhood to adult personality traits is that it is not the *actual* events which are important but rather the child's *interpretation* of the events. Freud himself sanctioned this reasoning in a 1905 essay, "My Views on the Part Played by Sexuality in the Etiology of the Neuroses," in which he reported:

> *Investigation into the mental life of normal persons . . . yielded the unexpected discovery that their infantile history in regard to sexual matters was not necessarily different in essentials from that of the neurotic. . . . The important thing, therefore, was evidently not the sexual stimulation that the person had experienced during childhood; what mattered was, above all, how he had reacted to these experiences.*

This revealing admission indicates that by 1905 Freud was aware that there are no differences in the actual sexual events of childhood between adults with neuroses and adults without neuroses; thus there was in fact no scientific basis for his theory of personality development. Such a realization was undoubtedly a major reason for Freud's tepid interest in scientific validation of his theory. As Philip Rieff noted, Freud "came to view the imagined as having no less power than the actual." It is precisely such reasoning that has led psychoanalysis to its discredited status within the scientific community. As Eysenck succinctly summarized it: "No other discipline claiming our attention has ever so clearly and decisively cut itself off from experimental testing of its theories—even astrology and phrenology make claims that are empirically testable."

Evidence for Genetic Determinants of Personality

At the same time as direct and indirect tests of Freud's theory have consistently suggested that the theory is incorrect, evidence has emerged that genetic factors are the single most important determinants for many, if not most, personality traits. Among humans such evidence has come predominantly from studies of twins and from adoption studies.

Older investigations of twins reared apart included studies of 20 pairs in America, 12 pairs in Denmark, and 44 pairs in England. The latter, for example, estimated that 61 percent of the personality trait labeled as "extraversion" is determined by genetics and described an identical twin pair raised apart, "who, though they knew nothing of one another, both kept many pets, played comic parts in amateur dramatics and liked to take on jobs as door to door saleswomen . . ." Another earlier study of 850 high school twin pairs (514 identical, 336 fraternal) reported that approximately 50 percent of the variation in personality traits as measured by questionnaires appear to be inherited. Similarly, a carefully controlled study of newborn twins showed that identical twins are much more similar than fraternal twins, especially on such measures as social awareness, fearfulness, and their tendency to smile and make sounds.

Recent studies of twins and adoptees have provided stronger evidence for the importance of genetic determinants of personality traits. The most highly publicized of these studies has been the Minnesota Study of Twins Reared Apart, which has compared 86 twin pairs reared apart (56 identical, 30 fraternal) with 331 pairs reared together

(217 identical, 114 fraternal). The media has popularized the stories of several pairs reared apart who, when finally brought together, were found to be extraordinarily similar (such as Bridget and Dorothy, who each wore seven rings and three bracelets, or Jim and Jim, who each smoked the same brand of cigarettes, drove the same kind of car, and were both part-time deputy sheriffs).

Although data analysis from the Minnesota study has not been completed, that which has been published has, like earlier studies, shown that "genetic factors exert a pronounced and pervasive influence on behavioral variability." Regarding specific personality traits, the study found "that, on the average, about 50% of measured personality diversity [i.e., the amount of variation within a population] can be attributed to genetic diversity." Some personality traits such as "negative emotionality" (stressed, harassed, anxious, angry) appeared to be more than 50 percent inherited, while others such as "social closeness" (prefers emotional intimacy, turns to others for help) were somewhat less than 50 percent inherited. Most surprising was the finding that variation in some social attitudes such as religiosity (religious interests and attitudes) and traditionalism (tendency to follow rules and moral standards) also appeared to be approximately 50 percent inherited.

In addition to the Minnesota twin study, other studies of twins reared apart are presently under way in both Sweden and Finland. The Swedish study, numerically larger than the Minnesota study, includes a large number of twin pairs over age 50, and its results suggest that genetic determinants of personality are important, but become less important with age. Even among such older twin pairs, however, it was found that one-third of their attitudes about responsibility for misfortune and control of their lives was attributable to genetic factors. In the Finnish twin study, it was found "that twins reared apart were in some traits more alike than twins reared together," a consequence of conscious efforts made by twins reared together to differentiate themselves and establish separate identities. In another study of twins, the "rough estimate of broad heritability" for altruism was reported to be 56 percent, for empathy 68 percent, and for nurturance 72 percent. Shyness is another personality trait that is thought to be highly hereditary and which has been well studied both in the Colorado Adoption Project and at Harvard University.

The personality trait that has been the most controversial in terms of heritability has been criminality. A twin study based on Denmark's national twin registry found that the chances of a second male twin being arrested if the first one had been arrested was 18 percent among same-sex fraternal twins but 34 percent among identical twins. A similar study in Norway reported lower concordance rates (15 percent for fraternals, 26 percent for identicals), but the Norwegian study examined a broader range of crimes, including driving offenses and treason.

An extensive adoption study carried out by Dr. Sarnoff A. Mednick and his colleagues in Denmark found that adopted boys had a 13.5 percent chance of being convicted of a crime as adults if neither their adopted nor biological parents had ever been convicted; this increased only slightly to 14.7 percent if their adoptive (but not biological) parents had been convicted and increased to 20 percent if their biological (but not adoptive) parents had been convicted. If both adoptive and biological parents had been convicted, then 24.5 percent of the boys were later convicted of a crime. Dr. Mednick and his colleagues concluded that "some factor transmitted by criminal parents increases the likelihood that their children will engage in criminal activity."

Although it does appear that genetics play some role in determining criminality, the magnitude of that role is very much in dispute. As Christopher Jencks noted in a review of such studies: "While the adoption studies certainly suggest that genes matter, they do not suggest that genes matter very much." The adoption studies have also been criticized because of possible selective placement of sons from criminal homes into similar homes. Nor is it clear precisely what is inherited that leads to criminal behavior; it may simply be personality traits such as a lack of empathy, impulsiveness, the inability to delay gratification, narcissism, or other similar traits that make criminal behavior more likely when a person is exposed to certain social conditions or milieus.

Unfortunately, arguments about the heritability of criminality have tended to obscure emerging data which suggest that many personality traits are highly heritable. The other controversial measure that has generated many times more heat than light has been the intelligence quotient, the nucleus of the nature-nurture controversy for three-quarters of a century. That there are individual differences in intelligence is indisputable; whether an IQ test measures such differences, as opposed to measuring scholastic aptitude, classroom learning, or exposure to language, is very much open to question. There are many better measures of brain function that have been studied and that have high heritability (e.g., perception speed, spatial visualization ability, perfect musical pitch); it is more productive to study such brain functions than to investigate IQ, which is so obviously contaminated by social and cultural factors.

The findings that genetic factors are important determinants of human behavior have come as no surprise to veterinarians and breeders of dogs and horses; indeed, even individuals whose pet dog or cat has had a litter have noted marked personality differences among the offspring from birth onwards. It has been known for many years at the laboratory level that mammals such as mice and rats can be inbred to produce strains that are more or less active or timid; in one study of mice, for example, "after 30 generations of selection a 30-fold difference exists between the activity of the high and the low lines." Similarly, extensive work has shown that "heredity is an important quantitative determiner of behavior in dogs and that genetic differences in behavior can be as reliably measured and analyzed as can heredity differences in physical size." In field research among chimpanzees (one of whom she named Freud), Jane Goodall reported "pronounced individual variation in behavior that gives each chimpanzee his or her unique personality," although neither Goodall nor other researchers have yet studied the relationship of such personality differences to personality characteristics of the chimps' parents and grandparents.

In summary, evidence is rapidly accumulating that genetic factors play a major role in determining many personality characteristics. As noted by Drs. Alexander Thomas and Stella Chess in their landmark study of 133 subjects from infancy to adulthood: "Continuity over time of one or another temperamental characteristics from infancy to early adult life has been strikingly evident in a number of our subjects." In a recent review of behavioral genetics, Dr. Robert Plomin observed that over the past decade the "increasing acceptance of hereditary influence on individual differences in development represents one of the most remarkable changes in the field of psychology." This is simply formalizing for the scientific community what parents of more than one child have known all along.

Evidence for Non-Genetic Determinants of Personality

If genetic factors account for approximately 50 percent of many personality traits, then non-genetic factors must account for the remaining 50 percent. The non-genetic canopy covers a broad array of possible influences, ranging from diminished oxygen supply to a child while the child is in the womb to birth trauma, infections, early familial childhood experiences, accidents, peer pressure, school experiences, and cultural influences. Is there any evidence suggesting the relative importance of early familial childhood experiences within this spectrum of non-genetic determinants?

Studies done to date suggest that early familial childhood experiences constitute a very small percentage of the total non-genetic influence on personality traits. The strongest evidence has come from adoption studies in which, according to one review of these studies, it has been shown that "pairs of unrelated children reared together in the same households show no more resemblance in personality than pairs of children chosen at random." Despite the fact that adopted children are raised together and experience the same father and mother, their personality characteristics are no more similar than two individuals who have never met. One study of extraversion, the personality trait which has probably been studied most extensively, concluded that familial experiences "could account for no more than 5% of the variance" for this personality trait. Studies of identical twins reared apart have come to conclusions similar to those of adoption studies; researchers in the Minnesota study reported that "the effect of being reared in the same home is negligible for many psychological traits," and the Swedish study concluded that "the influence of rearing in the same family (shared family environment) is negligible. . . . Estimates of [the shared environment] were consistently low, accounting for less than 10% of the total phenotypic variance."

This does not mean that early childhood experiences play *no* role in shaping personality traits. By imparting values to young children, the parents can affect the *expression* of personality even if they cannot affect the child's personality per se. For example, an intelligent, extroverted, aggressive boy may become a missionary if raised in a deeply religious family, a corporate executive if raised in an upper class business-oriented family, or a major drug dealer if raised in a family in which drugs and alcohol are prevalent. The boy's underlying personality is basically the same, although the way it is expressed is quite different. It also seems clear that some personality traits are more genetically determined than others.

It is also apparent that *grossly* aberrant childhood experiences may permanently alter a child's personality traits. For example, in a recent article, "Childhood Traumas," in the *American Journal of Psychiatry*, a case was cited where a six-year-old girl was attacked at a circus by a runaway lion; she sustained severe bites to her face and scalp, which led to multiple surgical procedures and disfiguration. The fact that the girl, "previously outgoing and friendly, stuck close to home and rarely ventured out into her neighborhood" is hardly surprising. Other grossly aberrant childhood experiences that might permanently affect personality traits include severe and disabling illnesses such as polio, severe physical abuse, sexual abuse (which may lead to dissociative states or "multiple personalities" in adulthood), and witnessing or participating in human or natural disasters, such as a flood or a murder. Even in these circumstances, however, it is impressive how many children emerge from such experiences with little or no apparent effects on their personality.

Freud's followers, in their attempts to prove the importance of childhood experiences, are fond of citing the studies of Konrad Lorenz and Harry Harlow and their colleagues. Lorenz "imprinted" baby geese by using himself as their mother early in development, and the geese thereafter followed him. As one psychoanalyst explained: ". . . the Oedipal period—roughly three and a half to six years—is like Lorenz standing in front of the chick [sic], is the most formative, significant, molding experience of human life, is the source of all subsequent adult behaviors." Similarly, Harlow et al. deprived infant monkeys of their mothers and contact with other monkeys, feeding them instead from bottles held by wire-and-cloth "surrogate" mothers. These experiences are certainly as aberrant for the geese and monkeys as was the runaway lion for the girl cited above. The surprising finding was not that the behavior of the geese and monkeys were affected by these experiences but rather how little they were affected; the monkeys, for example, were found to make a relatively good social adjustment if they were exposed to *any* other monkeys even if they had had no monkey mother.

The other work put forth by Freud's followers to prove the importance of childhood experiences are the studies of psychoanalysts René Spitz and John Bowlby. Spitz recorded the effects of maternal deprivation by studying infants in English orphanages during World War II; those who were deprived of all but minimum human contact became withdrawn and depressed, were more sickly, and had a higher mortality rate. The circumstances in these orphanages were grossly aberrant compared with normal homes and resembled conditions described more recently for orphanages in Romania; the fact that the infants were affected by such conditions should therefore be expected. More controversial were the post-war reports of John Bowlby on the effects of being orphaned. His oft-quoted reports to the World Health Organization claimed to have proven that the loss of one's mother in infancy leads almost inevitably to severe problems later in life. Many child development specialists have questioned Bowlby's conclusions, even including psychoanalysts such as Anna Freud, who maintained that "how the child reacted to the loss of the mother depended on the strength of his ego structure."

More recent studies have suggested that Bowlby's conclusions were fundamentally flawed and that the effects of separation of an infant from its mother are not necessarily harmful. One group of researchers studied children who lived in an institution during their first years and were later adopted or returned to their biological parents. The institution was noteworthy for its high turnover of staff, so that by age 4 each child had had an average of 50 different caretakers. When these children were evaluated at ages 8 and 16, however, it was found that "the formerly institutionalized children had no more serious health, behavior or emotional problems" than children who had not been institutionalized. Another study compared children who had been in orphanages or foster homes with those who had been adopted and found no significant difference between the two groups at age 35. Still another study, which followed children who had been "badly mistreated" as young children and then adopted, found the vast majority to be doing well as adolescents. Dr. Wagner H. Bridger, in a recent review of such studies, summarized the findings: "Short-term events usually do not have long-term effects. Even the worst early experience may do little permanent harm if the child's environment later improves."

Except for grossly aberrant experiences, then, there is no evidence that the normal developmental events of childhood shape later personality traits to any significant degree, although they may affect the expression of these traits. Even though Americans have become conditioned through Freudian influence to thinking that each minor event of childhood plays a crucial role in determining adult personality, the truth appears to be quite different. Several studies in fact have shown that most parents treat their children remarkably similarly at the same stage of development. In one longitudinal study, 50 mothers were videotaped interacting with two of their children when each was 1 year of age. The results indicate that "the mothers were remarkably consistent in their behavior toward their two children at the same age. . . . These data suggest that differential maternal treatment of their children in infancy does not appear to be a major source of the marked individual differences within pairs of siblings." Another study found that only 9 percent of children reported that there was "much difference" in the way their parents had treated them compared to their siblings, while another 35 percent reported "a bit of difference." Two

studies of twins also found that parents treated them both very similarly.

If early childhood experiences are comparatively unimportant
causes of personality traits except in children subjected to grossly aberrant circumstances, then what *are* the nongenetic factors that constitute
the other approximately 50 percent of personality trait variance which
is not due to genetic factors? Peer pressure, school experience, cultural
learning and teachers, scout leaders, and coaches almost certainly play
major roles, although these influences are difficult to quantify. It has
been noted in one study that "in families in which the parents speak a
language different from the larger community, or have a variant accent,
children invariably grow up speaking the language and accent of the
community and not that of the home. This again suggests the extent to
which we tend to overestimate the importance of parental influence on
the cognitive development of children."

It is also becoming increasingly clear that nongenetic biological
factors may be important antecedents of personality traits in some
individuals. These include instances in which a child has a diminished blood supply to the brain during pregnancy, is subjected to
certain chemicals at crucial stages of development (e.g., cocaine or
alcohol), has minor brain damage from birth trauma or as the result
of a childhood infection which affects the brain, is severely malnour-
ished as a child, or has trauma to the brain later in life. These aspects
of personality trait development, which have been relatively
neglected, are likely to become important areas for research in the
coming decade.

The Nature-Nurture Question Revisited

It is axiomatic that personality traits and human behavior are determined by the interaction of both nature and nurture, genes and culture
broadly defined. Geneticists invariably pay lip service to cultural concerns just as culturists recognize that genes play some role. Even Freud
acknowledged that "each ego is endowed from the first with individual
dispositions and trends [and] before the ego has come into existence
the lines of development, trends and reactions that it will later exhibit
are already laid down for it." Similarly, Margaret Mead, for whom cultural antecedents of behavior were all-powerful, wrote: "One of the
most fascinating preoccupations when one has a child of one's own is
watching for the appearance of hereditary traits and predispositions that
can be attributed to—or blamed upon—one side of the family or the

other. . . . The newborn baby does somehow embody the personality it will have."

The question, then, is not *whether* nature and nurture interact to determine personality traits and behavior but rather *how* they interact. Some light has been shed on this question in recent years by the twin studies of the late Dr. Ronald S. Wilson at the University of Louisville. By following 74 pairs of identical and fraternal twins from 3 months until 6 years of age, Wilson showed that the physical and mental development of identical twins has remarkably congruent spurts and lags compared with fraternal twins. It appears that individuals have, in Wilson's words, "a genetic blueprint" that not only determines *which* mental capacities and personality traits they will have but also *when* these capacities and traits will become manifest. Genes apparently are not static, one-time gifts of nature but rather are ongoing influences. Wilson summarized his observations by citing the following quotation from another researcher:

> It is most important to appreciate that the influence of genes is not manifested only at conception or at birth or at any other single time in the individual's life history. Developmental processes are subject to continuing genetic influence, and different genes are effective at different times.

What this means in practical terms is that how an individual is influenced by nongenetic factors, such as family, peers, culture, is even more complex than was previously suspected. There has been much discussion in recent years about reciprocal relationships and suitability between a child's personality traits and its parents' traits. For example, a shy or low-energy child may seem to be very "difficult" for parents who have the opposite characteristics. It is now clear that a temporal dimension must be added to this reciprocity; a child who is shy early in life may become prematurely so labeled by parents and peers in a way which makes the emergence of its later less-shy personality more difficult. Human personality traits apparently do not emerge as a single event in time, like a moth from a cocoon, but rather as an ongoing series of genetic events which continue over a person's lifetime.

What, then, is the role of parents, peers, culture, and other aspects of a child's environment? These provide the milieu in which genetic traits emerge and develop. Dr. Sandra Scarr, a behavioral geneticist, likened the role of a child's environment to the role of nutrition in determining height. A person with "medium-tall" genes will grow to be slightly under or slightly over average depending on that person's nutrition in childhood. However, "no matter how well-fed, someone with 'short' genes will never be taller than average." All inherited personality traits are so influenced by the environment; altruism or aggression may be encouraged and rewarded in one family, neighborhood, or culture, while the opposite may be true in another. But there is a limit to how much influence non-genetic factors can have on personality traits—you cannot make a silk purse out of a sow's ear but neither can you make a sow's ear out of a silk purse.

The suitability between a child's genetic endowment and the expectations of the child's environment has also been noted by Drs. Thomas and Chess in their longitudinal study of 133 children:

> *Parents may make the same demands on two children with different temperaments, and the effects on the children will be different. Thus, for example, parents may expect their child to adjust quickly and easily to beginning school but this may only be possible for the child who responds to new situations positively and adapts quickly to change. Successful adjustment will meet with parental approval and praise, making the parent-child interaction a positive one. If the child instead responds to new situations uneasily, tends to try to withdraw from them, and adapts slowly to change, the child will find the initial adjustment to school difficult. If the parents do not recognize this response as normal for the child, given the child's temperament, they may criticize and demand a quickness of adjustment of which the child is not capable. The parent-child interaction will be negative, in contrast to the positive interchange with the first case. In effect, the two children are experiencing the same parental attitudes and expectations, yet the effects are different, and the two childen are experiencing a radically different nonshared environment.*

Such observations have led some researchers to speculate that a significant portion of what is said to be the "environmental" influence on personality traits is in fact "due to genotype-environment interaction rather than representing a pure environmental source of variance." It is not the expectations of family, peers, or culture per se that are important but rather the expectations of family, peers, or culture *in the light of* the individual's genetic endowment.

The major impediment to further resolution of the nature-nurture question is resistance to research in this field. The assertions that genes are the single most important determinant of personality traits and that parental child-rearing practices are, except for gross aberrations, comparatively unimportant elicit cries of "Nazi" and "racist" because the nature-nurture question continues to be so politicized. When Dr. Irving Gottesman, a leading genetics researcher, lectured on behavioral genetics at the University of Texas, swastikas were painted over posters announcing the lecture. When Dr. Edward O. Wilson, a Harvard zoologist, gave a lecture on sociobiology at the meeting of the American Association for the Advancement of Science, a group of students poured water on him and called him a racist. At Berkeley, when anthropologist Dr. Vincent Sarich lectured on human genetic differences, his class was disrupted by "more than fifty students" who called him a racist.

Since genetics research has historically been associated with eugenics and with conservative or reactionary worldviews, much of the resistance to examining genetic aspects of personality traits has been associated with individuals with liberal political beliefs. One of the most prominent among such individuals has been Harvard University geneticist Dr. Richard C. Lewontin, who has also been called "an outspoken and idiosyncratic Marxist." Lewontin, in *Not in Our Genes* (1984), which was coauthored with neurobiologist Dr. Steven Rose and psychologist Dr. Leon J. Kamin, criticizes research on the genetics of human traits and abilities as "biological determinism," which the authors claim is "part of the attempt to preserve the inequalities of our society and to shape human nature in their own image." Lewontin, Rose, and Kamin say that they, by contrast, "share a commitment to the prospect of the creation of a more socially just—a socialist—society." Prior to genetics the privileged and wealthy classes of the world rationalized their position with religious myths, but with the discovery of genetics a new pseudoscientific rationalization became available. Lewontin et al. are continuing the tradition of Franz Boas, Ruth Benedict, Margaret Mead, and others in opposing genetic research on human personality traits because of its misuse.

To misuse the findings of genetics, however, does not make them incorrect. It is incontrovertible that there are genetic differences between individuals both in terms of physical characteristics (e.g., height), mental abilities (e.g., visual perception), and personality traits (e.g., shyness). The Declaration of Independence declares that "all men are created equal"; as Dr. Irving Gottesman has noted, this did not mean that all men have equal abilities but rather that all men should have "equal protection of the law and the right to equal opportunity for developing talents to the fullest extent compatible with the rights of others."

Just as there are genetic differences between individuals, so also it seems incontrovertible that there are genetic differences between groups of individuals. Such differences may involve physical (e.g., running speed) or mental (e.g., perception) abilities, personality traits (e.g., aggression), physiological characteristics (e.g., reaction to alcohol), or predisposition to specific diseases (e.g., phenylketonuria). For example, in the United States black athletes constitute 17 percent of major league baseball players, 60 percent of players in the National Football League, 75 percent of players in the National Basketball Association, and 98 percent of the top 50 sprinters in the 100-meter dash, yet **blacks constitute only 12 percent of the total population. There are probably no more than half a dozen sports fans in America who believe that this concentration of talent is purely nongenetic, although it is impolitic to say so. The reason is that any discussion of genetic differences between groups of individuals is equated with racism.**

For many people, the concept of racism implies that discrete groups called "races" still remain in the world. This was probably true thousands of years ago, but it is not true today—even isolated groups like the Australian Aborigines have been found to be surprisingly genetically heterogenous. As Lewontin et al. have pointed out, "no group is more hybrid in its origin than the present-day Europeans, who are a mixture of Huns, Ostrogoths, and Vandals from the east, Arabs from the south, and Indo–Europeans from the Caucasus." Franz Boas made the same point 60 years earlier: "In practically every nation there is a mixture of different types that in some cases intermingle and scatter through the whole country. . . . We have just as little right to say there is a Jewish race as that there is a French race or a German race or a Spanish race."

In studies of the distribution of human proteins in the world, it has been found that 85 percent of all differences exist "within the same local population, tribe, or nation; a further 8 percent is between tribes or nations within a major 'race'; and the remaining 7 percent is between major 'races.' " It is therefore scientifically incorrect to refer to races as still existing. Attempts to characterize a genetic trait which may exist among a group of people as a racial trait simply indicates ignorance, because 85 percent of human genetic variation exists within any group. Genetic differences between individuals overwhelm all other attempts to differentiate mankind, while genetic differences between groups have no predictive value for any single individual; for

example, anyone stupid enough to select a basketball team based on skin color alone may have Bill Cosby and Ray Charles on their team, with Larry Bird on the other team.

Another reason for contemporary resistance to examining the role of genetics in determining personality traits is that "genetic" has been popularly equated with "inevitable." As Lewontin et al. stated: "The determinists would have it, then, that human nature is fixed by our genes." Human nature is not "fixed" by genes but rather is confined within certain boundaries. The challenge for society is to provide all individuals with maximum opportunities to develop whatever genetic potential they have. This must begin by minimizing adverse environmental influences, such as poor nutrition or drug or alcohol intake during the prenatal period. Following birth, all infants and young children should have an intellectually stimulating environment [reports] such as might be available under an expanded Headstart and daycare program, immunizations and medical care guaranteed for all children, and learning levels in schools that allow all children—those with greater as well as lesser intellectual potential—to develop most fully. Research has shown that, given the great diversity in the timing of individual development, it is important to structure schools in such a way that late developers can be detected and moved to their proper level of ability. An aspect of democracy is equality of opportunity; as Dr. Theodosius Dobzhansky summarized it in his lucid book *Genetic Diversity and Human Equality*: "People can be made equal or unequal by the societies in which they live; they cannot be made genetically or biologically identical, even if this were desirable."

Freudian Theory and Phrenology

In order to put the scientific basis of Freud's theory into proper perspective, it is useful to compare it to a similar movement from the 19th century. Phrenology was a theory of human personality and behavior that was purported to be scientifically based, and also became very popular in America. The parallels between the phrenological movement and the Freudian movement a century later are striking.

Phrenology evolved from the writings of Franz Joseph Gall, who, like Freud, completed his medical training at the University of Vienna and set up a fashionable private practice in that city. Gall utilized his private patients to develop a theory that specific mental functions were not only associated with specific areas of the brain—an idea which has since been proven—but that these functions could be measured by bulges on the skull overlying the brain areas. Gall described 37 separate brain functions that he claimed could be physically identified on the skull, including self-esteem, mirthfulness, benevolence, spirituality, tune, and order. As in the case of Freud, Gall's ideas were not well accepted in Vienna, so he went to Paris and other European cities to lecture. He was accompanied by a devoted follower, Johann Spurzheim, who carried on the work after Gall's death in 1828. Four years later Spurzheim went to New England, where he gave a series of lectures in Boston and New Haven, thereby initiating the rapid spread of phrenology in America.

The strongest support for phrenology in America was initially among educated groups in New York and Boston. The noted educator Horace Mann "regarded phrenology as the greatest discovery of the ages and built all his theories of mental and moral improvement upon the ideas which it had furnished him." Samuel Gridley Howe, philanthropist and reformer, was president of the Boston Phrenological Society and his wife, author and educator Julia Ward Howe, used phrenological principles in her teaching. Henry Ward Beecher recommended "a practical knowledge of the human mind as is given by phrenology" as the best training to be a Christian. Horace Greeley popularized phrenology in his *New York Tribune* columns and suggested "that railroad trainmen be selected by the shape of their heads." Prominent individuals ranging from P. T. Barnum, Clara Barton, Amelia Bloomer, and John Brown to Andrew Carnegie, Thomas Edison, James A. Garfield, Ulysses S. Grant, and Brigham Young, all submitted their skulls for examination, and the results were published in the *Phrenological Journal*.

Phrenology also had an impact on American authors. Edgar Allen Poe, who said that phrenology "has assumed the majesty of a science and as a science ranks among the most important which can engage the attention of thinking beings," frequently referred to it in his stories. Walt Whitman included his own phrenological chart in early editions of *Leaves of Grass*. William Cullen Bryant and Mark Twain were both phrenological enthusiasts, but Herman Melville satirized it in *Moby Dick* when he described bulges on the whale's head. Popular magazines such as the *Ladies' Magazine* and the *Boston Literary Magazine* carried columns on it. The *Phrenological Magazine and New York Literary Review* combined phrenology with literature, similar to the *New York Review of Books'* amalgamation of Freudian theory and literature a century later.

Phrenology maintained a solid base among American physicians with the establishment of between 40 and 50 local Phrenological Societies, journals such as the *American Phrenological Journal* and the *Annals of Phrenology*, and a training course at the American Institute of Phrenology. By 1837 the chair in the department of mental and moral philosophy at the University of Michigan had been offered to George Combe, America's best known phrenologist. Some of the nation's leading psychiatrists, including Drs. Amariah Brigham and Isaac Ray, were profoundly influenced by phrenology and attempted to use its principles to treat mentally ill patients. The *American Journal of Insanity* ran numerous articles about it.

Phrenology's real popularity, however, came from its application to problems of living, such as "health, temperance, the raising of children, memory training, eugenics, matrimony, and religion." Despite the fact that Gall, phrenology's founder, had himself been politically conservative and socially elitist, his movement in America became associated with liberal politics and social reform. According to one historian, "in some ways its acceptance was a shibboleth of liberalism." Child rearing was especially affected, with much responsibility being placed on mothers to develop the brain's proper faculties while their children were still young. Criminology was also strongly influenced: "Phrenologists came to the eminently reasonable if unconventional conclusion that the vast majority of criminals are swayed by impulse and do not have sufficient moral sense to be inhibited by possible retribution." Punishment therefore would not work and "prisons should rather be rehabilitation centers." George Combe did a phrenological study of prisoners in Sing Sing Prison, the same prison in which Bernard Glueck would do research on Freud's theory 70 years later.

As phrenology spread throughout America it was increasingly criticized on a variety of grounds. It was said to be unscientific, "a quackery which succeeds by boldness"; indeed, after Gall did his original studies, his followers made virtually no effort to do further research on it. Phrenology was also said to subvert religion and to promote "atheism, materialism and determinism; the real casualties were moral responsibility and free will." In *The Humbugs of New York* (1839), a physician advised: "Avoid phrenologists as worse than the French infidels."

The farther phrenology spread, the less respectability it had. By the 1860s, it had become incorporated as a regular act in traveling carnivals, with phrenologists sometimes performing readings blindfolded. One phrenologist alone performed more than 200,000 examinations, and phrenology became increasingly big—and lucrative—business. As this popularization and vulgarization took place, the elite of Boston and New York, its original pillars of support, gradually lost interest.

By the 1890s phrenology had lost its elegance but was still occasionally used by individuals who were searching for a means for understanding human behavior. One such individual was anarchist Emma Goldman, who submitted to a cranial examination shortly after her release from New York's Blackwell Prison. The results of the examination, published in the *Phrenological Journal* in February 1895, were said to illustrate that "the facial signs of destructiveness are very pronounced. . . . Approbativeness and firmness are especially strong. . . . Almost utter lack of reverence and faith." Most importantly, the examiner reported that "her upper forehead is beautifully developed. . . . However her lower forehead is almost as defective as the upper portion is fine. . . . This shows a want of observation, precision, accuracy and specification in her collection or application of data. In other words, she will reason profoundly but often upon insufficient evidence."

Phrenology apparently did not provide Emma Goldman with what she sought. Eight months later, in October 1895, she did find it in Vienna. The lecturer was Sigmund Freud, who had assumed the mantle that had been worn by Franz Josph Gall in the same lecture halls a century earlier; the "Freudian Century" had begun.

10

An Audit of Freud's American Account

*Does not every science come in the end to a kind
of mythology?*
—SIGMUND FREUD, IN A LETTER TO ALBERT EINSTEIN

On May 1, 1889, Sigmund Freud treated his first patient, "Emmy von N.," with a new treatment that had been suggested to him by Dr. Joseph Breuer. Over the following decade, Freud continued developing this treatment method and named it "psychoanalysis." He also elaborated his belief that childhood sexual experiences cause most adult problems, and he explored dreams and the unconscious as means for retrieving childhood memories that could then be used to understand the origins of the problems. One hundred years have passed since Freud embarked on these peregrinations into the human psyche, an appropriate amount of time in which to assess the effects of his work. In doing so it is also worth noting the ironies that have evolved during Freud's century in America.

The Ironies of Freudian America

More than any country in the world, America accepted Freud's theory and popularized his name, and yet America was the one country that Freud truly hated. He often referred to it as "a gigantic mistake." When Max Eastman asked him why he hated it, Freud replied:

*"Hate America?" he said. "I don't hate America, I regret it!" He
threw back his head again and laughed hilariously. "I regret that
Columbus ever discovered it! . . . America is a bad experiment conducted by Providence. At least I think it must have been Providence. I
at least should hate to be held responsible for it." More laughter.*

On another occasion he said that the discovery of tobacco was "the
only excuse I know for Columbus's misdeed." Freud was never able to
reconcile himself to the fact that the one country he detested was the
country which had accepted his ideas most enthusiastically.

There were many reasons for Freud's antipathy to America. In correspondence he referred to Americans as "savages" and said "that [psy-
cho]analysis suits Americans as a white shirt suits a raven." In one letter
Freud likened the use of Viennese-inspired psychoanalysis in America
to the degrading employment of Viennese women as servants in other
countries. He also claimed that his brief visit to America in 1909 had
exacerbated his intestinal problems despite the fact that he had been
complaining of similar symptoms for two decades prior to his visit. In a
bizarre addenda, Freud alleged that the American trip had also caused
his handwriting to deteriorate.

Freud's principal contact with Americans, except for his brief visit,
was with individuals who came to Vienna to be psychoanalyzed by
him. For many years American and English patients constituted more
than half of Freud's practice. These individuals brought with them a
wide assortment of problems, ranging from poet Hilda Doolittle's concern about her bisexuality to banking heir James Loeb's seclusiveness
(after completing psychoanalysis with Freud, Loeb "built a large house
on a deeply wooded estate . . . where he lived as a recluse"). American
patients also brought with them considerable sums of money—an
important reason why Freud treated them. "What is the use of Americans if they bring no money?" Freud asked in 1924. "They are not
good for anything else." On another occasion he said that "America is
useful for nothing else but to supply money." Freud resented the fact
that most of the income from the sale of his books was coming from
America. "Is it not sad," he said, "that we are materially dependent on
these savages who are not better-class human beings?"

Freud also alluded to political reasons for disliking America.
According to one scholar, "Freud disliked American ideals of equality,
and in particular egalitarianism between the sexes." He referred to
American "petticoat government" and said that "American women are
an anti-cultural phenomenon." During his American visit, Freud had

also been repelled by the equality accorded to blacks and said that America "is already threatened by the black race." It should also be recalled that Freud sometimes identified with German causes, and during World War I he had "continued to identify the cause of the Central Powers as his own." During that war all three of Freud's sons fought against the Allies; his eldest son, Martin, was captured, and Freud was uncertain for several weeks whether or not he had been killed. In addition, the only son of Freud's favorite sister, Rosa, was killed, and several of his closest associates, such as Abraham, Eitingon, Ferenczi, Rank, as well as his son-in-law, all fought against the Allies. Thus for Freud America had been, and apparently continued to be, the enemy.

Another irony of Freudian America has been the association of Freud's name with American social reform and liberal politics. Freud himself was apolitical in his younger years and rarely voted, but he grew conservative as he aged. On one occasion Max Eastman asked Freud about his political beliefs and Freud replied: "Politically I am just nothing." In 1918 Freud expressed skepticism about the Russian Revolution, and in 1930 said, "I have no hope that the road pursued by the Soviets will lead to improvement." Freud also argued that "the psychological premises on which the [Communist] system is based are an untenable illusion," and he was described as listening to his brother "discuss the evils of socialism with an understanding smile."

When the reactionary Dollfuss regime took power in Austria in 1932, Freud supported it even as Dollfuss abolished free speech, banned the Communist Party, and brutally crushed a strike by the liberal Social Democrats. Like many conservative intellectuals of the period, Freud was sympathetic to Benito Mussolini's attempts to bring order to Italy and, when asked to inscribe a book for him, wrote: "Benito Mussolini with the respectful greeting of an old man who recognizes in the ruler the cultural hero." Freud's opposition to Hitler, however, was unequivocal because he understood the dangers posed by the Fuhrer's anti-Semitism.

Equally ironic is the association of Freud's name with humanism and egalitarianism in America. According to David Riesman, Freud was "a believer in the theory of elites: that society was inevitably divided between a small class of leaders and a large class of the led. . . . As one gets glimpses of [Freud's] own behavior towards the few lower-class people he came in contact with, such as maids and cabdrivers, one finds a tendency in him to be exploitative, even mean." Freud acknowledged this trait as one of his major shortcomings. To one friend he wrote: "In the depths of my heart I can't help being convinced that my dear fellow men, with a few exceptions, are worthless." And to another: "I have found little that is 'good' about human beings on the whole. In my experience most of them are trash, no matter whether they publicly subscribe to this or that ethical doctrine or to none at all." Freud's derogation of fellow humans did not stop with patients but extended to his followers as well: "The unworthiness of human beings, even of analysts, has always made a deep impression on me. . ."

Freud's belief in social and political elitism led logically to the development of psychoanalysis as a treatment method for the elite. Freud's initial patient, "Emmy von N.," was a wealthy and well-educated widow, and affluent patients subsequently constituted the vast majority of patients in his practice. According to one study only 3 percent of Freud's patients were poor. Economist Peter F. Drucker, whose parents knew Freud, claimed that much of the resistance to Freud in Vienna did not come from Freud's sexual ideas but rather from his elitist practice:

> *Freud did not accept charity patients, but taught instead that the psycho-analyst must not treat a patient for free, and that the patient will benefit from treatment only if made to pay handsomely. . . . Medical Vienna did not ignore or neglect Freud, it rejected him. It rejected him as a person because it held him to be in gross violation of the ethics of healer.*

The development of psychoanalysis in America has continued the elitist tradition of Freud, while at the same time it has tried to integrate this tradition with liberal and socialist political beliefs. The result has been a Janus-headed creature with one side showing Midas and the other side Marx. The inherent contradictions of Freudian therapy in America were illustrated as early as 1915 when Max Eastman, a fervent Marxist, extolled the merits of psychoanalysis, which he said then cost $200 to $500 per month [$2,000 to $5,000 in today's dollars]. Such contradictions were also visible in 1926 when Sandor Ferenczi attracted to his New York lecture series a large number of Marxists as well as a "fleet of limousines." In the 1920s Freud charged his American patients $20 per hour [$200 in today's dollars], which was the same fee which Dr. Abraham Brill was charging Mabel Dodge in New York.

Over the years occasional psychoanalysts have distinguished themselves among their peers by charging whatever the traffic will bear. One noteworthy example was Dr. Gregory Zilboorg, who in 1917 was a Socialist member of the Kerensky government in Leningrad but who in 1940 was charging $100 per hour [$770 in today's dollars] as a New York psychoanalyst. Such fees, especially when juxtaposed with the liberal political rhetoric of the profession, inevitably led to cynicism among many practitioners, as when Dr. Karl Menninger observed that "in our actual work we labor with a few rich individuals . . . whose personal salvation or lack of salvation will not make very much difference to the world." Such cynicism also spilled over to the general public, as in this 1935 assessment in the *American Mercury*: "If the Freudian doctrine is true, and the salvation of man's conflicts lies only in universal and catholic psychoanalysis, then the Freudians stand accused of being the most avaricious men in the world."

Despite their liberal political rhetoric, traditional practitioners of Freud's therapy in America have perpetuated the elitism of Freud himself. A 1966 survey of psychoanalysts reported that the average analyst treated only 28 patients per year and saw another 23 in consultation. Among patients seen in psychoanalysis, 83 percent had college degrees and 50 percent had completed graduate school. This demographic description of patients seen in the 1960s echos an earlier satire on psychoanalytic patients published in 1913:

> *Brilliant results are sometimes obtained in a given case by a single hour's use of the method every day for three years. This is particularly true where both the patient and the physician are versed in Greek and Roman mythology, classical Greek poetry, legend and tragedy, folk-lore and song.*

It also foreshadowed an assessment of Freudian therapy in 1990 by Dr. Robert Michels, a trained psychoanalyst:

> *If we regard it as a medical treatment, this limited availability is a public health crisis and an ethical scandal. Seen as a way of illuminating and enriching life, however, psychoanalysis resembles world travel, a season ticket to the opera, or higher education at a great private university.*

Freud's Credits: The Unconscious, Humanism, and Psychotherapy

In assessing the contributions of Freudian theory to American thought and culture, the role that Freud played in post-Victorian sexual reform should not be forgotten. Although disputes over birth control and homosexuality have been rendered substantially moot by the availability of contraceptives and by gay liberation, Freudian theory played an important role in extending the frontiers of morality at a time when abstinence was still a virtue. As Havelock Ellis noted retrospectively in 1939: "Whatever we may ultimately come to think of psychoanalysis as a technical method, it supplied an immense emphasis to the general recognition and acceptance of the place of sex in life." As succinctly stated by another observer: "Freud found sex an outcast in the outhouse, and he left it in the living room, an honored guest."

Freud's contribution to interest in dreams and the unconscious will probably be regarded in historical perspective as his most important legacy. Although Freud viewed both primarily as means for eliciting childhood memories and thereby proving his theory that childhood sexual traumas underlay adult psychopathology, his synthesis of existing ideas into a form which could be understood by a wider audience transformed slips of the tongue and phallic symbols into a universal language. Recent research on dreams suggests that Freud's representation was somewhat simplistic; dreams may not in fact be a royal road to the unconscious but they remain at least a serviceable secondary road. Similarly, new research on the unconscious has shown it to be much more complex than Freud portrayed. Memories, now known to be of several kinds, are assigned to different compartments of the brain, while emotions have been shown to be neurophysiologically independent of conscious thought processes. G. Stanley Hall's characterization of the intellect as "a speck afloat on a sea of feeling" is proving to be accurate. Freud's popularization of dreams and the unconscious has also produced a wealth of material for use by artists and writers, which is probably the principal reason why this group has been so strongly attracted to Freudian theory.

Another major Freudian contribution has been the nidus which his theory has provided for the growth of humanistic and egalitarian thought in America. Although Freud personally shared neither of these attitudes, his theory became grafted onto existing egalitarianism and evolved into a clarion call for social reform and the improvement of man. Reuben Fine emphasized this aspect of Freudian thought in his *History of Psychoanalysis*, calling it "a grand vision of mankind. . . . The vision of what mankind could become. . . . Daily life is a mere caricature of what people might be." Similarly, Laura Fermi noted that "in America it was the humanitarian tradition that drew psychiatrists and psychiatric social workers into psychoanalysis." Critic Walter Kaufmann, in a 1960 essay comparing Freud to Abraham Lincoln, was also drawn to this aspect of his thought:

> *Like no man before him, [Freud] lent substance to the notion that all men are brothers. Criminals and madmen are not devils in disguise, but men and women who have problems similar to our own; and there, but for one experience or another, go you and I.*

It is presumably this facet of Freud that has also attracted many socially conscious and humanistic Americans, including in recent years such individuals as Michael Harrington and Robert Coles.

Perhaps the most fundamental change in American culture wrought by Freud's legacy has been the popularization of counseling and psychotherapy. In no other nation in the world have these become so much a part of the culture as they have in America. While it is acknowledged that an increasing proportion of counseling and psychotherapy is not directly Freudian, most of it is indirectly Freudian insofar as it assumes that adult problems are a consequence of childhood experiences and that talking about or understanding the childhood experiences will help solve the adult problems. It seems likely that the immense popularity of counseling and psychotherapy in America is a direct consequence of Freud's greater popularity here compared to other countries.

There is a laudatory aspect to the widespread use of counseling and psychotherapy. It has promoted a more inner-directed culture in which intrapersonal feelings and interpersonal relations are accorded greater importance and life is said to be more than merely the accumulation of material possessions. Large numbers of Americans derive great satisfaction from pursuing in myriad ways their psychic nirvana and will argue that the pursuit itself has greatly enhanced the quality of their lives.

It should be noted, however, that the particular method of psychotherapy that Freud promoted—psychoanalysis—has proven to be neither more nor less effective than other brands of psychotherapy. Fisher and Greenberg, in an exhaustive review of research done on the efficacy of psychoanalysis, concluded that "there is very little evidence in the experimental literature even suggesting that the results of therapies called 'psychoanalysis' are in any way different from the results obtained by treatments given other labels." This finding is in accord with a large body of research on psychotherapy that has emerged in the past three decades, which strongly suggests that various psychotherapies are effective—i.e., the person undergoing the psychotherapy feels better—not because of the particular method of psychotherapy being used but rather because of generic principles of psychotherapy that underlie virtually all forms of it.

The generic principles of psychotherapy were clearly described by Dr. Jerome Frank in his widely read book, *Persuasion and Healing* (1961). They have continued to be refined over the years and are thought to include at least the following: personal qualities of the therapist (genuineness, empathy, warmth), mechanisms for raising the expectations of feeling better in the person seeking therapy (e.g., when the therapist hangs a diploma from a very prestigious university on the office wall), the ability of the therapist to put a name on what is wrong (which has been called the principle of Rumpelstiltskin), and finally the ability of the therapist to persuade the person that he/she has achieved mastery over the problem.

Virtually all techniques of psychotherapy utilize these underlying generic principles, and it is these principles which make the psychotherapies effective. Thus it is not surprising that, when studies are done comparing Freudian psychoanalysis, Rogerian therapy, primal scream, Gestalt therapy, or other forms of psychotherapy, all of them achieve remarkably similar results. One researcher has likened the situation to the making of a stew using one moose and one rabbit. The moose represents the generic principles underlying all psychotherapies, while the rabbit represents the particular method. It is evident that one may vary the type of rabbit used, but the stew will still taste the same. Thus one of Freud's contributions to American culture has been the popularization of counseling and psychotherapy in general, but Freud's psychoanalysis has not proven to be more efficacious than any other technique.

Freud's Debits: Narcissism, Irresponsibility, Denigration of Women, and Misallocated Resources

While Freud's theory and its ensuing movement have made some important contributions to American thought and culture, they have also brought with them many liabilities. One of the most important of these has been Freud's contribution to the "Me" generation, a suzerainty of self that focuses one's primary attention on personal happiness. The core of Freud's theory and therapy are both fundamentally narcissistic in assuming that one's happiness is the greatest good. This was clearly described in 1956 by Alfred Kazin, who wrote, "The overwhelming success of Freudianism in America lies in the general insistence on individual fulfillment, satisfaction and happiness. . . . The insistence on personal happiness represents the most revolutionary force in modern times."

The narcissism inherent in Freudian theory and therapy is equally applicable to many of its psychotherapeutic offshoots. From Adlerian psychotherapy to Zaraleya psychoenergetic technique, the main ingredient in the alphabet soup of American psychotherapies is "Me." This is well illustrated by Fredrick Perl's "Gestalt Prayer," the mantra of the "Me" generation:

> *I do my thing, and you do your thing,*
> *I am not in this world to live up to*
> *   your expectations,*
> *And you are not in this world to live up*
> *   to mine.*
> *You are you and I am I.*
> *And if by chance we find each other,*
> *   it's beautiful.*
> *If not, it can't be helped.*

Indeed, the focus of virtually all psychotherapy systems, many of which come and go in America like cerebral fall fashions, is not some higher ideal, not one's fellow man, but merely oneself.

Several observers have commented on America's increasing preoccupation with self in the 20th century. Historian Christopher Lasch, in *The Culture of Narcissism,* depicted an apocalyptic picture of American civilization collapsing around its ego: "In a dying culture, narcissism appears to embody—in the guise of personal 'growth' and 'awareness'—the highest attainment of spiritual enlightenment." Lasch himself, however, relied heavily on Freudian theory for his formulations of narcissism and failed to understand that Freudian theory has been a major contributor to the problem he described. Columnist Charles Krauthammer also noted a slide toward national narcissism, in which the new commandment is: "Love thyself, then thy neighbor. The formulation is a license for unremitting self-indulgence, since the quest for self-love is never finished and since the obligation to love others must be deferred while the search continues."

There have always been narcissistic individuals, but 20th-century America may be the first culture in which the quest for self and happiness has been equated with the greater good. These quixotic purblind seekers migrate from therapist to therapist in a perpetual search for a psychic Rosetta stone which they believe will rid them of inhibitions and thereby liberate them. Kazin referred to this as "the lap-dog psychology of Americans . . . the myth of universal 'creativity,' the assumption that every idle housewife was meant to be a painter and that every sexual deviant is really a poet." Kazin adds that under such circumstances it is "easy for the hack and the quack to get together!" An even harsher assessment of the effects of Freudian therapy was rendered by Edward and Cathey Pinckney, who wrote that "the one most outstanding accomplishment of psychoanalysis seems to be that it takes someone who is covertly obnoxious and makes him overtly obnoxious." One cannot help but wonder what would happen to many of the individuals in long-term psychotherapy if they spent the same hours working in community service or volunteering their time for the benefit of others that they now spend in the eructation of childhood trivia.

A second major liability of Freud's theory has been its promotion of irresponsibility. This evolved logically from the belief that individuals are governed by powerful unconscious forces, arising from early childhood experiences, which thereby usurp their freedom of action. It has led, in the words of one writer, to "the golden age of exoneration. . . . Almost nobody can really be held accountable. . . . Bonnie and Clyde came along too soon. Nowadays they could settle for a year at the Betty Ford Clinic as victims of compulsive bank-robbing addiction."

It is the areas of child rearing and criminal behavior in which Freudian theory has had the most profound effect and in which traditional concepts of responsibility have been challenged. In the Freudian scheme, men and women are seen increasingly as puppets of their psyches governed primarily by the edicts of their egos. The corollary of "don't blame me" is "blame my parents," expressed clearly from the earliest days of the Freudian movement, as in this 1926 rendition:

> *Mental hygienists are stressing one great point, namely, that in most cases of nervousness, in many cases of delinquency, in some cases of insanity, and almost all cases of child behavior or conduct disorders, the trail leads inevitably and directly back to the* home *and the* parents. *[Emphasis in original.]*

In the Freudian schema, mother, father, family, social circumstances, and culture become the causal agents for whatever is wrong. The ripple of personal irresponsibility spreads slowly outward to cover ever-greater areas until the very terms "good" and "bad" seem to no longer have meaning.

A third important effect of Freud's theory on American culture has been the denigration of women. Freud admitted that he did not understand women—he called them "a dark continent"—but at the same time he was remarkably dependent on them. For most of his life Freud surrounded himself with women—a wife, daughter, sister-in-law, and female followers—who venerated him and looked after his every need; Freud's wife is even said to regularly have put toothpaste on his toothbrush for him. There are also suggestions that Freud had difficulties in his relationships with women; one example of this was his failure to attend the funeral of his own mother, despite the fact that he was vacationing only two hours away at the time.

Freudian theory is inherently misogynistic and patronizing. Freud said that women are more narcissistic than men and have "little sense of justice." He called girls "the little creature without a penis" and viewed women as men manqués, relegated to a perpetually inferior status by this lack of a penis. According to Fisher and Greenberg, "Freud theorized that the female never fully accepts her lack of a penis. He consequenty portrayed her as unable to shake a chronic sense of body inferiority, envious of those who do possess a penis, and motivated to find substitutes." One such substitute, according to Freud, was the wish to have babies, which he viewed as an attempt to compensate for not having a penis. For Freud, women were not merely handmaidens for men but anatomically and intellectually inferior beings; as he once phrased it, women "have come into the world for something better than to become wise." Even Ernest Jones, whose biography portrays Freud as a saint, acknowledged that Freud's view of women was to have "as their main function to be ministering angels to the needs and comforts of men."

A few of Freud's followers, most notably Karen Horney, disagreed with this denigration of women and publicly said so. The majority, however, accepted Freud's evaluation of women and conveyed this assessment, implicitly if not explicitly, to their patients. When Karl Menninger wrote in 1939 about "the deepest hurt of all—'Mother failed me,' " he was reflecting a view of mothers that is inherent in Freudian theory. This theory has led logically to an epidemic of mother-blaming and women-bashing among mental health professionals. For example, a study of 125 articles in professional journals published between 1970 and 1982 reported that "mothers were held responsible for 72 different kinds of psychological disorders in their children . . . not a single mother was ever described as emotionally healthy, although some fathers were, and no mother-child relationship was said to be healthy." A social worker recalled that during her training, "We took it for granted that mothers caused much pathology. . . . If a patient is in trouble, the underlying assumption is that the mother must have done something wrong. . . . If [the problem] is not viral or bacterial, it must be maternal." The concept of pathogenic mothering has come to permeate almost all forms of psychotherapy and is used to explain most individual and family pathology.

Juxtaposed against the relative roles men and women play in causing individual and social pathology, the Freudian tradition of mother-blaming and women-bashing appears absurd. Although women constitute 51 percent of the American population, they are responsible for only 13 percent of assaults, 9 percent of driving-related felonies, 6 percent of robberies and burglaries, and 5 percent of murders. Only 1 percent of individuals on death row are women and "in most of these cases a co-defendant [a man] committed the actual crime." Women are conspicuously absent from accounts of public corruption and financial swindles and have historically played a very small role in initiating wars. And yet, according to Freudian theory, women rather than men are said to be responsible for many of the world's problems.

The increasing denigration of women as Freudian theory spread across America has produced an extraordinary amount of guilt and anguish. Women have been caught between the Scylla of restricting children, thereby damaging their fragile Freudian egos or the Charybdis of letting children do what they want, thereby spoiling them. Freudian theory, originally advertised as the cornerstone for modern women's child rearing, has been a millstone instead. Women have been made to feel guilty not merely for the indiscretions of their children but, according to Freudians such as Karl Menninger, for many of the world's woes as well. To assuage their guilt, women often undertake psychotherapy with the same mental health professionals who have been promoting Freud's theory. From the professionals' point of view this is good business: Create a problem and then get paid for solving it. From another point of view, however, one is reminded of Karl Kraus's observation: "Psychoanalysis is the malady of which it considers itself the remedy."

Another dismaying effect that Freudian theory has brought about in American thought and culture has been the misallocation of resources. This has been most prominent in the field of mental illness, where the vast majority of the nation's 200,000 psychiatrists, psychologists, and social workers spend their time doing counseling and psychotherapy based, directly or indirectly, on Freudian theory. Talking to one's therapist about your mother has become virtually a national pastime and continues to usurp an extraordinarily large share of professional resources that should be going to the seriously mentally ill.

The losers in this misallocation of resources have been individuals with serious mental illnesses such as schizophrenia and manic-depressiveness. Such individuals are both poor candidates for Freudian-inspired counseling and psychotherapy because their brain dysfunction often preclude logical thinking and because such counseling and psychotherapy are ineffective modes of treatment for them. Individuals with serious mental illnesses require medication and rehabilitation rather than discussions about early childhood experiences. Freud himself virtually ignored patients with serious mental illnesses, writing, "I do not like these patients." He believed that such patients were not appropriate candidates for his treatment and most of Freud's followers have believed likewise. One result is the sad spectacle of approximately 200,000 untreated mentally ill individuals among the nation's homeless population, despite the fact that America has more mental health professonals than any other country; the scene is one more legacy of Freudian theory.

Another aspect of this misallocation of resources has been the loss of professionals whose energy has been diverted toward psychotherapy and away from social change. This is seen most prominently in social work, whose practitioners were once at the forefront of efforts to improve society through combating poverty and improving working conditions. In recent decades social workers have become almost exclusively psychotherapists, tinkering with individual egos rather than trying to transform social systems. The consequence of this change was recently articulated by a senior social worker:

> When we sever the historical precedents that distinguish social work from other professions—working with the poor and disenfranchised—we risk becoming the same as psychology and psychiatry practitioners, except we charge less. . . . If we renounce our responsibility to the poor and if social change again becomes popular, we can only hope the poor will forgive us.

Despite the historical association of Freudian theory with social reform in America, the two in practice are antithetical. Freudian theory is inherently passive and self-absorbed and assumes that change must begin within individuals rather than within social systems. An individual who is a reformer and who undertakes psychoanalysis quickly learns that protestations against social inequities are mere displacements of one's anger against father, another manifestation of neurosis. In a similar vein, Arthur Miller likened psychoanalysis to "being bled white by a gratifying yet sterile objectivity," and Norman Mailer criticized it as follows:

> *In practice, psychoanalysis has by now become all too often no more than a psychic blood-letting. The patient is not so much changed as aged, and the infantile fantasies which he is encouraged to express are condemned to exhaust themselves against the analysts' nonresponsive reactions. The result for all too many patients is a diminution, a "tranquilizing" of their most interesting qualities and vices. The patient is indeed not so much altered as worn out—less bad, less good, less bright, less willful, less destructive, less creative.*

For social activists the analysis of their unconscious motives for wanting to improve the world has frequently transformed zeal into pabulum. This has been true not only for reformers like Jerry Rubin who, following his opposition to the Vietnam War, "experienced Est, Gestalt therapy, bioenergetics, Rolfing, massage, jogging, health foods, t'ai chi, Esalen, hypnotism, modern dance, meditation, Silva Mind Control, Arica, acupuncture, sex therapy, Reichian therapy, and More House—a smorgasbord course in New Consciousness." It is also true for thousands of young, socially conscious activists and potential reformers who, once they begin Freudian-inspired psychotherapy, embark on an endless rotation around their psychic navels and disappear forever from the political scene. Philip Rieff noted this "indifference to politics of many educated Americans of liberal political persuasion who have been instructed by Freudianism." The loss of these individuals from the struggle for social change has not only been a misallocation of resources but has also been one of the most important negative effects of Freudian theory on American culture.

Freudian Theory as Religion

The similarities of Freudian theory and therapy to a religion have been noted and commented upon from its earliest inception in Vienna. Indeed, even the first psychoanalytic patient, Bertha Pappenheim, who was treated by Freud's mentor, Dr. Josef Breuer, noted that "psychoanalysis in the hands of the physician is what confession is in the hands of the Catholic priest." This parallel was also clearly drawn by Max

Graf, one of the original members of Freud's Wednesday Society, whose son, Little Hans, was used to explore phobias. Graf recalled:

> *There was an atmosphere of the foundation of a religion in that room. Freud himself was its new prophet who made the theretofore prevailing methods of psychological investigation appear superficial. Freud's pupils—all inspired and convinced—were his apostles.*

When Freud excluded Alfred Adler because he minimized the importance of sex and thus deviated from Freudian theory, Graf observed: "Freud—as head of a church—banished Adler; he ejected him from the official church. Within the space of a few years I lived through the whole development of a church history. . . ."

Wilhelm Steckel also remarked on religious aspects of the Freudian movement. Originally one of Freud's patients, Steckel was a founder of the Wednesday Society and a devoted follower of Freud. In his biography, Steckel described himself as "the apostle of Freud who was my Christ." Freud apparently encouraged such fealty; in 1913 he "anointed" five of his most trusted followers—Hanns Sachs, Otto Rank, Ernest Jones, Karl Abraham, and Sandor Ferenczi—as a secret committee and gave each of them gold rings mounted with an antique Greek intaglio like his own.

Steckel's use of Christ was not precisely the right image for Freud and his early followers—from its inception in 1902 until 1906, all 17 members of the Wednesday Society were Jewish. Prior to the formation of the society, Freud had used the Viennese lodge of B'nai B'rith, which he had joined in 1897, as a forum for his ideas, and between 1897 and 1902 he presented several papers on his theory at their meetings. Dennis B. Klein, in *Jewish Origins of the Psychoanalytic Movement*, showed how profoundly Freud's ideas were influenced by Jewish Talmudic traditions, while David Bakan, in *Sigmund Freud and the Jewish Mystical Traditions*, maintained that many of Freud's ideas about sexuality were "startlingly close" to the mystical Judaism of the cabala. Given this traditional religious ethos, it is hardly surprising that when a psychoanalytic society began meeting in Zurich, a rumor circulated that it would have to pay "special taxes on the grounds of being a religious body."

When the Freudian movement arrived in New York, it retained a superficial aura of secular Judaism, but had broadened into a psychoanalytic ecumenicism. Freud offered an understanding of the unknown, an explanation for evil, and in later years (in *Totem and Taboo*) even a myth of creation. Irrational behavior and evil could be explained by contorted libidos and errant ids—products of childhood experiences. Freud's postulates were appealing to men and women who had set aside formal religion but who still craved the certainty of an Oedipal catechism. This theological appeal of Freudian theory was well described by Hanns Sachs. After reading *The Interpretation of Dreams*, Sachs wrote: "I had found the one thing worthwhile for me to live for; many years later I discovered that it was also the only thing I could live by."

Sachs also compared the training of psychoanalysts to that of the clergy:

> *Religions have always demanded a trial period, a novitiate, of those among their devotees who desired to give their entire life into the service of the supermundane and the supernatural, those, in other words, who were to become monks or priests. . . . It can be seen that analysis needs something corresponding to the novitiate of the Church.*

Inherent in any understanding of what is wrong with an individual is the possiblity of making it right. Under Freudian theory one's errors do not carry personal blame, for they are inevitable sequelae of childhood experiences. The confession of an individual undergoing Freudian therapy is not an act of absolution but rather an act of understanding and enlightenment. Redemption within a Freudian framework comes about through linking neuroses of the present with events of the past; it is intellectual salvation, but theoretically no less effective for being so.

The redemptive appeal of Freud's message was extensively explored in Klein's *Jewish Origins of the Psychoanalytic Movement*. Klein claimed that there was an "interpenetration of the Jewish redemptive vision with the psychoanalytic movement's hope for the eradication of neurosis." Insofar as evil originated in childhood experiences, it could be banished if childhood experiences were modified. Fritz Wittels, another of Freud's early followers, recalled: "Some of us believed that psychoanalysis would change the surface of the earth, believed that the Victorian age in which we then lived would by way of a psychoanalytic revolution be followed by a golden age in which there would be no room for neuroses any more. We felt like great men. . . ." Freud was the prophet who would lead mankind out of the land of Id and into the Promised Land—a contemporary Moses with a couch.

The religious appeal of Freud's theory was a particular reason for its acceptance in America in the early years of this century. A revolt was occurring not only against Victorian morality but against traditional Christianity as well. Freud's theory proffered an explanation of the unknown, a means of redemption, and a vision for a better future. For scientific men who had been deeply religious, men such as James Putnam and William A. White, the religious aspect of Freud's thinking was very important, for it promised a scientific base for their spiritual interests. As one historian noted, Putnam and White "found in 'mental health' a new basis for contemplating the purposes of human existence which they could reconcile with their scientific values." Skeptics of Freud's theory also noted the religious aspect of the movement and criticized his followers, who were said to "accept Freudism as a religion [and] not admit any qualification to the least of its tenents. . . ."

To many inhabitants of Greenwich Village, Freud's theory appealed less as a scientific soul than as a socially respectable mysticism. The same Max Eastman who spent three months undergoing a psychic cure at Doctor Sahler's New Thought Sanitarium would write of Freud's ideas: "They were all plausible to me—homosexuality, mother fixation, Oedipus, Electra and inferiority complexes, narcissism, exhibitionism, autoeroticism, the 'masculine protest.' " The same Mabel Dodge who regularly consulted faith healers would undergo psychoanalysis with Drs. Brill and Jelliffe while continuing to visit her faith healers between analytic sessions. Such mysticism merged easily with advocacy for sexual freedom and social reform; in 1922 more than 400 women attended a "psychic tea" sponsored by the National Opera Club, at which Freud's theory was sandwiched between a defense of spiritualism and an impassioned plea for psychic liberation by discarding one's corsets. A belief in Freud, then, was not a formal religion in Greenwich Village, but served the same informal function; *Three Essays on the Theory of Sexuality* was the Village Vulgate, and scripture was regularly cited according to Freud. For some individuals Freudian theory continues today to serve these same religious functions.

Beyond its religious value, will Freud's theory have any lasting importance? Nobel Laureate Peter Medawar suggested that Freud's theory will be seen in retrospect as "the most stupendous intellectual confidence trick of the twentieth century; and a terminal product as well—something akin to a dinosaur or a zeppelin in the history of ideas, a vast structure of radically unsound design and with no posterity." In a similar view, Vladimir Nabokov predicted that "our grandsons no doubt will regard today's psychoanalysts with the same amused contempt as we do astrology and phrenology."

These assessments, however, are too harsh. Freudian theory will more likely be recalled as having played an important role in the sexual liberation of America, as having familiarized the concept of the unconscious, and as having helped our culture turn inwards and focus more on the quality of our intrapersonal and interpersonal lives. And yet there are indications that Freud's time in America has come and gone. His debits—narcissism, irresponsibility, the denigration of women, and the misallocation of resources—far outweigh any remaining assets which Freudian theory might have to offer. The challenge for the 21st century is to place human behavior on a more solidly scientific foundation and to ensure that all children have the maximum opportunity to **develop the potentials with which they have been born. Freudian theory would appear to have no role in this endeavor since it has no scientific base. It will slowly fade from view, therefore, just as the Cheshire cat once did, except that in this case the grin will fade first, and the genitals last of all.**

Appendix A:

An Analysis of Freudian Influence on America's Intellectual Elite

As discussed in chapter 8, in *The American Intellectual Elite*, Charles Kadushin asked 110 American intellectuals to rate each other's influence "on cultural or socio-political issues." From these ratings, Kadushin derived a list of the 70 "elders" who had been rated as the most influential by their peers. In order to assess Freudian influence on this group, I took the two highest categories, consisting of 21 individuals, and examined their writings. In addition, I sent a brief questionnaire to the 13 of these who were still alive in 1990, asking them: "How extensively have you read the works of Freud or other psychoanalytic writers?" and "How important have Freudian (and other psychoanalytic) ideas been in the development of your own thought?" Replies were received from 8 of the 13 who were sent questionnaires.

The following is an analysis of Freudian influence on these 21 intellectuals. In addition, I noted the political inclinations of each in order to determine how closely Freudian influence was associated with liberal political belief among these intellectuals. (For a summary of this data, see chapter 8.)

Hannah Arendt: Philosopher and author of *The Origins of Totalitarianism*, Arendt seems not to have been influenced by Freudian theory as evidenced in her work.

Daniel Bell: Prominent sociologist and author of *The End of Ideology*, Bell "read extensively in the Freudian canon. . . . I would say that I have probably read all of the meta-psychological works of Freud with some care," as well as some works of Jung, Horney, Reich, and Adler. He also underwent personal psychoanalysis "for about five years." While teaching at the University of Chicago (with David Riesman) and Columbia University, Bell included Freud's work in some of his courses. There are occasional references to Freudian theory in his later work (e.g., a chapter on national character in *The Winding Passage*), but Bell says he lost interest in psychoanalysis after 1960 and now finds Freudian theory "when applied to history and sociology . . . badly deficient." Politically, Bell "delivered street-corner speeches for Socialist candidates while still a teenager" and was identified with liberal anti-communism in the postwar period. He currently characterizes himself as being a "socialist in economics, a liberal in politics, and a conservative in culture."

Saul Bellow: Winner of the Nobel Prize for literature in 1976, Bellow has shown a significant influence of Freudian and especially Reichian ideas in his novels. According to one biographer, "*Henderson the Rain King* . . . is so completely charged with Reichianism that almost every page, every narrative sequence, every piece of description is full of it. . . . One cannot, I feel, understand or evaluate *Henderson the Rain King* without first coming to terms with its Reichianism." *Seize the Day* and *Herzog* are also said to be "saturated with Reichianism." "One can assume," the biographer concludes, "that Reich was an important influence on Bellow all through the fifties and early sixties." Another critic has speculated that the two African tribes in *Henderson the Rain King* were set up by Bellow as "a choice between Reichian and Freudian understandings of the human situation" and that "Bellow sees the world much as Freud was beginning to see it when he outlined his bleak vision of the human predicament in *Civilization and Its Discontents*." Politically, Bellow is said to have "refused to fall in with the radicalism of the *Partisan* crowd in the 1950s . . . [and] to resist identification with the extremes of both right and left."

Noam Chomsky: Linguist, prominent opponent of the war in Vietnam, and an intellectual leader of the New Left, Chomsky says he "read Freud, Horney, Sullivan and others as a teenager and found [their work] fascinating," but that Freudian theory has had little, if any, influence on the development of his thought.

John Kenneth Galbraith: Regarded as "one of the leading liberal economists of the postwar era," Galbraith says that his personal acquaintance with Freudian theory "has been extraordinarily haphazard and casual," although he regards it as an "extremely important and fascinating line of thought."

Paul Goodman: As discussed in chapter 5, Goodman was an ardent and militant supporter of the theories of both Sigmund Freud and Wilhelm Reich. Politically, he was an anarchist and "chief spokesman for the non–Marxist tradition of western radicalism."

Richard Hofstadter: Historian and author of *The Age of Reform*, Hofstadter seems not to have been significantly influenced by Freudian therapy as evidenced in his work.

Irving Howe: Editor of *Dissent*, there is insufficient information available with which to assess the influence of Freudian theory on his thought.

Irving Kristol: Editor of *Public Interest*, Kristol says he has "read extensively" the writings of Freud and other psychoanalytic theorists. However, he adds, that "my reading only served to increase an original skepticism" and that Freudian theory has not been important in the development of his thought.

Dwight Macdonald: As discussed in chapter 4, many of Macdonald's writings "abounded with psychological and psychiatric terminology," and he was very interested in the theories of Wilhelm Reich. The precise extent of Macdonald's influence by Freudian and other psychoanalytic theories is not clear but was probably significant. Politically, he was a Marxist and leading defender of Trotsky in the 1930s.

Norman Mailer: Novelist and journalist, Mailer has been strongly influenced by psychoanalytic theory at the same time as he has ridiculed its practice. According to one critic, "Mailer began as an ardent Freudian." Mailer called Freud "a genius, an incredible mighty discoverer of secrets, mysteries and new questions." Later he became "profoundly infuenced by Wilhelm Reich" and Reich's belief in sexual energy; Mailer even built his own Reichian orgone box in which to absorb psychic energy, and he appears to have accepted the Reichian belief "of the good orgasm as the cure-all for mental and physical ailments." At the same time Mailer has portrayed psychoanalysts themselves as instruments of conformity and social control. Politically, Mailer has ranged from advocating anarchism and socialism to being a supporter of Henry Wallace, John Kennedy, and Jimmy Carter.

Herbert Marcuse: As discussed in chapter 8, Marcuse was a philosopher whose work was profoundly influenced by Freudian theory. Politically, he was a Marxist and an intellectual leader of the sixties New Left.

Mary McCarthy: As discussed in chapter 4, McCarthy underwent psychoanalysis three separate times and used Freudian themes prominently in her novels. In the 1930s, she was an active supporter of Leon Trotsky and other left-wing political causes.

Daniel P. Moynihan: Author and liberal United States Senator, Moynihan was one of the architects of the community mental health center movement and has continued to be one of the leading supporters of social programs. Insufficient information is available, however, with which to assess the influence of Freudian theory on his thought.

Norman Podhoretz: Editor of *Commentary*, Podhoretz has "read most of Freud's works but my reading in the other psychoanalytic writers has been spotty." In the 1960s, Podhoretz was clearly influenced by Freudian theory. He serialized Goodman's *Growing Up Absurd* and helped get it published. He also promoted Norman O. Brown's *Life Against Death* as "a great book by a major thinker. . . . I also went around trumpeting its virtues to everyone in town." Politically, Podhoretz "tilted toward the Left" early in his career, but then became a member of "the neoconservative brain trust."

David Riesman: Sociologist and author of *The Lonely Crowd* (1950), Riesman was strongly influenced by Freudian theory. His mother had been "an early and ardent admirer of Freud" and had been analyzed by Karen Horney. Riesman recalled: "There was a time in my life when I prided myself on having read everything translated into English that Freud wrote." Riesman was personally analyzed by Erich Fromm. In the late 1940s, Riesman helped develop an undergraduate course in the social sciences at the University of Chicago, in which several works of Freud were among the required readings; in the four lectures on Freud that he gave in this course, he characterized Freud as "one of the great intellectual heroes of all time." Riesman believes that in his later work "psychoanalytic ideas have shaped my approach only in the most general way." Politically, Riesman has been generally regarded as a liberal; Norman Podhoretz called *The Lonely Crowd* "perhaps the single most important expression of [postwar liberalism] to have come out of the field of sociology."

Arthur Schlesinger, Jr.: Author of *The Vital Center* (1949) ("the manifesto of postwar liberalism"), Schlesinger is regarded as one of the nation's leading liberal historians. He has published biographies of Andrew Jackson and Franklin D. Roosevelt and worked as an adviser and speechwriter for both Adlai Stevenson and John F. Kennedy. He acknowledges having read "a fair amount of Freud and a certain amount of Jung, Fromm and other writers of the psychoanalytic school" and at one time "had an interest in the possible applicability of Freudian insights to the study of history." He doubts that Freudian theory has been vital in the development of his thought, but adds, "like everyone in my generation, I use Freudian patter and metaphor in conversation."

Robert Silvers: Since 1963 Silvers has been co-editor of the *New York Review of Books* (designated by Tom Wolfe as "the chief theoretical organ of radical chic"). When asked about the influence of Freudian theory on the development of his thought, Silvers replied: "I have tried to understand Freudian theories and the various criticisms of them and have published a great many essays and reviews about them—some of them sympathetic to Freudian theories, some unsympathetic, and some that seem to me neither sympathetic nor unsympathetic." Since it began publishing, the *New York Review* has probably included more reviews of Freud's ideas than any non-psychiatric publication. Silvers's political leanings appear to be to the left, both because of the liberal politics of the *New York Review* (it has been called a "platform of the radical Left") and because of Silvers's public support for liberal causes.

Susan Sontag: Author of *Play It as It Lays*, Sontag has called Freud "the most influential mind of our culture" and a "revolutionary mind." As a seventeen-year-old university student she audited a course on Freud and married the course's instructor, Philip Rieff, ten days after meeting him. According to a Sontag critic, "Sontag helped her then husband, Philip Rieff, with [his book] *Freud, The Mind of the Moralist*," a book which is highly flattering to Freud. Sontag was a prominent supporter of leftist political causes but publicly denounced communism in 1982.

Lionel Trilling: As discussed in chapter 4, Trilling had a personal psychoanalysis and thereafter was one of the most active literary prose-lytizers for Freud's theory. Politically, he was a Marxist and active sup-porter of Leon Trotsky in the 1930s.

Edmund Wilson: As discussed in chapter 4, Wilson became inter-ested in Freud's theory as early as 1915 and used psychoanalytic themes prominently in his literary criticism. Politically, he was an active sup-porter of Socialist and Communist party candidates in the 1930s.

Appendix B:

Is Toilet Training Related to "Anal" Personality Traits?
A Summary of 26 Research Studies

The core of Freud's theory of human behavior is a belief that early childhood experiences are crucial in determining adult personality traits. Freud designated the important periods of childhood development as the oral, anal, and Oedipal stages. The major event during the anal stage is toilet training; Freud said that "the way in which this training is carried out determines whether or not anal fixations result."

The following are summaries of 26 research studies that have been cited as directly bearing on Freud's theory about toilet training and "anal" personality traits. These studies are summarized in chapter 9.

Gilbert V. Hamilton, *A Research on Marriage* (New York: A. and C. Boni, 1929).

> As part of an interview of 200 married men and women, the subjects were asked: "Do you recall whether or not, as a child, you were fond of prolonging the act of moving your bowels for the sake of pleasant thrills which it gave you?" Those who answered "yes" (39 subjects), "no, but it brought some sense of relief" (10 subjects), and "no, but I was troubled with constipation" (10 subjects) were grouped together and called "anal erotic." This group was then compared with the others who answered "no" without qualifications (not "anal erotic") on self-reported personality traits of stinginess or extravagance, hoarding of inanimate objects, personal appearance, sadistic fantasies, and masochistic fantasies. The author used no statistical analysis but concluded the "the evidence is uniformly on the side of Freud's general contention that stinginess, extravagance, sadism, masochism, a tendency to hoard comparatively useless things, orderliness, and carelessness sustain important dynamic relationships to anal erotic tendencies." In fact

when the author's numbers are analyzed statistically using a chi square, the only statistically significant associations are found for sadistic and masochistic fantasies.

It should be noted that the researcher was a psychoanalytically oriented psychiatrist who acknowledged being "dependent at every turn on Freud's insight." The subjects were recruited from among his friends in New York City from the upper socio-economic class and were mostly of liberal and radical political persuasion; thus it is a reasonable assumption that the majority of subjects were, or had been, psychoanalyzed and thus were aware of Freud's postulated relationship between childhood attitudes toward bowel movements and later personality traits. The fact that the research found no significant associations except for sadistic and masochistic fantasies is remarkable given the self-selected subjects and strong psychoanalytic bias of the questions. Methodologically it also seems absurd to call individuals "anal erotic" who recall bowel movements as giving them a sense of relief or who were constipated in childhood. Despite the scientifically pathetic quality of this study, it was still being cited by researchers in the 1950s as having shown that "these men and women, as adults, showed a higher frequency of reported stinginess or extravagance than the non-anal ones. . . . These differences are all in line with the theory of the anal character" (E. R. Hilgard et al., *Psychoanalysis As Science*, 1952, p. 16).

Mabel Huschka, "The Child's Response to Coercive Bowel Training," *Psychosomatic Medicine* 4(1942):301–08.

This frequently cited study is a survey of the toilet-training history of 213 children ages 1 to 13 who had been referred to the Child Psychiatry Clinic at Cornell University Medical Center in New York City. The children had been sent to the clinic for a varity of problems including motor disturbances (e.g., tics, speech disturbances), "physical symptoms for which no organic basis could be found," mental symptoms (e.g., pathological fear), and conduct disorders. Over half (104 out of 169) of the children for which definitive histories could be ascertained had had toilet training begun prior to 8 months of age, which the author defined as coercive. The implication of the study is that coercive toilet training causes problems as proven by the referral of such children to a child psychiatry clinic.

The study is virtually useless. First, the author presented no rationale for assuming that the problems with which the child presented should be in any way related to prior toilet training. Even more importantly, the author utilized no control group to ascertain toilet training practices in the general population, despite citing the most recent edition of the United States Children's Bureau publication, which advised that toilet training "should always be begun by the third month and may be completed during the eighth month." Thus the finding that approximately half of the parents had *begun* training prior to 8 months of age can hardly be surprising, and the lack of a control group renders the study meaningless.

Robert R. Sears, *Survey of Objective Studies of Psychoanalytic Concepts.* Bulletin of the Social Sciences Research Council, no. 51. New York, 1943.

In a college fraternity, 27 men rated each other on personality traits of stinginess, orderliness, and obstinancy. There was a small but significant correlation between these traits in some individuals, leading the author to conclude that "there is in a sample of young males a slight tendency for the anal character traits to form the kind of constellation observed by Freud." Data on toilet training was not collected in this study, so it merely provided some support for the existence of personality traits, which have been labeled as "anal" in some individuals.

Gerald S. Blum, "A Study of the Psychoanalytic Theory of Psychosexual Development," *Genetic Psychology Monographs* 39(1949):3–99.

This study utilized a projective test that has been widely discussed in the psychology literature since it was published. A series of 12 cartoon pictures depicting a dog named Blacky are presented to a subject who is asked to tell a story about each picture. The pictures were created to elicit psychoanalytic themes and are not subtle; for example, a picture of Blacky watching another dog getting his tail chopped off is supposed to elicit themes of castration. The picture that is supposed to elicit themes connected with the anal stage of development depicts Blacky defecating between the kennels of his mother and father.

Blum's original study was done with Stanford University students who had chosen to take a general psychology course. His results showed that students who responded to one picture with psychoanalytic themes tended to respond to other pictures in the

same way. There are major problems with both validity and relia-
bility of the test (*see* Paul Kline, *Fact and Fantasy in Freudian Theory*,
1972, pp. 81–83). More seriously, according to Hilgard et al., "the
pictures themselves give such strong suggestions that the data are
corrupted by the theory. . . . the Blacky test provides a degree of
validation of psychoanalytic theory with regard to internal consistency but, because it remains within the framework of analytic theory, does not provide 'objective' validation of the underlying
assumptions" (Ernest R. Hilgard et al., *Psychoanalysis As Science*,
1952, pp. 19 and 155.). The Blacky test is essentially a test of the
internal validity of a belief system and would be analagous to
showing subjects a series of pictures with themes of Catholicism or
Republicanism; students who accept that belief system would
respond to many of the pictures in a similar way.

Amy R. Holway, "Early Self-Regulation of Infants and Later Behavior
in Play Interviews," *American Journal of Orthopsychiatry*
19(1949):612–23.

Seventeen middle- and upper-class children ages 3 to 5 were
evaluated during a 30-minute interview regarding their play habits.
Play was recorded either as "realistic," "fantasy," "hostile aggression," or "tangential." These scores were then compared with the
age and method of toilet training, using information obtained from
an interview with the mother. The author reported that the combined score of play rated as "fantasy," "hostile aggression," and
"tangential" correlated "above the 5 percent level" with "the
method of bowel control education used by the parents."

Two major problems render this study worthless. First, the
number of subjects (17) is too few. More importantly, it may be
questioned why "fantasy" play, "hostile aggression," and "tangential" play together should be expected to correlate with toilet-
training experiences. As an example of "fantasy" play, which was
interpreted by the researchers as indicative of "dealing with a
painful reality situation," the author cited: "child piles up the furni-
ture to make a beanstalk."

William H. Sewell, "Infant Training and the Personality of the Child,"
American Journal of Sociology 58(1952–53):150–59.

Information on bowel training was obtained from the mothers
of "162 farm children of old American stock" from stable homes.

The children were ages 5 and 6. This data was compared with personality characteristics obtained by personality tests and by ratings obtained from the children's mothers and teachers. No correlation was found between "the personality adjustment and traits of the children" and bowel training measured either by the time of induction (early or late) or whether the child was punished for toilet-training accidents. The author concluded that "the results of this study cast serious doubts on the validity of the psychoanalytic claims regarding the importance of the infant disciplines and on the efficacy of prescriptions based on them."

Robert R. Sears, John W. M. Whiting, Vincent Nowlis, and Pauline S. Sears, "Some Child-Rearing Antecedents of Aggression and Dependency in Young Children," *Genetic Psychology Monographs* 47(1953):135–234.

Information on child-rearing practices were obtained from the mothers of 40 preschool children in Iowa. This was then correlated with measures of aggression as computed from direct observation, doll play, and preschool teachers' ratings. No significant correlations were found, and the authors concluded that "it appears that the severity of food scheduling, weaning, and toilet training are unrelated to any of several forms of later preschool aggression."

John W. Whiting and Irvin L. Child, *Child Training and Personality: A Cross-Cultural Study* (New Haven: Yale University Press, 1953).

The authors selected anthropological materials available from 75 "primitive societies" and rated them for various child-training procedures, including toilet training. This was then compared with "customs relating to illness and the threat of death . . . as indices of personality characteristics of the typical members of a given society" (p. 161). One major problem with this study was the assumption that such customs should relate to child-rearing practices at more than a random level. Another problem was that some of the anthropological studies had been carried out by researchers who were firm believers in psychoanalytic theory and thus might have reported the data within such a construct. Finally, the attempt to correlate child-rearing practices to "customs relating to illness and the threat of death" was done utilizing "psychoanalytic theory [which] has suggested that extreme frustration or extreme indulgence of a particular form of behavior in childhood may produce a continuing fixation of interest on that particular form of behavior" (p. 315). Despite having loaded the research dice strongly in favor of psychoanalytic theory, the results showed virtually no correlation between toilet-training practices and customs relating to illness and death.

Thelma G. Alper, Howard T. Blane, and Barbara K. Abrams, "Reaction of Middle and Lower Class Children to Finger Paints as a Function of Class Differences in Child-Rearing Practices," *Journal of Abnormal and Social Psychology* 51(1955):439–85.

This study compared behavior during finger painting in 18 middle-class and 18 lower-class nursery school children. It was assumed, with no supporting data, that the middle-class children had had more severe toilet training. It was further hypothesized, using psychoanalytic reasoning, that the middle-class children would be more uncomfortable getting messy during the finger painting. The middle-class children did indeed "show a lower tolerance for getting dirty, for staying dirty, and for the products they produce while dirty . . ." However, since data on actual toilet-training practices was not obtained, and since the authors failed to consider alternative explanations such as middle-class parents telling their children not to get dirty more often than lower-class parents, the study is worthless as a test of Freud's theory.

Arnold Bernstein, "Some Relations Between Techniques of Feeding and Training During Infancy and Certain Behavior in Childhood," *Genetic Psychology Monographs* 51(1955):3–44.

In a well-baby clinic in New York City, 47 children ages 4 to 6 were selected at random and asked to play with finger paints, cold cream, and dolls. Data was also collected from the mothers on the history of toilet training and constipation in the children. "No relationship was found between coercive toilet training and collecting, constipation or response to smearing tests."

Robert R. Sears, Eleanor E. Maccoby, and Harry Levin, *Patterns of Child Rearing* (Evanston, Ill.: Row, Peterson and Co., 1957).

A group of 379 Boston-area children were closely followed from birth to kindergarten. Extensive histories were obtained from their mothers, and the children were rated on aggressive behavior. A slight correlation was found between severe toilet training and aggressive behavior. The mothers, however, who used more severe toilet training also were found to be stricter in other ways, including more use of physical punishment, less tolerance of disobedience, greater demand for neatness and orderliness, et cetera. The authors thus concluded that a mother who used severe toilet training was also likely to demand orderliness; the strict mother would thus be the cause of both severe toilet training *and* orderliness as a personality characteristic rather than the toilet training per se causing the orderliness. "The so-called anal character," the authors noted, "could be a product of direct training."

A follow-up study was done on these children at age 12. There were found to be "minor indications of a positive relation between severe toilet training and later nursery school aggression" (Sears et al., *Identification and Child Rearing*, 1965, p. 129). Unfortunately, anal personality traits were not assessed. Furthermore, it may be questioned whether aggression is a valid measure of the effects of toilet training, since aggression is a personality trait said by Freudian theory to be intimately related to the oral, not the anal, stage of development.

Halla Beloff, "The Structure and Origin of the Anal Character," *Genetic Psychology Monographs* 55(1957):141–72.

In Northern Ireland 43 graduate students filled out a questionnaire designed to elicit personality characteristics thought to be related to the anal phase of development. Their mothers were then interviewed regarding the age at which toilet training was completed. It was found that some students did indeed have a cluster of personality characteristics designated as an "anal character"; however, this was "not related to bowel training experiences" but rather to "the degree to which the same components [anal character traits] are exhibited by the mother." The author concluded "that although the anal character is a meaningful dimension of variation for the description of our subjects' attitudes and behavior, it is not related to toilet training experiences, but strongly to the degree of anal character exhibited by the mother."

M. A. Straus, "Anal and Oral Frustration in Relation to Sinhalese Personality," *Sociometry* 20(1957):21–31.

In Sri Lanka 73 third-grade children were given personality tests (including the Rorschach). Information on toilet-training procedures was then obtained from their mothers. No significant correlations were found between the children's personality traits and toilet-training procedures. The study is of questionable validity, however, because the personality tests had not been validated for children in Sri Lanka.

Joseph Adelson and Joan Redmond, "Personality Differences in the Capacity for Verbal Recall," *Journal of Abnormal and Social Psychology* 57(1958):244–48.

Two tests of recall were given to 61 Bennington College freshmen. One test involved a passage about Freud and his theorized stages of psychosexual development; the other test was a passage about New England history. The students were also given the Blacky test (*see* Blum, 1949, as described previously) and were divided into "anal retentives" and "anal expulsives" on the basis of their response to the picture of Blacky defecating between the kennels of his mother and father. It is not at all clear, as Kline noted, "how they derive their hypothesis from psychoanalytic theory" (P. Kline, *Fact and Fantasy in Freudian Theory*, 1972, p. 85). All that can really be concluded from this study is that individuals who can recall a passage about Freudian theory more clearly (quite possibly because they are more interested or had prior exposure to the theory) tend to see the psychoanalytically derived Blacky pictures in a somewhat different way than students who are less interested in Freud but more interested in history. The study would appear to have no bearing on the question of whether toilet-training practices are related to adult personality characteristics.

Mary E. Durrett, "The Relationship Between Reported Early Infant Regulation and Later Behavior in Play Interviews," *Child Development* 30(1959):211–216.

The frequency of aggression was measured in two standard doll play interviews in 60 4- and 5-year-old upper-middle-class children in Tallahassee, Florida. Information on toilet training of the children was obtained from the mothers. The measures of aggression were then compared to the history of toilet training. "None of the correlation coefficients between aggression and early regulation was significant," and in fact none of those involving toilet training were even in the direction predicted by Freudian theory.

As noted above, however, aggression is not a valid measure of anal traits according to Freudian theory.

Daniel R. Miller and Guy E. Swanson, *Inner Conflict and Defense* (New York: Henry Holt and Co., 1960).

In Detroit 104 junior high school boys were assessed on a variety of measures to identify the defense mechanisms which they used. Information was obtained from their mothers on child-rearing practices, including toilet training. No correlations were noted between the use of defense mechanisms (e.g., denial) and toilet training. Even if correlations had been found, it is unclear how it would have supported Freud's theory.

Frank Pederson and David Marlowe, "Capacity and Motivational Differences in Verbal Recall," *Journal of Clinical Psychology* 16(1960):219–22.

Using 70 Ohio State University students, this study was an attempt to replicate the Adelson and Redmond (1958) study described above and, as such, incorporated all the shortcomings of that study. The results it obtained were the opposite of the Adelson and Redmond study, lending further support to the impression that both studies are worthless.

Joseph C. Finney, "Maternal Influences on Anal or Compulsive Character in Children," *Journal of Genetic Psychology* 103(1963):351–67.

In Hawaii 31 boys referred to a Child Guidance Clinic were rated on "anal character," stubborness, and submissiveness by clinic staff and teachers. The study was seriously flawed methodologically and did not gather information on toilet training. The results confirmed the validity of a constellation of traits referred to as "anal character." A strong association was found between these traits in the boys and rigidity in their mothers.

E. M. Hetherington and Yvonne Brackbill, "Etiology and Covariation of Obstinacy, Orderliness and Parsimony in Young Children," *Child Development* 34(1963):919–43.

In Newark 35 kindergarten children were given 10 behavioral tests to assess personality traits of obstinacy, orderliness, and parsimony. Information on toilet training (age of starting, age of completion, and severity) was obtained from their mothers. No correlations were found and the authors concluded "that analytic theory is not correct in maintaining that too early, too late, or too severe training leads to high degrees of obstinacy, orderliness, and parsimony. . . . It is time for psychoanalysis to reconsider its adamant perpetuation of this aspect of its theory." Information was also gathered on these same personality traits in the children's mothers and fathers, and the authors found a "remarkable similarity in pattern between boys and fathers and between girls and mothers."

Robert R. Sears, Lucy Raul, and Richard Alpert, *Identification and Child Rearing* (Stanford: Stanford University Press, 1965).

In yet another attempt to correlate children's aggression with experiences of infancy, 40 children at Stanford University were assessed for aggression and the data was compared with information obtained from their mothers on infant experiences, including "severe and demanding toilet training." For boys only there was a correlation (-.47) between severity of toilet training and "tattling" as one of the 14 measures of aggression. The authors concluded that "the actual number of significant correlations [in the study] is hardly beyond what can be accounted for by chance. . . . The evidence is certainly heavily against any interpretation that infant frustration per se is an important antecedent [of aggression]."

Edward Gottheil and George C. Stone, "Factor Analysis of Orality and Anality," *Journal of Nervous and Mental Disease* 146(1968):1–17.

A group of 179 young men (ages 17 to 26) were given a questionnaire as they were being processed into military service. The questionnaire contained items about bowel habits as well as items designed to measure "anal character" traits. No information on actual toilet training was obtained from the mothers in this study. Using factor analysis it was found that there was a clustering of personality traits consistent with the "anal character," but it was rather weak; "the oral and anal trait factors together accounted for only 5.3 percent of the total variance." No association was found between anal traits and bowel habits. The authors concluded: "On the basis of these findings, it would seem reasonable to wonder about the place of the mouth and the anus in the concepts and theory of the oral and anal character types."

Jerry S. Wiggins, Nancy Wiggins, and Judith C. Conger, "Correlates of Heterosexual Somatic Preference," *Journal of Personality and Social Psychology* 10(1968):82–90.

At the University of Illinois 95 male students in an introductory psychology course filled out questionnaires "as a course requirement." One questionnaire asked them to rate their preference for nude silhouettes of women with varying sized breasts, buttocks, and legs. Other questionnaires rated personality traits and habits. No information was collected on toilet training or other childhood experiences. Preference for various anatomical regions on the women's silhouettes were then correlated with personality traits and habits.

The results showed some self-explanatory correlations (e.g., students who date frequently and read *Playboy* preferred women with large breasts) as well as many which appeared to be random (e.g., business majors preferred women with large buttocks). Size of buttocks elicited fewer correlations than size of breasts or legs. Students who scored high on the trait of orderliness on the Edwards Personal Preference Schedule preferred women with large buttocks; the authors state that this correlation is consistent with Freudian theory, but it appears to be no more than chance. The author's conclusion from this study surely ranks among the classic scientific endeavors of psychoanalysis: "Within the limitations of the present design, it may be concluded that the female body parts of breasts, buttocks and legs are important determinants of male heterosexual somatic preference."

Paul Kline, "Obsessional Traits, Obsessional Symptoms and Anal Erotism," *British Journal of Medical Psychology* 41(1968):299–305.

In England 46 "volunteer students" (not otherwise described) were administered 2 questionnaires measuring obsessional personality traits, 2 questionnaires measuring anal personality traits, and then were asked to react to the Blacky cartoon (*see* Blum, 1949, as described previously) of a dog defecating between the kennels of its father and mother. No information on toilet training was obtained because the author considered such data unreliable. The author found some correlation between obsessional personality traits and anal personality traits (orderliness, parsimony, and obstinacy) as well as correlations with reactions to the Blacky cartoon. Since questions on one of the anal trait questionnaires included such things as

"Do you regard the keeping of household dogs as unhygienic?" such correlations are hardly surprising and have no bearing on Freud's hypothesized relationship of personality traits to toilet training. The most that can be said for this study is that it demonstrates that students who are concerned with cleanliness are also concerned with orderliness and react more strongly to a cartoon of a dog defecating where it should not defecate.

Paul Kline, "The Anal Character: A Cross-Cultural Study in Ghana," *British Journal of Social and Clinical Psychology* 8(1969):201–10.

In Ghana 123 university students were given a questionnaire designed to measure "anal character" (e.g., "Do you keep careful accounts of the money you spend?"). They scored significantly higher than counterparts in British universities, which the author admits was probably due to the more stringent university entrance requirements in Ghana—that is, only the hardest working and best organized students are likely to be admitted. Information on toilet training in Ghana proved to be impossible to obtain, and in a later publication the author says that his investigation "must be regarded as a failure" (P. Kline, *Fact and Fantasy in Freudian Theory*, 1972, p. 59).

Tupper F. Pettit, "Anality and Time," *Journal of Consulting and Clinical Psychology* 33(1969):170–74.

At New York University 91 students in a psychology course were asked to fill out 3 questionnaires measuring anal personality traits and one questionnaire measuring concerns about time and punctuality. Since the anal personality questionnaires included items about punctuality and the time questionnaire contained items pertaining to anal personality traits (e.g., "I feel I am more conscientious about being on time than most people"), a correlation of the time questionnaire with the questionnaires measuring anal personality was inevitable. The author concluded that "this study gives considerable support to the theoretical relations between time and anality," which is simply to say that orderliness and punctuality are related.

Seymour Fisher, *Body Experience in Fantasy and Behavior* (New York: Appleton-Century-Crofts, 1970).

Male college students were given a Body Focus Questionnaire designed to ascertain what parts of their body they were most consciously aware of. They were also given a series of personality tests (including Blum's Blacky Test, previously discussed) to measure personality characteristics. Fisher found that the students who showed a high degree of "back awareness" showed greater "sensitivity to stimuli with anal connotations, negative attitudes toward dirt, [and] measures of self-control and orderliness." No information on toilet training was obtained. It would appear, therefore, that Fisher merely verified a cluster of personality traits consistent with the "anal character" and that college students with these traits had a higher "back awareness," possibly because they were aware of Freudian theory and responded to the Body Focus Questionnaire in a manner which they believed was expected of them. Since no **information on toilet training was obtained, the research has no direct bearing on Freud's theory.**

Notes

Preface

"The treatment of the id by the odd." E. M. Thornton, *The Freudian Fallacy: An Alternative View of Freudian Theory* (Garden City, N.Y.: Dial Press, 1984), 244.

"biographical truth is not to be had." Paul Roazen, *Freud and His Followers* (New York: New York University Press, 1984), 12.

1.
Freud, Social Reform and Sexual Freedom

"If I had my life . . . psychoanalysis." Ernest Jones, *The Life and Work of Sigmund Freud*, 3 vols. (New York: Basic Books, 1957), 3:392. The three volumes of the Jones biography were published in 1953, 1955 and 1957, respectively.

Emma Goldman. Unless otherwise noted, the information on Emma Goldman is quoted from her autobiography *Living My Life* (New York: Alfred E. Knopf, 1931) and from two recent biographies: Alice Wexler, *Emma Goldman, An Intimate Life* (New York: Pantheon

Books, 1984) and Candace Falk, *Love, Anarchy and Emma Goldman* (New York: Holt, Rinehart and Winston, 1984).

"an eminent young professor." Goldman, *Living My Life*, 173.

"something new and wonderful . . . my own." Wexler, *Emma Goldman, An Intimate Life*, 36.

"the killing . . . of a life." Ibid., 63.

"It was Freud . . . homosexuality." Ibid., 295; quoting a 1929 letter from Goldman to Beckman.

"For the first time . . . my own needs." Goldman, *Living My Life*, 173.

"I always felt . . . revulsion." Wexler, *Emma Goldman, An Intimate Life*, 22.

"His simplicity and earnestness . . . into broad daylight." Goldman, *Living My Life*, 173. It is unclear from Goldman's biography whether she actually attended the lectures given by Freud on October 14, 21, and 28, 1895 or heard him give a subsequent lecture in 1896 as stated by Alice Wexler, one of Goldman's biographers (Wexler, 48).

"Only people . . . as Freud." Ibid.

"plump, demure . . . in white."
Nathan Hale, *Freud and the Americans* (New York: Oxford University Press, 1971), 5, 269. The fact that Goldman occupied a front-row seat is mentioned by Fred H. Matthews, "Freud Comes to America" (M.A. thesis, University of California at Berkeley, 1957), 10.

"he stood out . . . among pygmies." Goldman, *Living My Life*, 455. There is no evidence that Goldman ever underwent a personal psychoanalysis; in fact, she once labeled the process as "nothing but the old confessional" (Wexler, 295); however, many of her closest friends did so, and one of them, André Tridon, practiced as a lay analyst (Hale, 327).

title of Professor. Jones, *Freud*, 1:339.

"in those days . . . name was mentioned." Max Graf, "Reminiscences of Professor Sigmund Freud," *Psychoanalytic Quarterly* 11 (1942):465–76.

"voluntary . . . coitus interruptus." Sigmund Freud, "My Reply to Criticisms on the Anxiety Neurosis" (1895) in *The Complete Psychological Works of Sigmund Freud*, James Strachey, ed., (London: Hogarth Press, 1966) 1:108. For additional Freud quotations from his 1895 lectures, see Percival Bailey, *Sigmund the Unserene* (Springfield: Charles Thomas, 1965), 17.

"the unique significance . . . established." Sigmund Freud, "My Views on the Part Played by Sexuality in the Aetiology of the Neuroses" (1905), in *Collected*

Papers (New York: International Psycho-analytic Press, 1924) 1:272–83.

"The energy of the sexual . . . in the neurosis." Sigmund Freud, *Three Contributions to the Theory of Sex* in *The Basic Writings of Sigmund Freud*, Abraham A. Brill, ed., (New York: Modern Library, 1938), 573. Freud's essay was first published in 1905.

"no neurosis . . . *vita sexualis*." Freud, "My Views on the Part Played by Sexuality in the Aetiology of the Neuroses" (1905), in *Collected Papers*, 1:272.

"I stand for . . . sexual life." Freud to James J. Putnam, 1915, in J. C. Burnham, "Psychoanalysis and American Medicine, 1894–1918," *Psychological Issues* 5 (1967): 1–249.

"the misuse of the sexual function." Jeffery M. Masson, *The Assault on Truth: Freud's Suppression of the Seduction Theory* (New York: Farrar, Straus and Giroux, 1984), 74, quoting Wilhelm Fliess, "The Nasal Reflex Neurosis," published in 1893.

"The Nose and Female Sexuality." Freud to Wilhelm Fliess, 8 October 1895 in *The Complete Letters of Sigmund Freud to Wilhelm Fliess, 1887–1904* Jeffery M. Masson, ed. (Cambridge: Harvard University Press, 1985), 141.

"one-sided facial spasm." Freud to Fliess, 12 June 1895, Ibid., 131.

"neuralgic stomach pain." Masson, *Assault on Truth*, 77.

"stomach ailments and menstrual problems." Ibid., 57.

"At least half a meter of gauze."

Ibid., 62.

"her face . . . in." Ibid., 70.

"hemorrhages were . . . longing." Ibid., 67.

"holding in your hands . . . anything." Freud to Fleiss, 22 June 1895, in *The Complete Letters*, 133.

"great clinical secret . . . memories." Freud to Fliess, 15 October 1895, Ibid., 144. See also the letter of October 8, 1895.

"I consider the two neuroses . . . conquered." Freud to Fliess, 16 October 1895, Ibid., 145.

"*topical* around . . . several decades." Lancelot L. Whyte, *The Unconscious Before Freud* (New York: Basic Books, 1960), 169–70.

"Every extension of knowledge . . . the unconscious." Ibid., 176. Whyte cites several of Nietzsche's thoughts on the unconscious written between 1876 and 1888.

Freud belonged to a Reading Society. Frank J. Sulloway, *Freud, Biologist of the Mind: Beyond the Psychoanalytic Legend* (New York: Basic Books, 1979), 468.

[Freud] was indebted to Nietzsche. Whyte, *The Unconscious Before Freud*, 175.

"the imaginary fulfillment of wishes." Sulloway, *Freud, Biologist of the Mind*, 324.

Karl Scherner. Ibid., 325.

"the claim that dreams . . . successive personalities." Sulloway, *Freud, Biologist of the Mind*, 322–23. See also Henri Ellenberger, *The Discovery of the Unconscious* (New York: Basic Books, 1970), 303–11.

"my views concerning . . . infantilism." Freud, *Collected Papers*, 1:272–83.

"he dreamed . . . most mighty." Lionel Trilling, *Freud and the Crisis of Our Culture* (Boston: Beacon Press, 1955), 31.

"Freud was continually preoccupied . . . ten years." Sulloway, *Freud, Biologist of the Mind*, 25.

"I have almost finished . . . seeing them go astray." Freud to Martha Bernays, 28 April 1885 in *The Letters of Sigmund Freud*, ed. Ernst L. Freud (New York: McGraw Hill, 1964), 140–41.

"the myth of the hero" and ". . . heroic destiny." Sulloway, *Freud, Biologist of the Mind*, 36, 42, and 476 ff. This aspect of Freud's character is also explored by Percival Bailey in *Sigmund the Unserene* (e.g., p. 36 where Bailey says that Freud "was not always careful to tell the whole truth").

"We are certainly . . . from afar." Freud to Carl Jung, 17 January 1909 in *The Freud/Jung Letters* ed. William McGuire (Princeton: Princeton University Press, 1974), 195–96.

replica of the statue. Ronald W. Clark, *Freud: The Man and the Cause* (New York: Random House, 1980), 358.

"every day . . . drawing it." Ibid.

"Sometimes I have crept . . . the Ten Commandments." Earl A. Grollman, *Judaism in Sigmund Freud's World* (New York: Bloch Publishing Co., 1965), 110, quoting Freud's essay "The Moses of Michelangelo," published anonymously in 1914 in *Imago*.

"had emotional reasons for identifying . . . predecessor." Jones, *Freud*, 3:368.

"there can be little doubt . . . with Moses." Reuben Fine, *A History of Psychoanalysis* (New York: Columbia University Press, 1979), 266.

"one cannot avoid . . . Moses." Peter Gay, *Freud: A Life For Our Time* (New York: W. W. Norton, 1988), 317 f.

"a kind of psychical counterpart to wireless telegraphy." Sigmund Freud, "Dreams and Occultism," *New Introductory Lectures on Psychoanalysis, Standard Edition*, vol. 22 (London: Hogarth Press, 1964), 36.

"often heard . . . her voice." Jones, *Freud*, 3:380.

"a quite peculiarly intimate . . . daughter." Jones, *Freud*, 3:224.

"he referred to . . . correspondence." Paul Roazen, *Freud and His Followers* (New York: New York University Press, 1984), 390.

"magical actions . . . averting disaster." Ibid., 382.

"found himself . . . his child's life." Ibid.

"a revolutionary difference . . . of psychoanalysis." Ibid., 386.

"brought a telepathist to a meeting." Ibid., 233.

"if I had my life . . . psychoanalysis." Freud to Hereward Carrington, 1921, quoted in Jones, 3:392.

his experiments with cocaine. Robert Byck, ed., *Cocaine Papers: Sigmund Freud* (New York: New American Library, 1974).

"a magical drug . . . lasting euphoria." Jones, *Freud* 1:78–97.

"fortified himself with . . . cocaine." Gay, *Freud: A Life for Our Time*, 50.

"Woe to you, my Princess . . . this magical substance." Freud to Martha Bernays, 2 June 1884, in Byck, 10–11.

"the third scourge of humanity." Jones, 1:94, quoting Erlenmeyer.

Freud–Fliess letters. The following quotations are from Masson, *The Complete Letters*.

E. M. Thornton. *The Freudian Fallacy: An Alternative View of Freudian Theory* (Garden City, N.Y.: Dial Press, 1984).

Studies of cocaine abusers. Dale D. Chitwood, "Patterns and Consequences of Cocaine Use," in *Cocaine Use in America: Epidemiologic and Clinical Perspectives*, eds. Nicholas J. Kozel and Edgar H. Adams, NIDA Research Monograph 61 (Washington: Government Printing Office, 1985), 125.

"now suffering painfully . . . in Breslau." Freud to Fliess, 12 December 1897, in Masson, *The Complete Letters*, 285.

"cardiac weakness . . . pain." Freud to Fliess, 27 September 1899. Ibid.

Freud scholar Peter Swales: Peter J. Swales, "Freud, Cocaine and Sexual Chemistry: The Role of Cocaine in Freud's Conception of the Libido," in *Sigmund Freud: Critical Assessments*, vol. 1, Laurence Spurling, ed. (London: Routledge, 1989).

"I don't think he . . . for 15 years." Ernest Jones to Siegfried Bernfeld, 3 May 1952. This letter and the two letters following are in the Siegfried Bernfeld Collection in the Library of Congress and are quoted with the kind permission of Mr. Mervyn Jones.

"I am . . . not mentioning that." Jones to Siegfried Bernfeld, 28 April 1952.

"Before . . . everyone he met." Jones to Siegfried Bernfeld, 9 May 1952.

"by the time he published . . . in place." Gay, *Freud: A Life for Our Time*, 103–4.

"an historic moment." Jones, *Freud*, 3:354.

"dream specimen." Fine, 28, quoting Erik Erikson.

"Since, however . . . . such discoveries." Sigmund Freud, *The History of the Psychoanalytic Movement* (New York: Collier Books, 1963), 55. First published in 1914.

"a certain lewd . . . obscene label." Hale, *Freud and the Americans*, 260.

"only to physicians and lawyers." Ibid.

"In reality the new-born . . . childhood." Havelock Ellis, *Studies in the Psychology of Sex*, vol. 6 (Philadelphia: F. A. Davis, 1910), 36.

"Immoral . . . time and place." Hale, *Freud and the Americans*, 62.

"semi-detached" marriages. Oscar Cargill, *Intellectual America* (New York: Macmillan, 1941), 620.

"laissez-faire sexual morality." Cargill, *Intellectual America*, 617.

Margaret Sanger. David M. Kennedy, *Birth Control in America: The Career of Margaret Sanger* (New Haven: Yale University Press, 1970), 13.

Dr. William J. Robinson. See Hale, *Freud and the Americans*, 271, quoting Robinson's essay in his *American Journal of Urology and*

Sexology, 1915.

"The chief sin of the world . . . literally true." Hale, *Freud and the Americans*, 251.

"but he had to . . . at the door." John C. Burnham, "Psychoanalysis in American Civilization Before 1918," Ph.D. diss., Stanford University, 1958, 317. Quoted with the kind permission of Professor John C. Burnham.

"barbaric and bestial proclivities." Ibid., 214.

"Viennese libertine." *New York Times*, undated citation by Catherine L. Covert, "Freud on the Front Page: Transmission of Freudian Ideas in the American Newspaper of the 1920's," (Ph.D. diss., Syracuse University, 1975), 271. Quoted with the kind permission of Ms. Carolyn S. Holmes, executrix of estate of Catherine L. Covert, 1991.

"worship of Venus and Priapus." Hale, *Freud and the Americans*, 300.

"a direct invitation." Ibid.

"I am beginning . . . medicine." William S. Sadler, *Worry and Nervousness* (Chicago: McClurg, 1914), 357.

"I also think . . . too great." Freud letter to Carl Jung, 17 January 1909, in McGuire, *The Freud/Jung Letters*, 195.

"The true meaning of a dream . . . to digestion." H. L. Mencken in *Smart Set* 35 (1911): 153–55, quoted in Matthews, 43.

"To Freudian writers . . . generally speaking." J. V. Haberman, "A Criticism of Psychoanalysis," *Journal of Abnormal Psychology* 9 (1914): 265–80.

"an ingeniously obscene imagination." Warner Fite, "Psycho-Analysis and Sex Psychology," *Nation* 10 Aug. 1916, 127–29.

"peddler of pornography." Hale, *Freud and the Americans*, 310.

"But this is . . . police court." C. Ladd Franklin, "Freudian Doctrines," *Nation* 19 Oct. 1916, 373–74.

"of such vile nature . . . its value." Charles Bruehl, "Psychoanalysis," *Catholic World* Feb. 1923, 577–89.

"lovers took to urging . . . neurotic." Cargill, *Intellectual America*, 608.

"people gave freer . . . to inhibit." Sidney Ditzion, *Marriage, Morals and Sex in America* (New York: Bookman Associates, 1953), 361.

"the idea . . . and the illicit." Edwin B. Holt, *The Freudian Wish and Its Place in Ethics* (New York: Henry Holt and Company, 1915), vi.

"psychoanalysis and sex . . . identical." Clarence P. Oberndorf, *A History of Psychoanalysis in America* (New York: Grune and Stratton, 1953), 134–35.

"the element of sex . . . the sex element." Robert S. Woodworth, "Some Criticisms of Freudian Psychology," *Journal of Abnormal Psychology* 12 (1917): 174–94.

"I have devoured . . . this sort." Ibid.

"Yes, I am of course . . . against repressed desires." Hale, *Freud and the Americans*, 300.

Puritans were "sexually abnormal." Covert, 273, quoting André Tridon.

"that she could not tolerate . . . herself." Sigmund Freud, "Wild Psychoanalysis" (1910) in Strachey, *Works of Sigmund Freud*, 221–227.

"psychoanalysis . . . disorders." Ibid.

"higher and nobler knowledge." Hale, *Freud and the Americans*, 416, quoting the *New York Times* 12 Feb. 1916, 6 and 14 Feb. 1916, 12.

"psychoanalysis was . . . morals." Ibid.

"I thought it . . . tender love." Lavinia Edmunds, "His Master's Choice," *Johns Hopkins Magazine*, April 1988, 40–49. The case involved Dr. Horace Frink, one of Freud's favorite followers and who had clear evidence of manic-depressive psychosis at the time Freud was urging him to marry his former patient.

"to have electrified many physicians." Barbara Sicherman, "The Quest for Mental Health in America, 1880–1917" (Ph.D. diss., Columbia University, 1967), 196.

medical dissertation. Burnham, "Psychoanalysis in American Civilization Before 1918." In 1894 William James had noticed a paper on hysteria by Freud and Josef Breuer, but between then and 1906 there was virtually no mention of Freud in the American medical literature.

"a nobler self." Matthews, "Freud Comes to America," 57.

"idealist ethic . . . Divine Purpose." Ibid., 54.

"man's spiritual . . . Infinite." Ibid., 55.

"a decorative . . . touch." Ibid.

"glowing . . . bewilderment." Ibid., 56.

"with an attempt . . . of God." Burnham, "Psychoanalysis in American Civilization Before 1918," 303.

"The National Society for the . . . Insane." Clifford Beers to William James, 16 April 1907 in *The Inner World of American Psychiatry 1890–1940: Selected Correspondence*, ed. Gerald N. Grob, (New Brunswick: Rutgers University Press, 1985), 143.

"levied one . . . ever been heard." Gerald N. Grob, *The State and the Mentally Ill* (Chapel Hill: University of North Carolina Press, 1966), 265.

"mental hygiene . . . justify physical hygiene." Sicherman, 332.

"the moral . . . ventilation." Ibid., 365.

"all forms of . . . unhappiness." Ibid., 333.

White had urged reforms. Ibid., 364.

"psychic infections." Ibid., 367, quoting Salmon in 1912.

"dependent . . . basis for psychoses." Ibid.

"practically all the hopeful . . . schools." Ibid., 279.

"an ardent . . . Catholic priest." Abraham A. Brill, "Professor Freud and Psychiatry," *Psychoanalytic Review* 18 (1931): 241–46.

"a barren . . . haphazard therapy." Abraham A. Brill, A Psychoanalyst Scans His Past," *Journal of Nervous and Mental Disease* 95 (1942): 537–49.

"they are doing that Freud stuff." Abraham A. Brill, "The Introduction and Development of Freud's Work in the United States," *American Journal of Sociology* 45 (1939): 318–25.

"worked heart and soul . . . to psychiatry." Ibid.

"a fine collection . . . of sex practices." "Brands for the Burning," *Time*, 27 Jan. 1941, 30.

"seemed even more . . . American analysts." Burnham, "Psychoanalysis in American Civilization Before 1918," 149.

"invariably." Ibid., 68.

"no neurosis . . . *vita sexualis*." Freud, *Collected Papers* 1:272.

"pure air and food." Hale, *Freud and the Americans*, 396.

"the urge is there . . . itself." John D'Emilio and Estelle B. Freedman, *Intimate Matters: A History of Sexuality in America* (New York: Harper and Row, 1988), 223.

"Brill conspicuously . . . to be sexual." Burnham, "Psychoanalysis and American Medicine, 1894–1918," 110.

"frank insistence . . . in risqué jokes." Hale, *Freud and the Americans*, 392.

"the most important . . . Declaration of Independence." Justin Kaplan, *Lincoln Steffens: A Biography* (New York: Simon and Schuster, 1974), 199.

a minimum of six inches. Henry F. May, *The End of American Innocence: The First Years of Our Own Time* (New York: Oxford University Press, 1959), 338.

"cut off from the Village . . . the Id." Milton Klonsky, "Greenwich Village: Decline and Fall," *Commentary* 6 (1948): 461. Quoted with the kind permission of *Commentary*.

Dadaists occasionally picnicked: Raymond Nelson, *Van Wyck Brooks: A Writers Life* (New York: E. P. Dutton, 1981), 95.

"bourgeois pigs." Ibid.

"serenely happy." Max Eastman, *Enjoyment of Living* (New York: Harper and Brothers, 1948), 253.

"delivered through a serrated gold crown." Hale, *Freud and the Americans*, 246.

"I wanted to believe . . . in mental healing." Eastman, *Enjoyment of Living*, 242.

"read Freud . . ." and ". . . in my make-up." Ibid., 491.

his sister and his mother. Ibid., 317, 344, and 356.

"which I believe . . ." and ". . . free and energetic." Quotes are from Max Eastman's two articles, "Exploring the Soul and Healing the Body" and "Mr. Er-Er-Er-Oh! What's His Name?" *Everybody's Magazine* June 1915, 741–50 and July 1915, 95–103.

"mental cancers . . . will disappear." Ibid.

"The attitude . . . the world." Ibid.

Freud Dell: James B. Gilbert, *Writers and Partisans: A History of Literary Radicalism in America* (New York: Wiley, 1967), 56.

"busy analyzing . . . his hands." Burnham, "Psychoanalysis in American Civilization Before 1918," 266, quoting Sherwood Anderson.

"everyone at that time . . . about it." Frederick J. Hoffman, *Freudianism and the Literary Mind* (Baton Rouge: Louisiana State University Press, 1957), 58.

"frustrated excitement." Hale, *Freud and the Americans*, 309.

"change a thin . . . contented mate." Ibid.

"immensely indebted to psychoanalysis." Frankwood E. Williams, *Proceedings of the First International Congress on Mental Hygiene* (New York: The International Committee for Mental Hygiene, 1930), 620.

"a new view . . . supplementing it." Floyd Dell, *Homecoming* (New York: Farrar and Rinehart, 1933), 293–94.

"prophet of the new liberalism." Charles Forcey, *Crossroads of Liberalism* (New York: Oxford University Press, 1961), 108. For other accounts of Lippmann's early years, see Edward L. Schapsmeier and Frederick H. Schapsmeier, *Walter Lippmann: Philosopher-Journalist* (Washington: Public Affairs Press, 1969) and Stow Persons, *American Minds* (New York: Henry Holt and Co., 1958).

"as . . . *The Origin of the Species.*" Gay, *Freud: A Life for Our Time*, 458.

"The greatest advance . . . of human character." Walter Lippmann, *A Preface to Politics* (New York: Macmillan, 1913), 85.

"the suffrage movement . . . in the human psyche." Forcey, *Crossroads of Liberalism*, 112.

"I cannot help feeling . . . to thought." Walter Lippmann, "Freud and the Layman," *New Republic* 2 (1915): 9–10.

"a few years . . . of the Psychoanalytic Society." Ronald Steel, *Walter Lippmann and the American Century* (Boston: Little Brown, 1980), 48.

"I wish Walter Lippmann . . . a little." Ibid., 173.

"in continental Europe . . . may be dispensed with." Abraham A. Brill, "The Psychopathology of the New Dances," *New York Medical Journal* 49 (1914): 834–37.

"Socialists . . . Modern Artists." Gilbert, *Writers and Partisans*, 26.

"display . . . psychology of Sigmund Freud." Schapsmeier, *Walter Lippman*, 4.

"aroused a . . . lively discussion." Brill, "The Introduction and Development of Freud's Work," 318–25.

"several . . . give-aways." Mabel D. Luhan, *Movers and Shakers* (New York: Harcourt Brace and Co., 1936), 142.

". . . a large, soft, overripe Buddha . . . at his feet." Joseph R. Conlin, *Big Bill Haywood and the Radical Union Movement* (Syracuse: Syracuse University Press, 1969), 87.

"The questions . . . by medical men." Brill, "The Introduction and Development of Freud's Work," 318–25.

"a pioneer in the cult of the orgasm." Christopher Lasch, *The New Radicalism in America: The Intellectual as a Social Type* (New York: Alfred A. Knopf, 1965), 118. See also 124.

"the woman . . . young clerks." Luhan, *Movers and Shakers*, 509.

"revolution . . ." and ". . . to pathology." "Dreams of the Insane," *New York Times*, 2 March 1913, sec. 5:10. The article is not signed but is identical in style and content to the one which Kuttner did sign the following year.

"Freud and . . . Revealing them." Ibid.

A year later Kuttner wrote. Alfred Kuttner, "What Causes Slips of the Tongue," *New York Times*, 18 Oct. 1914, sec. 5:10.

"mind was . . . as a neurotic." Peter C. MacFarlane, "Diagnosis by Dreams," *Good Housekeeping*, Feb. 1915, 125–33, March 1915, 278–86.

"This . . . about sex." Ben Hecht, *Gaily, Gaily* (New York: Doubleday and Company, 1963), 65.

"to liberate . . . our literature." Bernard DeVoto, "Freud in American Literature," *Psychoanalytic Quarterly* 9 (1940): 236–45.

"came to me . . . about it." Burnham, "Psychoanalysis in American Civilization Before 1918," 268.

"I gave them . . . or imagined." Brill, "The Introduction and Development of Freud's Work," 318–25.

Dreiser read books recommended by Brill. For a description of the Dreiser-Brill friendship see W. A. Swanberg, *Dreiser* (New York: Charles Scribner's Sons, 1965), 234, 271, and 287.

"began a series . . . by a stenographer." Ibid., 399.

he asked Brill for a letter. Ibid., 311–12.

"was always entangled . . . with courage and adventure." Ibid., 246.

"that the conflicts . . . and creativity." Ibid., 295.

"strong, revealing light . . . and my work." W. David Sievers, *Freud on Broadway: A History of Psychoanalysis and the American Drama* (New

York: Hermitage House, 1955), 68.

objected to one of his affairs. Swanberg, *Dreiser*, 320.

"The novel . . . as a baby." May, *The End of American Innocence*, 294.

"sickened and . . . grew perverse." Frederick J. Hoffman, *The Twenties* (New York: The Free Press, 1949), 30–31.

"one of the heroes of modern thought." Hoffman, *Freudianism and the Literary Mind,* 256.

"as a kind of therapy for Oppenheim." See May, 325 and James Hoopes, *Van Wyck Brooks: In Search of American Culture* (Amherst: University of Massachusetts Press, 1977), 136.

Oppenheim himself later became. Hale, *Freud and the Americans*, 327.

"Three-quarters of the poetry . . . Freudian in content." DeVoto, "Freud in American Literature," *Psychoanalytic Quarterly* 9 (1940): 236–45.

"decided very early . . . as I could." Hoffman, *Freudianism and the Literary Mind*, 280.

"Oh, just one . . . soupçon of jealousy." Frederick L. Allen, *Only Yesterday: An Informal History of the 1920s* (New York: Harper and Row, 1931), 76, quoting F. Scott Fitzgerald's *This Side of Paradise*.

Arthur Hopkins. Sievers, *Freud on Broadway*, 46–47.

"a minor character . . . 'love-cracked.' " Ibid., 50.

"an ingenious . . . giddy faddist." Ibid., 53–54.

"You could not go . . . hearing of someone's complex." Albert Parry, *Garrets and Pretenders: A History of*

Bohemianism in America (New York: Dover Publications, 1960), 278. First published in 1933.

"I beg your pardon . . . in such a way." Hale, *Freud and the Americans*, 429.

"the psychoanalytic era." Sievers, *Freud on Broadway*, 65.

"dominating . . . his sexual partner." Ibid., 77.

"With flasks . . . and individual freedom." Ibid., 79.

"a great force." Ibid., 67.

"If there is . . . Dr. Freud." Alfred Kazin, *On Native Grounds* (New York: Reynal and Hitchcock, 1942), 194.

O'Neill was in psychoanalysis. Sievers, *Freud on Broadway*, 116. O'Neill's analyst is said to have been Dr. Gilbert V. Hamilton who published a study of sexual practices in marriage called *A Research on Marriage* in 1929 (*see* Appendix B); O'Neill was therefore almost certainly one of the subjects in this study.

"O Oedipus . . . is adopting you." Sievers, *Freud on Broadway*, 117.

"As soon as I . . . speak on problems of sex." Brill, "The Introduction and Development of Freud's Work," 318–25.

"we could reduce . . . smallpox and typhoid." Hale, *Freud and the Americans*, 351.

"unshackle our libido." Oberndorf, *A History of Psychoanalysis*, 134.

"in extension classes . . . New York area." Hale, *Freud and the Americans*, 400.

"smoking is . . . oral eroticism." Edward L. Bernays, *Biography of An Idea: Memoirs of Public Relations*

Counsel (New York: Simon and Schuster, 1965), 386–87. Bernays was also related to Freud through Freud's wife because Bernay's father was the brother of Freud's wife. Freud's sister and his wife's brother had married three years before Freud himself did and had then emigrated to New York City in 1892 when Edward Bernays was one year old.

"the first . . . to advertising." Ibid., 395.

Dr. Brill . . . and the legalization of abortions. See letter to Brill from Abraham L. Wolbarst, 21 December 1943, Brill Collection, Library of Congress.

"an important mode of . . . forbidden wishes." Hale, *Freud and the Americans*, 474.

"Puritan prudery . . . our present social system." Brill, "The Psychopathology of the New Dances," 834–37.

"two timid . . . matrimony." Ibid.

"sympathetic to the . . . pre-war period." Hale, *Freud and the Americans*, 348.

Trotsky "wrote intelligently about Freud." Daniel Aaron, *Writers on the Left* (New York: Harcourt Brace and World, 1961), 46.

"both contemplate . . . harmony." Fine, *History of Psychoanalysis*, 447, quoting Eastman's *Marx, Lenin and the Science of Revolution*, 101.

"smells . . . of sexual novelties." Louis Hartz, *The Liberal Tradition in America* (New York: Harcourt Brace Jovanovich, 1955), 243.

"determine the . . . guest list." Stephen J. Whitfield, *A Critical American: The Politics of Dwight*

Macdonald (Hamden, CT: Shoe String Press, 1984), 15.

guests included such notables. Margaret Sanger, *Margaret Sanger: An Autobiography* (New York: W. W. Norton, 1938), 70.

"a psychoanalytic revolution . . . Golden Age." Fritz Wittels, "Brill—the Pioneer," *Psychoanalytic Review* 35 (1948): 397.

"the emphasis placed." Burnham, "Psychoanalysis in American Civilization Before 1918," 348.

"grew quickly . . . among intellectuals." Hale, *Freud and the Americans*, 476.

"around 1916 . . . Havelock Ellis." Ibid., 475.

"the speakeasy usually . . . as well." Allen, *Only Yesterday*, 82.

"an air . . . rather dashing and desirable." Ibid., 95.

"The closed car . . . and chaperones." Ibid., 83.

"house of prostitution on wheels." Ibid.

"a record of . . . magazine publishing." Ibid., 84.

Pulitzer Prize juries. Ibid., 98.

"not only . . . should be continuous." Ibid., 194.

"sex o'clock." D'Emilio and Freedman, *Intimate Matters*, 234.

"Sex is . . . news is news." Frederick L. Allen, *Since Yesterday: The 1930's in America* (New York: Harper and Row, 1939), 134.

"like mah-jongg or miniature golf." Stephen D. Becker, *Marshall Field III* (New York: Simon and Schuster, 1964), 134.

"to preface . . . from Freud." Caroline Ware, *Greenwich Village, 1920–1930* (Boston: Houghton

Mifflin, 1935), 256.

"crested in the . . . Depression." Covert, "Freud on the Front Page," iii.

"the number . . . between 1925 and 1926." Hale, *Freud and the Americans*, 477.

In this author's study. I counted the number of non-duplicated articles about Freud or psychoanalysis in *The Reader's Guide to Periodical Literature* for each year from 1910 to 1988, then divided by the total number of pages for that year to yield a rate per 1000 pages. As reflected by such articles, the years of greatest interest in Freud and psychoanalysis were 1915–1922, 1939 (obituaries and recollections at the time of his death), 1956–1959, and 1974–1976. The periods of lowest interest, all roughly equivalent, were 1930–1935, 1940–1947, and 1980–1988.

"Every issue an Oedipus Complex." E. E. Cummings, "The Tabloid Newspaper," *Vanity Fair*, Dec. 1926, 86.

"who . . . turned on the gas." Hoffman, *Freudianism and the Literary Mind*, 68.

Psychic Psarah: Covert, "Freud on the Front Page," 51, quoting the Dearborn *Independent*, March 29, 1924.

"Doctor Paul Ehrich . . . from an Electra Complex." Hoffman, *The Twenties*, 231.

"a species of voodoo . . . human sacrifices." Morris Fishbein, *The New Medical Follies* (New York: Liveright, 1927), 197.

"find some . . . speak." Ibid., 199.

"as a candidate for oblivion." Covert, "Freud on the Front Page," 283.

"psychoanalysis has . . . exploitation." Anonymous, "Farewell to Freud," *Commonweal* 17 (1933): 452.

"the psychoanalytic . . . unintelligibility." W. Beran Wolfe, "The Twilight of Psychoanalysis," *American Mercury* Aug. 1935, 385–94.

major pulp magazines. Hale, *Freud and the Americans*, 545, note no. 21.

"touched only a small minority of psychiatrists." Ibid., 254.

only 16,250 copies. Ibid., 430.

"By 1930 the influence . . . of Americans." Ibid., 477.

"intellectual influence . . . in the twentieth century." Philip Rieff, *Freud: The Mind of the Moralist* (Chicago: University of Chicago Press, 1959), x–xi.

2.
Racism, Immigration and the Nature-Nurture Debate

"The racialists . . . ." Franz Boas, "An Anthropologist's Credo," originally published in the *Nation* Aug. 1938, 201 and reprinted without title in Clifton Fadiman, ed., *I Believe* (New York: Simon and Schuster, 1939), 19. Reprinted with the kind permission of the Nation Co., Inc.

"it is possible . . . in history." Thomas W. Gossett, *Race: The History of an Idea in America* (Dallas: Southern Methodist University Press, 1963), 418. Given this role it is extraordinary that a major biography of Boas has never been written, although one is said to be in progress.

"epoch-making discovery." Fred H. Matthews, "Freud Comes to America: The Influence of Freudian Ideas on American Thought 1909–1917" (M.A. thesis, University of California at Berkeley, 1957), 9.

he taught a seminar on Freud's theory. John C. Burnham, "Psychoanalysis in American Civilization Before 1918," 250.

totem poles were sex symbols. Melville J. Herskovits, *Franz Boas: The Science of Man in the Making* (Clifton, N.J.: Augustus M. Kelley, 1973), 91.

correspondence. For example see Boas' letter of 14 November 1912 to Smith E. Jelliffe in the Boas Collection, American Philosophical Society, Philadelphia.

first psychoanalytic journal. Nathan Hale, *Freud and the Americans: The Beginnings of Psychoanalysis in the United States* (New York: Oxford University Press, 1971), 329.

encourage his daughter. "The Reminiscences of Franziska Boas," transcript of taped interviews done in 1972 by the Oral History Research Office, Columbia University. Transcript in Boas Collection of the American Philosophical Society, Philadelphia. Quoted with the kind permission of the Oral History Research Office, Columbia University.

"in which the ideals . . . a living force." Boas, "An Anthropologist's Credo," 201.

resulted in two duels. See Alfred L. Kroeber, "Franz Boas: The Man," *American Anthropological Association Memoirs*, no. 45, 61 (1943), 7–8.

"So decisive . . . did not actuate them." Ibid. Reproduced by permission of the *American Anthropological Association Memoirs*, no. 61 (1943). Not for sale or further reproduction.

"I had seen . . . like ours." Herskovits, *Franz Boas*, 1. See also 7–8 for a statement of Boas' "two principles."

"These Jews . . . stupid camels." "The Reminiscences of Franziska Boas."

"within the decade . . . entered our ports." Francis A. Walker, "Restriction of Immigration," *Atlantic Monthly* June 1896, 822–29.

"Hungarians . . . Native people." Ibid.

"there is . . . our soil." Ibid.

"have none . . . existence." Ibid.

"specialized for . . . other nations." Allan Chase, *The Legacy of Malthus: The Social Costs of the New Scientific Racism* (Urbana, Ill.: University of Illinois Press, 1980), 14.

"all the Nordic people . . . organized government." Nicholas Pastore, *The Nature-Nurture Controversy* (New York: Garland, 1984), 64. Originally published in 1949.

Immigration Restriction League. John Higham, *Strangers in the Land: Patterns of American Nativism 1860–1924* (New Brunswick, N.J.: Rutgers University Press, 1955), 102–103 and Chase, 140–43. Both of these books provide excellent histories of this period.

characteristics are inherited. Chase, *Legacy of Malthus*, 154.

"annihilate . . . vicious protoplasm." Derek Freeman, *Margaret Mead and Samoa: The Making and Unmaking of an Anthropological Myth* (Cambridge, Mass.: Harvard University Press, 1983), 16, quoting Davenport.

"certain races from central Europe." Henry F. Osborn, "The Second International Congress of Eugenics Address of Welcome," *Science* 54 (1921): 311–313.

"a large class of vipers." Henry F. Osborn, "Can We Save America?" Manuscript in the Osborn Collection at the American Museum of Natural History, which was noted as having been prepared at the request of the editor of *McCall's* magazine. Quoted with the kind permission of the American Museum of Natural History.

"it is for . . . his spots." Ibid.

"fundamentally . . . differences." Ibid.

"that each . . . kind of soul." Henry F. Osborn to Charles E. Seashore, 30 December 1922, Osborn Collection, American Museum of Natural History. Quoted with the kind permission of the American Museum of Natural History.

"wretched outcasts," ". . . dominated this country," ". . . we call Russians." Madison Grant, "Restriction of Immigration: Racial Aspects," *Journal of National Institute of Social Sciences*, 6 (1921): 1–11.

Jesus Christ also had been Nordic. Grant to Henry Osborn, 3 May 1917, Osborn Collection, American Museum of Natural History. Quoted with the kind permission of the American Museum of Natural History.

friends with Theodore Roosevelt. For examples of this friendship see the Collection of the Immigration Restriction League, Houghton Library, Harvard University (e.g. Grant's letter to Prescott Hall 18 November 1915 and Roosevelt's letter to Grant 2 December 1919.

eugenics sermon contest. Daniel J. Kevles, *In the Name of Eugenics: Genetics and the Uses of Human Heredity* (New York: Alfred A. Knopf, 1985), 61. The description of Fitter Families' contests is also from this source.

"there was no ignoring . . . coast to coast." Higham, *Strangers in the Land*, 110.

if each Irishman would kill. Chase, *Legacy of Malthus*, 107.

In the 1890s. The incidents in Colorado, Louisiana and Pennsylvania are mentioned in Higham, *Strangers in the Land*, 90.

"night-riders burned dozens." Ibid., 92.

"while the cephalic index . . . anatomical relation." Franz Boas, "The Cephalic Index," *American Anthropologist* 1 (1899): 448–61.

". . . oust Boas." Kroeber, 17.

firing of Boas. The correspondence concerning Boas' ouster from the American Museum of Natural History is in the Boas Collection of the American Philosophical Society in Philadelphia and the

Osborn Collection of the American Museum of Natural History in New York. The relationship between Boas and Osborn was a complex one, superficially cordial even when they were attacking each other in the daily newspapers; despite their enmity regarding racial issues, Boas needed Osborn's help to insure that his students at Columbia had access to the Museum's resources while Osborn needed Boas to complete the reports of the Museum's Jesup North Pacific Expedition which Boas still had not finished in 1928.

"demonstrate . . . form of man." Franz Boas, *The Mind of Primitive Man* (New York: Macmillan, 1911), 53.

"the American-born . . . are increased." Ibid., 55.

"a Magna Carta of race equality." Leslie Spier, "Some Central Elements in the Legacy," *American Anthropological Association Memoirs* no. 89, 61 (1959): 146–55.

"astonishing conclusions." Grant to President William H. Taft, 22 November 1910, Taft Papers, Presidential Series no. 2, file 77, Division of Manuscripts, Library of Congress.

"made a most amazing report . . . of New York." Grant to Senator F. M. Simmons, 5 April 1912, Collection of the Immigration Restriction League, Houghton Library, Harvard University. Quoted by permission of the Houghton Library.

"current literature . . . led by Boas." Grant to Rev. Percy S. Grant, 8 April 1912, Collection of the Immigration Restriction League, Houghton Library, Harvard University. Quoted by permission of the Houghton Library.

"that great swamp . . . ," ". . . unsanitary surroundings," ". . . destroy higher types." Ibid.

courses in eugenics. Chase, *Legacy of Malthus*, 124.

"Love or Eugenics." Kevles, *In the Name of Eugenics*, 58.

"prevention of procreation . . . and degenerate persons." Stephan L. Chorover, *From Genetics to Genocide* (Cambridge: MIT Press, 1979), 42.

"orphans . . . and paupers." Chase, *Legacy of Malthus*, 16.

"chicken-stealing . . . theft of automobile." Ibid., 135.

8,500 individuals. Chorover, *From Genetics to Genocide*, 42.

"a work of solid merit." Frederick A. Woods, review of *The Passing of the Great Race*, by Madison Grant, *Science* 48 (1918): 419–20.

marked a turning point. Gossett, *Race*, 353.

"The Polish Jew . . . of the nation." Madison Grant, *The Passing of the Great Race* (New York: Charles Scribner's Sons, 1916), 14.

"the cross between . . . is a Jew." Ibid., 16.

"a round skull . . . was appreciably longer." Ibid., 15.

"be applied to . . . race types." Ibid., 46–47.

"Mistaken regard for . . . race." Ibid., 44.

by February 1921. Higham, *Strangers in the Land*, 308.

"bootlegging was half Jewish . . . and Irish." Paul Johnson, *Modern*

Times (New York: Harper and Row, 1983), 250.

Ku Klux Klan. See Chase, 645, f. 13. Their attacks on bootleggers is cited in Higham, 268.

In California. Higham, *Strangers in the Land*, 265.

In Alabama. Ibid.

"foreigners . . . set fire to their dwellings." Higham, *Strangers in the Land*, 264.

"race of people . . . gave them hospitality." Gossett, *Race*, 332.

among the five "Great Americans." Henry F. May, 137.

"abnormally twisted . . . in their habit." Higham, *Strangers in the Land*, 309.

"the mythical magical melting pot . . . into it." George H. Lorimer, "The Great American Myth," *Saturday Evening Post*, 7 May 1921, 20. See also the editorial of 30 April 1921. Quoted with the kind permission of the *Saturday Evening Post*.

"every American . . . immigration problem." Ibid.

"New York . . . unravel." Ibid.

"streams of undersized" and other quotes are from Kenneth L. Roberts, "Ports of Embarkation," *Saturday Evening Post*, 7 May 1921; "The Existence of an Emergency," *Saturday Evening Post*, 30 April 1921; and *Why Europe Leaves Home* (New York: Bobbs-Merrill Company, 1922), 10, 14, 22, and 230–31.

"the dangers . . . in racial differences." Gossett, *Strangers in the Land*, 404.

"our country . . . on both sides." Calvin Coolidge, "Whose Country is This?" *Good Housekeeping* 72 (1921): 13–107.

"rat-men." Gossett, *Strangers in the Land*, 405.

"certain races . . . ," ". . . bottom of the list." Henry F. Osborn, "Eugenics—The American and Norwegian Programs," *Science* 54 (1921): 482–84.

"We are engaged . . . well-founded government." Chase, 278, quoting Osborn.

"As science . . . society." Ibid.

"Rights and Wrongs on the Racial Question." Notes in the Osborn Collection, American Museum of Natural History. Listed with Boas as possible "opponents" were Robert Lowie of the University of California and Herbert Spinden of Harvard University.

"almost half . . . classed as morons." Cornelia J. Cannon, "American Misgivings," *Atlantic Monthly* Feb. 1922, 145–57.

"inferior men . . . demagogues." Ibid.

"the army tests . . . the Nordic race group." Chase, 271, quoting Carl C. Brigham, *A Study of American Intelligence* (Princeton: Princeton University Press, 1923), 207.

"only about fourteen." Walter Lippmann, "The Mental Age of Americans," *New Republic*, 25 Oct. 1922.

". . . the New Snobbery." Walter Lippmann, "A Future for the Tests," *New Republic*, 29 Nov. 1922.

"a dithyrambic praise . . . of his achievements." Franz Boas, "Inventing a Great Race," *New Republic* Jan. 1917, 305–307.

Quoted with the kind permission of the New Republic, Inc.

"the attempt to justify a prejudice." Franz Boas, review of *The Passing of the Great Race*, by Madison Grant, *American Journal of Physical Anthropology* 1 (1918): 363.

regular technical consultation. Kevles, *In the Name of Eugenics*, 135 and Hamilton Cravens, *The Triumph of Evolution: American Scientists and the Heredity-Environment Controversy, 1900–1941* (Philadelphia: University of Pennsylvania Press, 1978), 234.

Armenians were Mongoloid. Herskovits, *Franz Boas*, 28.

"to raise a race of supermen." Franz Boas, "Eugenics," *Scientific Monthly* 3 (1916): 471–78.

Boas' facial cancer. See "The Reminiscences of Franziska Boas." Boas' daughter later maintained that her father had not really had cancer and that the growth had been benign.

"I have rather suspected . . . camouflage." Grant to Henry Osborn, 29 October 1917, Osborn Collection, American Museum of Natural History. Quoted with the kind permission of the American Museum of Natural History.

"there are some distinct curiosities in it." J. M. Tanner, "Boas' Contribution to Knowledge of Human Growth and Form," *American Anthropological Association Memoirs*, no. 89, 61 (1959): 76–111.

"despite . . . certain anthropologists." Henry F. Osborn, "The Approach to the Immigration Problem Through Science." The paper was unpublished but was presented publicly and widely quoted by the newspapers. A version of the paper is in the Osborn Collection, American Museum of Natural History. See also Chase, 274–75.

"swayed not by . . . prejudice." Franz Boas, "The Question of Racial Purity," *American Mercury* Oct. 1924, 164.

"uncouth barbarian." Ibid.

"Would not . . . pour Nordic." Ibid.

"dubious comments about my own race." Henry F. Osborn, "Lo, the Poor Nordic!" *New York Times*, 18 April 1924, 13.

"entirely . . . of Madison Grant." Franz Boas, "Lo, the Poor Nordic!" *New York Times*, 18 April 1924, sec. 9:19.

"that Jews were excluded . . . from jobs." Higham, *Strangers in the Land*, 278.

Johnson-Reed Act: Chase, *Legacy of Malthus*, 300.

"America must be kept American." Gossett, *Race*, 407.

popular campus song. Ibid., 372.

"convenient jingle of words." Freeman, 9, quoting Francis Galton.

"the average citizen . . . ," ". . . capable of voting." Chase, *Legacy of Malthus*, 101, quoting Galton, 1894.

"so long as they . . . ," ". . . claims to kindness." Ibid., 100, quoting Galton, 1873.

"lust for equality . . . ," ". . . universal suffrage." Higham, *Strangers in the Land*, 272, quoting Hall.

"the true spirit . . . themselves and others." Henry F. Osborn, "The

Second International Congress of
Eugenics Address of Welcome."
"to bend . . . to the great god
Demos." Madison Grant, "Discus-
sion of Article on Democracy and
Heredity," *Journal of Heredity* 10
(1919): 164–65.
"In the democratic form . . . and
integrity." Grant, *The Passing of the
Great Race*, 5.
"the conduct . . . of the commu-
nity." Grant, "Democracy and
Heredity," 65.
"In America . . . the privilege of
wealth." Grant, *The Passing of the
Great Race*, 6.
"*Vox populi* . . . Chant for duty."
Ibid., 8.
"True aristocracy . . . in popula-
tion." Ibid., 7.
Boas himself was a Socialist. "The
Reminiscences of Franziska Boas."
"the greatest experiment . . . in
view." Margaret M. Caffrey, *Ruth
Benedict: Stranger in This Land*
(Austin: University of Texas Press,
1989), 289.
"was no more a Communist than
Muffin." "The Reminiscences of
Franziska Boas." A copy of the
FBI file on Boas is in the Boas
Collection, American Philosophi-
cal Society, Philadelphia.
"handed the poem . . . declared the
judge." Lillian Symes and Travers
Clement, *Rebel America* (New York:
Harper and Brothers, 1934), 307.
Senator Thomas R. Hardwick.
Frederick L. Allen, *Only Yesterday*,
42.
the house of . . . A. Mitchell
Palmer. Ibid.
"were herded . . . for a week."
Ibid., 48.

"authorities took . . . Communist
party." Ibid.
"fundamentally unnatural." Kevles,
In the Name of Eugenics, 88.
"deplored birth . . . indulgence."
Ibid., 52.
Dr. Abraham Jacobi. Margaret
Sanger, *An Autobiography*, 181.
features ridiculing Freud. See for
example Blanche Goodwin,
"Expression and the Fraudian
Complex," *Saturday Evening Post*,
13 June 1925, 70.

3.
The Sexual Politics of Ruth Benedict and Margaret Mead

"There is only . . . one burning
moment." Ruth Benedict,
undated journal fragment, in Mar-
garet Mead, *An Anthropologist at
Work: Writings of Ruth Benedict*
(Boston: Houghton Mifflin,
1959), 154.
Alfred Kroeber. Marvin Harris, *The
Rise of Anthropological Theory* (New
York: Crowell, 1968), 431.
Boas waived requirements. Judith
Modell, *Ruth Benedict: Patterns of a
Life* (Philadelphia: University of
Pennsylvania Press, 1983), 117.
Elsie Clews Parsons. See Peter H.
Hare, *A Woman's Quest for Science:
Portrait of Anthropologist Elsie Clews
Parsons* (Buffalo: Prometheus
Books, 1985). For references to
Parsons' interest in Freud, see
John C. Burnham, "Psychoanaly-
sis in American Civilization
Before 1918," 239–41.
"well known for . . . ideas." Mar-
garet M. Caffrey, *Ruth Benedict*,
95.

"sex relations should . . . by society." Henry F. May, *The End of American Innocence*, 309.

"passionate love . . . poets sing." Ibid.

"a kind of cultural fatalist." Edward Sapir to Ruth Benedict, 25 June 1922, in Mead, *An Anthropologist at Work*, 50.

"responses . . . into which he was born." Ruth Benedict, "Toward a Social Psychology," *Nature* 119 (1924): 51.

"The fundamental question . . . what to nature." Ruth Benedict, "Nature and Nurture," *Nation* 30 Jan. 1924, 118.

"world improver." Victor Barnouw, *Culture and Personality* (Homewood, Ill.: Dorsey Press, 1973), 110.

Boas' suggestion. See Modell, *Ruth Benedict: Patterns of a Life*, 248.

"How can we stop this epidemic of racism?" Ruth Benedict, foreword to 1945 edition of *Race: Science and Politics* (Westport, Conn.: Greenwood Press, 1982), x. Originally published in 1940.

"all men . . . available to all." Ibid., 160.

"Until the regulation of . . . racial groups." Ibid., 156–57.

supporter of Roosevelt's New Deal. Modell, *Ruth Benedict: Patterns of a Life*, 208.

"knew that some . . . be taken seriously." Mead, *An Anthropologist at Work*, 349.

"As for me . . . place you fill in my life." Benedict to Franz Boas, 26 December 1939, in Mead, *An Anthropologist at Work*, 417.

"coal black and piercing." Alfred L. Kroeber, "Franz Boas: The Man," *American Anthropological Association Memoirs* no. 61 (1943): 5–26.

"for Ruth . . . a better world." Modell, *Ruth Benedict: Patterns of a Life*, 66.

"apparently . . . one woman faculty member." Caffrey, *Ruth Benedict*, 190.

"that fire upon our flesh." Benedict, undated journal fragment, in Mead, *An Anthropologist at Work*, 154.

"We cast about . . . phantoms, shadows." Ibid.

"the more . . . maleness in herself." Modell, *Ruth Benedict: Patterns of a Life*, 89.

"the relativity of . . . so-called normal." Ruth Benedict, "Anthropology and the Abnormal," *Journal of General Psychology* 10 (1934): 59–82.

"The etiology of homosexuality . . . social." Ruth Benedict, "Sex in Primitive Society," *American Journal of Orthopsychiatry* 9 (1939): 570–73.

"our culturally discarded traits." Benedict, "Anthropology and the Abnormal," 59–82.

"A tendency . . . cultural." Ibid.

"There is something cruel . . . her terrible revenge." Edward Sapir to Ruth Benedict, 18 August 1925, quoted in Mead, *An Anthropologist at Work*, 85.

"She used to wonder . . . who is a cultural misfit." Margaret Mead, *Blackberry Winter: My Earlier Years* (New York: Simon and Schuster, 1972), 195–96.

Natalie Raymond and Ruth Valentine: For details of these relationships see Modell and Caffrey.

"exquisite responsiveness to literature." "Margaret Mead Answers," *Redbook*, May 1975, 64.

"spoke with an authority . . . in a teacher." Mead, *An Anthropologist at Work*, 4.

"very much a father figure . . . ," ". . . transmit the message." Jane Howard, *Margaret Mead: A Life* (New York: Simon and Schuster, 1984), 56–57.

"Mead considered . . . waste of time." Ibid., 248.

articles to psychoanalytic journals. See for example M. Mead, "An Ethnologist's Footnote to Totem and Taboo," *Psychoanalytic Review* 17 (1930): 297–301.

"perhaps the most fruitful attacks." Margaret Mead, *Growing Up in New Guinea* (London: Penguin Books, 1963), 174.

"the solution to The Oedipal Situation . . . ." Margaret Mead, *Male and Female: A Study of Sexes in a Changing World* (New York: Penguin Books, 1962), 132. First published in 1949.

"command of . . . professionals I ever knew." Howard, *Margaret Mead: A Life*, 332.

participated in a mass meeting. Mead, *Blackberry Winter*, 107.

"wearing red dresses . . . 'Internationale.' " Howard, *Margaret Mead: A Life*, 47.

support of the Communists. Melville Herskovits to Margaret Mead, 9 June 1923, Mead Collection, Library of Congress.

"he was . . . for the good of mankind." Margaret Mead, "Apprenticeship Under Boas,"

American Anthropological Association Memoirs no. 89, 61 (1959): 42.

"special and different." Mead, *Blackberry Winter*, 81.

"I had wondered . . . of the career-minded women I had met." Ibid., 196.

"I had my father's mind . . . his mother's mind." "Margaret Mead Answers Questions About," *Redbook*, Aug. 1975.

noted her masculine qualities. See for example, Bateson, 113.

"she . . . prided herself on her femininity." Robert Cassidy, *Margaret Mead: A Voice for the Century* (New York: Universe Books, 1982), 14.

"you just . . . fell in love with me." Bateson, *With a Daughter's Eye*, 24.

"Once a man wanted . . . a chance with her." Howard, *Margaret Mead: A Life*, 174.

Love letters to Mead. See the letters from Leah Josephson Hanna to Mead, Mead Collection, Library of Congress.

"Margaret . . . admired her very much." Howard, *Margaret Mead: A Life*, 44.

lesser interest in men. See for example Howard, 44, 50, and 113.

primarily as professional liaisons. See for example Mead's letters to Caroline Tennant Kelly in the Mead Collection, Library of Congress.

rented a room. Modell, *Ruth Benedict: Patterns of a Life*, 126.

"rests me like a . . . fireplace." Howard, *Margaret Mead: A Life*, 57.

"When touch seems . . . dignity in

living." Caffrey, *Ruth Benedict*, 192; Caffrey identified it merely as "Diary: 1923."

"perfect friendship . . . buying a new dress." Modell, *Ruth Benedict: Patterns of a Life*, 154–57.

lived together only once. Caffrey, *Ruth Benedict*, 201.

Mead was frigid. Howard, *Margaret Mead: A Life*, 62.

"Ruth and Gregory . . . and her remote beauty." Bateson, *With a Daughter's Eye*, 117.

"bisexual potentialities are normal." "Margaret Mead Answers," *Redbook*, July 1963, 24–29.

"we must recognize . . . human behavior." Margaret Mead, "Bisexuality: What's It All About?" *Redbook*, Jan. 1975, 29–31.

"a very large . . . capacity for love." Ibid.

"the individual . . . one human potentiality." M. Mead, "Cultural Determinants of Sexual Behavior," in *Sex and Internal Secretions*, ed. William C. Young, (Baltimore: Williams and Wilkins, 1961), 1433–79.

"whether they will become . . . life history." Mead, "Bisexuality: What's It All About?"

"the process by . . . sex attitudes." Mead, *Growing Up in New Guinea*, 174.

"and several . . . assumed to be homosexual." Howard, *Margaret Mead: A Life*, 50.

"a revolting theme." Jonathan Katz, *Gay American History* (New York: Avon Books, 1978), 128. For a history of the times, see also Deborah G. Wolf, *The Lesbian Com-*

munity (Berkeley: University of California Press, 1980).

"the conflicts . . . always exposed." Benedict, "Anthropology and the Abnormal," 59–82.

"the burden of nonconformity . . . and behavior." Mead, "Cultural Determinants of Sexual Behavior."

"upon the gifted . . . new worlds." Mead, *Growing Up in New Guinea*, 173.

she wrote sympathetically of auras. See the Mead monthly columns in *Redbook* in Feb. 1963, Jan. 1965, Oct. 1967, and March 1977.

American Society for Psychical Research. L. Lasagna, "Let Magic Cast Its Spell," *The Sciences* 24 (1984): 12.

"there *are* . . . flying objects." Margaret Mead, "UFOs—Visitors From Outer Space?" *Redbook*, Sept. 1974, 57–59.

"spirit guides." Howard, *Margaret Mead: A Life*, 412.

visited mediums. Howard, *Margaret Mead: A Life*, 187, 412–13.

"disciplined subjectivity." "A Conversation With Margaret Mead and T. George Harris on the Anthropological Age," *Psychology Today*, July 1970, 59–76.

her failure to use scientific methodology. For examples of criticism of Mead's failure to use scientific methodology, see Harris, *The Rise of Anthropological Theory*, 415 ff., and Sheila Johnson, "A Look at Margaret Mead," *Commentary* 55 (1973): 70–72. Another example of Mead's shortcomings as a scientist was her re-study of Manus published in 1956 as *New Lives for Old: Cultural Transformations—*

Manus, 1928–1953 (New York, William Morrow, 1960). Although she claimed to have collected 32,000 drawings, 20,000 photographs, and 299 reels of movie film as well as "a whole battery of modern tests: TATs, Mosaics, Bender-Gestalts, Stewart Ring Puzzles, Gesell Infant Development Tests, Caligor Eight-card Redrawing Test [and] Minnesota Paper Form Board," she published her book without utilizing virtually any of this material. At the end of the book she simply noted that "the test results have not yet been analyzed" but promised that they would be in the future (501).

to study Carl C. Brigham's book. Howard, *Margaret Mead: A Life*, 62.

occasional references to the nature-nurture debate. See, for example, the letter of Eleanor Phillips to Mead, 5 July 1923, Mead Collection, Library of Congress. It should be added, however, that the publicly available correspondence of both Mead and Benedict is remarkably free of references to either the eugenics controversy or to their personal relationship. A portion of the Ruth Benedict Collection at Vassar College will not be available for examination until 1999, a restriction dictated by Margaret Mead in her will in 1978.

"the problem . . . suggested by Boas." Mead, *Blackberry Winter*, 122.

". . . about black people today." "A Conversation With Margaret Mead."

"The home language was . . .

lower." Margaret Mead, "Sense and Nonsense About Race," *Redbook*, Sept. 1969, 35–42.

"extreme caution . . . habits of thought." Margaret Mead, "The Methodology of Racial Testing: Its Significance for Sociology," *American Journal of Sociology* 31 (1926): 657–67.

"degree of assimilation . . . of the English language." Franz Boas, "This Nordic Nonsense," *Forum* 74 (1925): 502–11.

"new and old elements of culture." Mead, "Apprenticeship Under Boas," 42.

Boas asked her instead. Mead, *Blackberry Winter*, 126–27. Boas had a special interest in this problem. Boas' boss during his days at Clark University had been G. Stanley Hall, whose 1905 book *Adolescence* had argued strongly that the stresses of the adolescent period were biological in origin. Boas had disagreed with Hall at the time of the book's publication; in the intervening years, Hall had supported the eugenicists, which made refutation of Hall even more desirable from Boas' point of view. Mead confirmed on several occasions that Boas had chosen the problem on which she was to work in Samoa. See for example *An Anthropologist at Work*, 14 and Margaret Mead, *Letters From the Field 1925–1975* (New York: Harper and Row, 1977), 19.

"always tailoring . . . priorities." Mead, "Apprenticeship Under Boas," 42.

"The far-reaching importance . . . of adult life." Edward Sapir, review of *Psychoanalysis as a Pathfinder*, by Oskar Pfister, *Dial* Sept. 1917: 267–69.

Sapir pressured Mead. Caffrey, *Ruth Benedict*, 198–99.

"Margaret described . . . each other." Bateson, *With a Daughter's Eye*, 125.

"Adolescence represented . . . satisfying ambitions." M. Mead, *Coming of Age in Samoa: A Study of Adolescence in Primitive Society* (New York: Mentor Books, 1949), 95–96. First published in 1928.

"that adolescence . . . make it so." Ibid., 137.

"the results . . . by our civilisation." Franz Boas, foreword to *Coming of Age in Samoa.*

"with the freedom of sexual life . . . ," ". . . crimes do not occur." Franz Boas, *Anthropology and Modern Life* (New York: Dover Publications, 1986), 125, 190. First published in 1928.

"The study of cultural forms." Franz Boas, *Encyclopedia of the Social Sciences*, 13 (1934): 34, quoted by Derek Freeman, *Margaret Mead and Samoa* (Cambridge: Harvard University Press, 1983), 101.

"heavy sense of responsibility." Mead, "Apprenticeship Under Boas," 42.

"a huge red hibiscus." Margaret Mead, *Letters From the Field 1925–1975*, 50. See also p. 40.

"These casual . . . with the salacious." Mead, *Coming of Age in Samoa*, 90.

"Familiarity with sex . . . as senility." Ibid., 92.

"The present problem . . . their consciences." Ibid., 142.

"The findings of the behaviourists . . . more conflicts." Ibid., 122.

"Illuminating . . . his pupils." Abraham A. Brill, comment on back cover of Mead's *Coming of Age in Samoa*, Mentor Books edition, 1949. It is not clear whether it was written for the cover or taken from a review.

"the young Margaret . . . adolescent informants." Freeman, *Anthropological Myth*, 240.

"There are . . . because she dug dirt." Nicholas von Hoffman, *Tales From the Margaret Mead Taproom* (Kansas City: Sheed and Ward, 1976), 97.

"I know . . . the position." Ibid., 101.

"probably no more . . . United States." Lowell D. Holmes, *Quest for the Real Samoa: The Mead/Freeman Controversy and Beyond* (South Hadley, Mass.: Bergin and Garvey, 1987), 78.

"the most common ground for divorce." Ibid., 82.

"homosexuals . . . without stigma or ridicule." Ibid., 78.

"Samoans very conservative about it." Ibid., 122.

"Even her most fervent . . . flair for languages." Howard, *Margaret Mead: A Life*, 207.

"They bully and chivvy . . . on head or heels." David Lipset, *Gregory Bateson: The Legacy of a Scientist* (Englewood Cliffs, N.J.: Prentice Hall, 1980), 136, quoting a Bateson letter.

"Margaret . . . wants to find." Holmes, *Quest for the Real Samoa*,

144, quoting his own 1967 letter.

Dadaism. One of the best sources of information on this movement in Paris is Robert Motherwell, ed., *The Dada Painters and Poets: An Anthology* (New York: Wittenborn Schultz, 1951).

"the Pope of Lesbos." George Wickes, *The Amazon of Letters: The Life and Loves of Natalie Barney* (New York: G. P. Putnam, 1976), 181.

"Ballet mécanique." A good account of this performance can be found in Hugh Ford, ed., *The Left Bank Revisited: Selections From the Paris Tribune 1917–1934* (University Park, Penn.: State University Press, 1972), 212–13, 220–21.

"Margaret . . . heard him come in." Howard, *Margaret Mead*, 98. The information on Cressman and Fortune is from Howard's interviews of Cressman.

"creative . . . practice of homosexuality." Mead, "Bisexuality: What's It All About?", 29.

"a culture that made the point so clearly." "A Conversation With Margaret Mead and T. George Harris," 59.

"a puritan society." Mead, *Growing Up in New Guinea*, 132.

"sex . . . inherently shameful." Ibid., 126.

"the Manus emphasized anality." Mead, *Blackberry Winter*, 200.

"In one, both men . . . unadorned partners." Margaret Mead, preface to the 1950 edition, *Sex and Temperament in Three Primitive Societies* (New York: Mentor Books, 1950), vi. The book was first published in 1935.

"We may say . . . to either sex." Mead, *Sex and Temperament*, 206.

"The differences . . . culturally determined." Ibid.

"I for one . . . she arrives at." Jessie Bernard, "Observations and Generalizations in Cultural Anthropology," *American Journal of Sociology* 50 (1945): 284–91.

"According to some . . . able to guess." Mead, 1950 preface to *Sex and Temperament in Three Primitive Societies*, vi.

"the individual . . . of his society." Ibid., 213.

"temperamental affinity for . . . the opposite sex." Ibid. 216.

"the pain of being born . . . ." Ibid., 215.

"in . . . ourselves and each other." Mead, *Blackberry Winter*, 216.

"Gregory and I were falling in love." Ibid., 217.

"Reo repudiated . . . a new romance." Bateson, *With a Daughter's Eye*, 138.

Fortune knocked Mead down. Howard, *Margaret Mead*, 160–61.

"the most important . . . profession." Harris, *The Rise of Anthropological Theory*, 406.

"become the willing . . . world." Mead, *An Anthropologist at Work*, 206.

"always had to work through interpreters." Ibid., 202.

"moderation is the first virtue." Ruth Benedict, *Patterns of Culture* (New York: Mentor Books, 1934), 101. Other quotations on the Zuni are taken from this chapter.

"Appollonian." Ibid., 158. Other quotations on the Kwakiutl are taken from this chapter.

Fortune approved Benedict's chapter. In later years Fortune claimed that he had not approved of Benedict's description of the Dobuans; see Lipset, *Gregory Bateson: The Legacy of a Scientist*, 137 f. Correspondence from the early 1930s between Benedict and Fortune, however, suggests that Fortune did, in fact, both review it and approve it.

"dour, prudish . . . as he asks none." Benedict, *Patterns of Culture*, 151. Other quotations on the Dobuans are taken from this chapter.

"Most people . . . are born." Ibid., 220–21.

"the biological bases . . . irrelevant." Ibid., 206.

"Man is not committed . . . germ plasms." Ibid., 27.

"racial heredity . . . in reality." Ibid., 28.

"The author demonstrates . . . the community." Mead, *An Anthropologist at Work*, 212, quoting Benedict.

"when the homosexual . . . always exposed." Benedict, *Patterns of Culture*, 229.

"an increased tolerance . . . usual types." Ibid., 235.

Natalie Raymond. See Modell, 188–90.

"spread word . . . a lesbian." Howard, *Margaret Mead*, 213. The incident concerned a divorce case in which Benedict had agreed to testify for the husband and his wife in turn threatened Benedict.

"was from the outset . . . criticism." Harris, *The Rise of Anthropological Theory*, 404.

"over-simplified" and "very misleading." Li An-Che, "Zuni: Some Observations and Queries," *American Anthropologist* 39 (1937): 62–77.

"Below the . . . other societies." Ibid.

"highly colored." Barnouw, *Culture and Personality*, 97.

"At any rate . . . overstatement." Ibid.

"I don't believe it." Mead, *Blackberry Winter*, 184.

"I know . . . without attacking it publicly." Ann Chowning, review of *Sorcerers of Dobu*, by Reo Fortune, *American Anthropologist* 66 (1964): 455–57.

"used no English . . . over me." Reo F. Fortune, *Sorcerers of Dobu* (New York: E. P. Dutton, 1932), xi.

threatened to spear him. Ibid., 104.

Mead pictured Reo Fortune. Mead, *Blackberry Winter*, e.g. 161, 191, and 211. See also Howard, 171, 267, and 314. Prior to Margaret Mead's death, I had conversations and correspondence with her regarding Fortune's mental state and how it might have affected his perception of the Dobuans. Mead acknowledged that Fortune may have had psychiatric problems, but adamantly denied that such problems had interfered with his perception of the Dobuans.

"cheerful, laughter-loving folk." Walter E. Bromilow, *Twenty Years Among Primitive Papuans* (London: Epworth Press, 1929), 95.

"There are others who . . . do not." Mead, *Blackberry Winter*, 184.

"a man who [gives] a counter-gift."

Géza Roheim, *Psychoanalysis and Anthropology* (New York: International Universities Press, 1950), 227.

Hogbin had told her. Mead letter to the author, 14 March 1974, and conversations with Mead about this period.

"the Dobuans . . . are not imaginary." Chowning, review of *Sorcerers of Dobu*, 455–57.

Margaret Mead acknowledged: Mead letter to the author, 14 March 1974.

Attempts by this author. Letter of 26 April 1983 sent on two separate occasions to Ann Chowning and accompanied by a telephone call. Chowning was living in Wellington, New Zealand, where many of Fortune's relatives were also living at that time.

"It was simple . . . and unyielding." Margaret Mead, *From the South Seas: Studies of Adolescence and Sex in Primitive Societies* (New York: William Morrow, 1939), x.

"Anthropologists had . . . measurable differences." "A Conversation with Margaret Mead and T. George Harris."

"his failure . . . such as Benedict." Verne F. Ray, review of *Franz Boas: The Science of Man in the Making*, by M. J. Herskovits, *American Anthropologist* 57 (1955): 140.

"Certainly . . . within the family." Melville J. Herskovits, *Franz Boas*, 71.

"It is simply . . . personality movement." Harris, *Anthropological Theory*, 407.

4.
Hitler's Resolution of the Nature-Nurture Debate

[Hitler] believed . . . the twentieth century. Paul Johnson, *Modern Times: The World from the Twenties to the Eighties* (New York: Harper and Row, 1983), 129.

"the kind of propaganda . . . among our sentimentalists." Madison Grant to Henry Osborn, 23 Feb. 1927, Osborn Collection, American Museum of Natural History.

"complete retraction . . . in such activities." John Higham, *Strangers in the Land: Patterns of American Nativism 1860–1924* (New Brunswick: Rutgers University Press, 1955), 327.

president of the Carnegie Institution. Hamilton Cravens, *The Triumph of Evolution: American Scientists and the Heredity-Environment Controversy, 1900–1941* (Philadelphia: University of Pennsylvania Press, 1978), 179–80.

"the army [IQ] tests . . . race group." Allan Chase, *The Legacy of Malthus: The Social Costs of the New Scientific Racism* (Urbana: University of Illinois Press, 1980), 271, quoting Brigham.

"one of the most agonizing . . . sciences." Ibid., 321.

"tests in . . . another tongue." Carl C. Brigham, "Intelligence Tests in Immigrant Groups," *Psychological Review* 37 (1930): 158–65.

"This review . . . foundation." Ibid.

"the project . . . could contribute." Chase, *The Legacy of Malthus*, 326.

"a definite race of . . . generations." Ibid., 328.

"the dominance of economics over eugenics." Ibid., 330.

"hopelessly perverted . . . and reactionaries generally." Daniel J. Kevles, *In the Name of Eugenics: Genetics and the Uses of Human Heredity* (New York: Alfred A. Knopf, 1985), 164.

"The Rich are Taller." Franz Boas, "The Rich are Taller," *New York Times*, 22 Nov. 1931, sec. 10:2.

"I do not believe . . . the author." Franz Boas, "Nordic Propaganda," *New Republic*, 7 March 1934, 106–108.

Boas promptly sent von Hindenburg. Caffrey, *Ruth Benedict*, 282.

"the crazy conditions in Germany." Ibid., 285.

"everything [else Boas] wrote . . . circulation." Melville J. Herskovits, *Franz Boas: The Science of Man in the Making* (Clifton: N.J.: Augustus M. Kelley, 1973), 117.

"the attempt . . . on a pseudo-science." Franz Boas, "Aryans and Non-Aryans," *American Mercury* June 1934, 219–23.

"Herr Hitler has . . . race problems." Boas, "Nordic Propaganda," 106–108.

Germans were the best educated. Johnson, *Modern Times*, 127, 130.

"Why is the white man . . . the gorilla." Robert Proctor, *Racial Hygiene: Medicine Under the Nazis* (Cambridge: Harvard University Press, 1988), 24.

Ploetz traveled to Iowa. Ibid., 98.

racial hygiene was being taught. Ibid., 38.

"should be . . . of the Eugenics Records Office." Ibid., 101.

"What we racial hygienists . . .

tested long ago." Ibid., 98.

"state sterilization authorities . . . ." Kevles, *In the Name of Eugenics*, 116.

"after the war . . . in the United States." Proctor, *Racial Hygiene*, 117.

Once the sterilization law. Ibid., 106–108.

German doctors experimented. Ibid., 110.

the falling birthrate. Ibid., 19.

"the government provided . . . to the fitter families." Kevles, *In the Name of Eugenics*, 117.

"a quality marriage." Ibid., 138.

"fit to marry." Ibid., 138–39.

"an enthusiastic trip." Geoffrey Hellman, *Bankers, Bones and Beetles: The First Century of the American Museum of Natural History* (Garden City, N.Y.: Natural History Press, 1969), 194.

"No one will be surprised . . . with it." Chase, *The Legacy of Malthus*, 343.

"cleansing process." Johnson, *Modern Times*, 342.

Davenport publicly defended. Chase, *The Legacy of Malthus*, 634, n. 9.

"Jews are . . . the sexual life." Proctor, *Racial Hygiene*, 53–54.

"sexual degeneration . . . is decent." Robert J. Lifton, *The Nazi Doctors: Medical Killing and the Psychology of Genocide* (New York: Basic Books, 1986), 42.

"associated sexual relations . . . on Jews." Marc Fisher, "Master of the Death Camps," *Washington Post Book World*, 9 June 1991, 12.

"Jewish science." Marie Jahoda, "The Migration of Psychoanalysis:

Its Impact on American Psychology," in *The Intellectual Migration*, eds. Donald Fleming and Bernard Bailyn, (Cambridge: Harvard University Press, 1969).

"non-Aryan physicians . . . insurance schemes." Proctor, *Racial Hygiene*, 151–52.

"What progress . . . burning my books." Peter Gay, *A Life for Our Time* (New York: W. W. Norton, 1988), 592.

One out of every five. Jahoda, "The Migration of Psychoanalysis," 282.

more than 200 German physicians. Ibid., 152.

"European psychoanalysis found . . . continental Europe." Laura Fermi, *Illustrious Immigrants: The Intellectual Migration from Europe 1930–41* (Chicago: University of Chicago Press, 1971), 142.

In Vienna. Ibid., 147.

13 percent of them emigrated. Proctor, *Racial Hygiene*, 15.

approximately 190 European psychoanalysts. Fermi, *Illustrious Immigrants*, 152.

The list of psychoanalysts who emigrated to America is compiled from Fermi, *Illustrious Immigrants*, 139–73.

New York had 76 members. Henry W. Brosin, "A Review of the Influence of Psychoanalysis on Current Thought" in *Dynamic Psychiatry*, eds. Franz Alexander and Helen Ross, (Chicago: University of Chicago Press, 1952), 525.

United States had more psychoanalysts. Reuben Fine, *A History of Psychoanalysis* (New York: Columbia University Press, 1979), 90.

Prior to 1929. Ibid., 110.

Alfred Adler's . . . "The Psychology of Marxism." Ibid., 438.

"in the service of Communist ideology." Walter Bromberg, *The Mind of Man: A History of Psychotherapy and Psychoanalysis* (New York: Harper and Row, 1954), 223.

the "Partisansky Review." Carol Gelderman, *Mary McCarthy: A Life* (New York: St. Martin's Press, 1988), 91.

"keep an open mind . . . in Freud." Isaac Deutscher, *The Prophet Armed: Trotsky 1879–1921* (New York: Oxford University Press, 1954), 193.

"We were all . . . contemplating it." William Barrett, *The Truants: Adventures Among the Intellectuals* (Garden City, N.Y.: Anchor Doubleday, 1982), 38.

"the most ambitious attempt . . . modernism." Christopher Lasch, *The Agony of the American Left* (New York: Alfred A. Knopf, 1969), 53.

"a New York intellectual . . . read *Partisan Review*." James Atlas, "The Changing World of New York Intellectuals," *New York Times Magazine*, August 25, 1985, 22–76.

"a Marxist purist." Stephen J. Whitfield, *A Critical American: The Politics of Dwight Macdonald* (Hamden, Conn.: Shoe String Press, 1984), 37.

"a heavier . . . insights of Freud." Leslie Fieldler, "The Ordeal of Criticism," *Commentary* 8 (1949): 504–506.

"Rahv had great respect . . . and

Literature." Alan Lelchuk, "Philip Rahv: The Last Years," in *Images and Ideas in American Culture*, Arthur Edelstein, ed. (Boston: Brandeis University Press, 1979), 210.

"a manic-impressive." Barrett, *The Truants*, 38–39.

"Most of us . . . great man." Ibid.

Julius Rosenberg, in alcove number 2. Alexander Bloom, *Prodigal Sons: The New York Intellectuals and Their World* (New York: Oxford University Press, 1986), 251.

began reading Freud. James B. Gilbert, *Writers and Partisans: A History of Literary Radicalism in America* (New York: John Wiley and Sons, 1967), 218.

William Phillips, ed., *Art and Psychoanalysis* (New York: Criterion Books, 1957).

Edith Kurzweil and William Phillips, eds., *Literature and Psychoanalysis* (New York: Columbia University Press, 1983).

"With . . . example of Chutzpah." Whitfield, *A Critical American*, 20.

"not a snob but a bit stupid." Ibid., 1.

"political career . . . by Jackson Pollack." Ronald Berman, *America in the Sixties: An Intellectual History* (New York: The Free Press, 1968), 4.

"his articles in *Politics* . . . ." Whitfield, *A Critical American*, 48.

the theories of Wilhelm Reich. Personal communication from Daniel Bell, 5 September 1990.

"flirted with radical politics." Bloom, *Prodigal Sons*, 78.

"vocabulary of psychoanalysis." Richard H. Pells, *The Liberal Mind in a Conservative Age: American Intellectuals in the 1940s and 1950s* (New York: Harper and Row, 1985), 190.

"nothing at all but mild, all-purpose left." Paul Johnson, *Intellectuals* (New York: Harper and Row, 1988), 260.

"astonished everyone . . . seeing Wilson." Gelderman, *Mary McCarthy: A Life*, 89.

McCarthy began psychoanalysis. Ibid., 91.

"that psychoanalysis . . . of myths." Ibid., 120.

"blend of avant-garde . . . like to become." Bloom, *Prodigal Sons*, 80, quoting Irving Howe.

Wilson had supported Foster. Ibid., 45; and Johnson, *Intellectuals*, 255.

he failed to file income tax returns. Johnson, *Intellectuals*, 266–67.

He recalled being introduced to Freudian theory. Edmund Wilson, *Classics and Commercials: A Literary Chronicle of the Forties* (New York: Farrar, Straus and Company, 1950), 58.

"a reminder that the lust for cruelty . . . they are." Charles P. Frank, *Edmund Wilson* (New York: Twayne Publishers, 1970), 191, quoting Wilson.

regularly beat Mary McCarthy. See Gelderman, 91 ff.

"a terrifying nervous breakdown." George H. Douglas, *Edmund Wilson's America* (Lexington: University of Kentucky Press, 1983), 47.

James's . . . attraction to little girls. Frank, *Edmund Wilson*, 54.

Jonson's . . . anal erotic tendencies. Ibid., 57. For a psychoanalytic view of Wilson's understanding of

Freudian theory, see Louis Fraiberg, *Psychoanalysis and American Literary Criticism* (Detroit: Wayne State University Press, 1960).

"Yeats, Freud . . . one's father." *Edmund Wilson Letters on Literature and Politics 1912–1972*, Elena Wilson, ed. (New York: Farrar, Straus and Giroux, 1977), 329.

"With the exception of Edmund Wilson . . . few decades." Irving Howe, "On Lionel Trilling," *New Republic*, 13 March 1976, 29.

supported William Foster. Irving Howe, *Socialism in America* (New York: Harcourt Brace Jovanovich, 1985), 60. See also Bloom, 47.

as members . . . for the Defense of Trotsky. Bloom, *Prodigal Sons*, 108.

friends with Whittaker Chambers. Ibid., 254.

"He had . . . something of a numinous glow." Barrett, *The Truants*, 175.

"brilliant" methodology. Lionel Trilling, "The Legacy of Sigmund Freud," *Kenyon Review* 2 (1940): 152–73.

Wordsworth's infantile narcissism. Barrett, *The Truant*, 178.

"the single . . . new liberalism." Norman Podhoretz, *Breaking Ranks: A Political Memoir* (New York: Harper and Row, 1979), 33.

"[He] has a grasp . . . into his criticism." Fraiberg, *Psychoanalysis*, 213,224.

Sidney Hook. For an assessment of his political activities see Bloom, 46, 108, and 254 and Whitfield, 39.

"a scientific mythology." Bloom, *Prodigal Sons*, 101.

"Freud's doctrines . . . of human behavior." Sidney Hook, *Out of Step* (New York: Harper and Row, 1987), 138.

"the influence . . . did not diminish." W. David Sievers, *Freud on Broadway: A History of Psychoanalysis and the American Drama* (New York: Hermitage House, 1955), 212.

"the conflicting emotions . . . the primal scene." Ibid., 215.

"perhaps the most original . . . in the thirties." Ibid., 261.

"the analytic situation . . . of 'transference.' " Ibid., 262.

Thornton Wilder. Ibid., 256–57.

"almost all of Freud." Ibid., 289.

"the first musical drama . . . ." Ibid., 291.

"reportedly written in tribute . . . ." Irving Schneider, "Images of the Mind: Psychiatry in the Commercial Film," *American Journal of Psychiatry* 134 (1977): 613–20.

"ran for 467 performances . . . ." Sievers, *Freud on Broadway*, 289.

"there is no department . . . not affected." Bernard DeVoto, "Freud's Influence on Literature," *Saturday Review of Literature*, 7 Oct. 1939, 10–11.

"a kind of magic show." Johnson, *Modern Times*, 12.

After undergoing psychoanalysis. Bruce L. Smith, "Intellectual History of Harold D. Lasswell" in *Politics, Personality and Social Science in the Twentieth Century: Essays in Honor of Harold D. Lasswell*, Arnold A. Rogow, ed. (Chicago: University of Chicago Press, 1969), 57.

Lasswell became a lay analyst. Roy R. Grinker, "Psychoanalysis and Autonomic Behavior," in Rogow, 108.

"that human motives . . . economic motives." Smith, "Intellectual History," 57.

Franz Boas . . . Spanish loyalists. Ruth Benedict to Margaret Mead, 1 Sept. 1937, Mead Collection, Library of Congress.

"carry on my research . . . going crazy." *Time*, 11 May 1937, 26.

"to its fanatical extreme . . . traditional culture." Franz Boas, "Race Prejudice from the Scientist's Angle," *Forum* 98 (1937): 90–94.

"the hysterical claims . . . scientific background." F. Boas, "An Anthropologist's Credo," 201.

Hitler invaded Poland. Johnson, *Modern Times*, 362.

"useless eaters" and "lives devoid of value." See Proctor, chap. 7, for a good description of this period in Germany.

"the natural and God-given inequality of men." Ibid., 181.

"for those hopeless ones . . . agony of living." Foster Kennedy, "The Problem of Social Control of the Congenital Defective: Education, Sterilization, Euthanasia," *American Journal of Psychiatry* 99 (1942): 13–16.

18,269 killed. Proctor, *Racial Hygiene*, 191.

275,000 mentally retarded. Stephen L. Chorover, *From Genetics to Genocide*, 101.

Wannsee conference. Lucy S. Dawidowicz, *The War Against the Jews* (New York: Holt, Rinehart and Winston, 1975), 136–39.

officials . . . possessed doctoral degrees. Robert N. Proctor, "Science and Nazism," *Science* 241 (1988): 730–31.

"Gas chambers . . . followed the equipment." Proctor, *Racial Hygiene*, 212.

"not much will remain . . . of the Jews." Dawidowicz, *The War Against the Jews*, 139.

"I can heartily recommend the Gestapo to anyone." Jones, *The Life and Work of Sigmund Freud*, vol. 3 (New York: Basic Books, 1957), 226.

"It is my last war." Otto Friedrich, *City of Nets: A Portrait of Hollywood in the 1940s* (New York: Harper and Row, 1986), 26.

"The battle . . . is now won." Margaret Mead, *From the South Seas: Studies of Adolescence and Sex in Primitive Societies* (New York: William Morrow, 1939), x–xi.

"We must do our share . . . all of us." Franz Boas, Radio broadcast of 27 Sept. 1941; published in Boas, *Race and Democratic Society* (New York: J. J. Augustin, 1945), 1–2.

"I have a new theory about race." Mead, *An Anthropologist at Work*, 355.

5.
Postwar Propagation of the Freudian Faith

"The concentration camp . . . of an earlier century." Stephen Whitfield, *A Critical American: The Politics of Dwight Macdonald* (Hamden, CT: Shoe String Press, 1984), 52.

"the crime without a name." Paul

Johnson, *Modern Times: The World from the Twenties to the Eighties* (New York: Harper and Row, 1983), 418.

"On the screen . . . many laughed." Alfred Kazin, *Starting Out in the Thirties* (Boston: Little Brown, 1965), 166.

"could imagine . . . us all up at once." Stephen J. Whitfield, "The Holocaust and the American Jewish Intellectual," *Judaism* (Fall 1979): 391–401, quoting Kazin. This article is an excellent account of the effects of the Holocaust on intellectuals.

"the horrible details . . . skin of prisoners." James McCawley, "Atrocities—World War II," *Catholic World* Aug. 1945, 378–84.

"the next day only 2,000 . . . and mouths." Emanual Myron, "Back-Page Story," *New Republic* 17 Feb. 1947, 12–15.

"The Germans are in many ways *like us*." Dorothy Thompson, "The Lesson of Dachau," *Ladies Home Journal* Sept. 1945, 6.

"Before what we now know . . . of men's suffering." Lionel Trilling, *The Liberal Imagination* (New York: Anchor Books, 1953), 256.

"at night the red sky . . . for miles." Lucy S. Dawidowicz, *The War Against the Jews* (New York: Holt, Rinehart and Winston, 1975), 148.

"an elegant figure . . . in posture." Robert J. Lifton, *The Nazi Doctors* (New York: Basic Books, 1986), 342–44.

"We are living . . . bars of soap." Irving Howe, "The New York Intellectuals: A Chronicle and a Critique," *Commentary* 46 (1968): 29–51. Reprinted with the kind permission of *Commentary* magazine.

four of his elder sisters. Paul Johnson, *A History of the Jews* (New York: Harper and Row, 1987), 511.

"The controversy regarding . . . ideologies." Calvin S. Hall, "Temperament: A Survey of Animal Studies," *Psychological Bulletin* 38 (1941): 909–43.

"ten millions . . . dealt with." Johnson, *Modern Times*, 271.

"probably the most . . . against its citizens." Ibid., quoting Leslek Kolakowski.

one million members. Ibid., 301.

"exceedingly wise and gentle . . . to him." Ibid., 276.

"a man enters prison . . . at all." Ibid., 275–76.

"By the end . . . begun to stink." Milton Klonsky, "Greenwich Village: Decline and Fall," *Commentary* 6 (1948): 461.

"The New York Intellectuals . . . discussions." Alexander Bloom, *Prodigal Sons*, 251–52.

"We have obviously . . . American life." Editorial, "Our Country and Our Culture," *Partisan Review* 19 (1952): 282–326, 420–50, and 562–97.

"The ideal of the workers' . . . in Thyestes." Lionel Trilling, "Our Country and Our Culture," 319.

"In the West . . . in the neck." Sidney Hook, "Our Country and Our Culture," Ibid., 570.

"In the chastened . . . the intellectuals." Philip Rahv, "Our Country and Our Culture," 304.

"Freud . . . as the prophet." Charles Kadushin, "The Friends and Supporters of Psychotherapy on Social Circles in Urban Life," *American Sociological Review* 31 (1966): 786–802.

"The demise of Marxism . . . the intellectual realm." Richard King, *The Party of Eros: Radical Social Thought and the Realm of Freedom* (Chapel Hill: University of North Carolina Press, 1972), 44.

"My difficulties were . . . how to love." Arthur Miller, *Timebends: A Life* (New York: Grove Press, 1987), 320–21.

" 'Being analyzed' . . . or racial issues." Melitta Schmideberg, "A Contribution to the History of the Psycho-Analytic Movement in Britain," *British Journal of Psychiatry* 118 (1971): 61–68.

Sigmund Freud "Dostoevski and Parricide," *Partisan Review* 12 (1945): 530–44.

"the way for the Report . . . sensible way." Lionel Trilling, "The Kinsey Report," *Partisan Review* 15 (1948): 460–76.

"one of the . . . twentieth century." Laura Fermi, *Illustrious Immigrants: The Intellectual Migration from Europe 1930–41* (Chicago: University of Chicago Press, 1971), 141, quoting psychoanalyst Henry W. Brosin.

"noblest . . . of Western culture." John Burnham, "From Avant-Garde to Specialism: Psychoanalysis in America," *Journal of the History of the Behavioral Sciences* 15 (1979): 128–34, quoting Lionel Trilling.

"those dialectical disciplines . . . the human race." Joel Kovel, "Psychoanalyst in New York," in *Creators and Disturbers: Reminiscences by Jewish Intellectuals of New York*, Bernard Rosenberg and Ernest Goldstein, eds. (New York: Columbia University Press, 1982), 238.

"We were . . . if you didn't." William Barrett, *The Truants: Adventures Among the Intellectuals* (Garden City, N.Y.: Anchor Doubleday, 1982), 230.

"When the political is the same." Klonsky, "Greenwich Village: Decline and Fall," 461.

"the full terror of Nazism . . . and 'race'. " Margaret Mead, *Blackberry Winter: My Earlier Years* (New York: Simon and Schuster, 1972), 220.

"The anthropologist is concerned with . . . total systems." Margaret Mead, *From the South Seas: Studies in Adolescence and Sex in Primitive Societies* (New York: William Morrow, 1939), xxv.

Mead had written in three weeks. Jane Howard, *Margaret Mead: A Life* (New York: Simon and Schuster, 1984), 236.

"We must see this war . . . half forged." Margaret Mead, *And Keep Your Powder Dry: An Anthropologist Looks at the American Character* (New York: William Morrow, 1942), 261.

"develop . . . a world built new." Ibid., 273, quoting Mead.

"two scientific approaches . . . Freudian psychology." Margaret Mead, "The Study of National Character," in *The Policy Sciences*, Daniel Lerner and Harold D. Lasswell, eds. (Stanford: Stanford University Press, 1951), 70–85.

"the most fruitful . . . psychoanalysis." Howard, *Margaret Mead: A Life*, 189, quoting Mead.

"that the child learns . . . practices." Margaret Mead, "The Study of Culture at a Distance," in *The Study of Culture at a Distance*, eds. Margaret Mead and Rhoda Metraux (Chicago: University of Chicago Press, 1953), 37.

"a large part . . . in any society." Ibid.

"rituals, myths, films, popular art." Ibid.

"that human cultures . . . in history." Mead, "The Study of National Character."

"to work seriously with the zones of the body." Margaret Mead, *Male and Female: A Study of Sexes in a Changing World* (New York: Penguin, 1962), 348. First published in 1949.

"her Balinese field trip . . . research frame." Marvin Harris, *The Rise of Anthropological Theory* (New York: Crowell, 1968), 434.

"the scenes in Shakespeare . . . was Oedipal." Mary C. Bateson, *With a Daughter's Eye: A Memoir of Margaret Mead and Gregory Bateson* (New York: William Morrow, 1984), 43.

"a coarsening of the whole intellectual approach." Howard, *Margaret Mead: A Life*, 332.

Menninger had approvingly cited. William C. Menninger, "Characterologic and Symptomatic Expressions Related to the Anal Phase of Psychosexual Development," *Psychoanalytic Quarterly* 12 (1943): 161–93.

"Wherever Mead went . . . her good friends." Howard, *Margaret Mead: A Life*, 332.

what she had dreamed. Ibid., 248.

"did not maintain . . . patients." Ibid., 259.

"The [office] atmosphere . . . had happened." Ibid., 260.

"for the upper ten percent . . . no analyst." Ibid.

Benedict had characterized kite-flying. Margaret M. Caffrey, *Ruth Benedict: Stranger in This Land* (Austin: University of Texas Press, 1989), 319.

"psychologically show . . . criminality." Judith S. Modell, *Ruth Benedict: Patterns of a Life* (Philadelphia: University of Pennsylvania Press, 1983), 270.

Romania's child-rearing methods. Ibid.

"an unborn mouse." Howard, *Margaret Mead: A Life*, 224.

"and quite happily, platonic." Ibid., 293.

"I have never been in Japan . . . at war." Geoffrey Gorer, "Themes in Japanese Culture," *Transactions of the New York Academy of Sciences* 5 (1943): 106–124. Quoted with the kind permission of Annals of The New York Academy of Sciences.

key to the Russian character. Geoffrey Gorer and John Rickman, *The People of Great Russia* (London: Cresset, 1949). See also Mead and Metraux, *The Study of Culture at a Distance*.

"Gorer attempted to show . . . swaddling." Harris, *The Rise of Anthropological Theory*, 445.

swaddling hypothesis from Mead. Ibid.

"the cradleboard [among Indians] . . . ." Harold Orlansky, "Infant Care and Personality," *Psychological Bulletin* 46 (1949): 1–48.

1961 survey of AAA. Weston Labarre, "Psychoanalysis and Anthropology," *Science and Psychoanalysis* 4 (1961): 10–20.

"varying . . . insight." Ibid.

"We are really getting." Caffrey, *Ruth Benedict*, 249.

"then learning to be." Ibid., 248.

Karen Horney's course "Culture and Neurosis." Ibid., 250.

"the culturalists." Reuben Fine, *A History of Psychoanalysis*, 139.

"the two disciplines . . . license." Harris, *The Rise of Anthropological Theory*, 448.

"shake off her psychic angularities." W. David Sievers, *Freud on Broadway: A History of Psychoanalysis and the American Drama* (New York: Hermitage House, 1955), 220. See also, Wilfrid Sheed, *Clare Boothe Luce* (New York: E. P. Dutton, 1982), 58.

"psychoanalysis becomes . . . malicious chatter." Sievers, *Freud on Broadway*, 221.

Luce hires Freud's nephew. W. A. Swanberg, *Luce and His Empire* (New York: Charles Scribner's Sons, 1972), 55, 87.

Clare Boothe Luce also hired Bernays. Ibid., 194–195.

"A boom has overtaken . . . psychoanalysis." Francis S. Wickware, "Psychoanalysis," *Life*, 3 Feb. 1947, 98–108.

"It merely . . . a cure." Ibid.

"repressed sexual . . . both sexes."

Ibid.

"these rates . . . ends." Ibid.

"already . . . trends." Ibid.

"The True Freudians." *Time* 10 Sept. 1945, 70–72.

"had some extraordinary results." "Are You Always Worrying?" *Time* 25 Oct. 1948, 64–72.

"of the same . . . mother-in-law." Ibid.

"The emphasis on two-bit psychiatry . . . ." Ezra Goodman, *The Fifty-Year Decline and Fall of Hollywood* (New York: Simon and Schuster, 1961), 248.

Gregory Zilboorg, "Psychoanalysis and Religion," *Atlantic Monthly* Jan. 1949, 47–50.

Erich Fromm, "Oedipus Myth," *Scientific American* Jan. 1949, 22–23.

"caused . . . to improve." Sid Caesar, "What Psychoanalysis Did For Me," *Look* 2 Oct. 1956, 48–49.

"freak bestseller." Brock Brower, "Who's In Among the Analysts," *Esquire* July 1961, 78–84.

more articles on Freud. See notes for chapter 1 regarding how these calculations were made from the *Reader's Guide to Periodical Literature*.

Hannah Lees, "How I Got Caught In My Husband's Analysis," *Good Housekeeping* Nov. 1957, 80–279.

Anonymous, "Psychoanalysis Broke Up My Marriage," *Cosmopolitan* Oct. 1958, 70–79.

Gerald Sykes, "Dialogue of Freud and Jung," *Harper's* May 1958, 66–71.

David Bakan, "Moses in the Thought of Freud," *Commentary* 26 (1958): 322–31.

Anonymous, "Psychoanalysis and Confession," *Commonweal* 69 (1959): 414–15.

"Because of . . . Sigmund Freud." Lucy Greenbaum, "Dreams— Fantasies or Revelations?" *New York Times Magazine*, 10 Nov. 1946, 15–61.

"never before . . . on a wider basis." William C. Menninger, "Analysis of Psychoanalysis," *New York Times Magazine*, 18 May 1947, 12–50.

"the gist of Freud's theory . . . ." Franz Alexander, "Wider Fields for Freud's Techniques," *New York Times Magazine*, 15 May 1949, 15–53.

"few men have had a greater influence on their age." Leonard Engel, "Analysis of Sigmund Freud," *New York Times Magazine*, 4 Oct. 1953, 12–22.

"In the same way . . . a man can have." Alfred Kazin, "The Freudian Revolution Analyzed," *New York Times Magazine*, 6 May 1956, 15–38.

"In attempting an overview . . . for source material." Sievers, *Freud on Broadway*, 400.

Tennessee Williams's psychoanalysis. Roger Boxill, *Tennessee Williams* (New York: St. Martin's Press, 1987), 130. It should be noted that a correction sheet inserted in the front of the book said that Williams "entered," not "underwent" psychoanalysis in 1957.

"the quintessence . . . id and ego-ideal." Sievers, *Freud on Broadway*, 376–77.

"he uses . . . reality." Esther M. Jackson, *The Broken World of Tennessee Williams* (Madison: University of Wisconsin Press, 1966), 60–61.

William Inge's psychoanalysis. Sievers, *Freud on Broadway*, 352.

"considerable Freudian influence." Ibid.

"draw upon Freudian insights . . . the trite." Ibid., 355.

"the feelings . . . exposed in us." Ibid., 452.

"the members of the audience . . . a psychoanalytic one." Lionel Abel, *The Intellectual Follies: A Memoir of the Literary Venture in New York and Paris* (New York: W. W. Norton and Company, 1984), 221–22.

90 percent patient frequency of plays. Kadushin, "The Friends and Supporters of Psychotherapy," 548.

"to . . . drinking and psychoanalysis." Abel, *The Intellectual Follies*, 222.

Eliot had been hospitalized. Jeffrey Berman, *The Talking Cure: Literary Representations of Psychoanalysis* (New York: New York University Press, 1987), 93–97.

"appears to . . . in 1921." Ibid., 93.

"as a tragic . . . anomaly." Sievers, *Freud on Broadway*, 281.

Hellman began psychoanalysis. William Wright, *Lillian Hellman: The Image, The Woman* (New York: Simon and Schuster, 1986), 170–72.

Hellman's second therapist. Ibid., 408.

"my good friend, Gregory Zilboorg." Sievers, *Freud on Broadway*, 283.

"plethora of Freudian . . . vibrato." Wright, *Lillian Hellman: The*

Image, The Woman, 208.

Hellman . . . joined the Communist Party. Carl E. Rollyson, *Lillian Hellman: Her Legend and Her Legacy* (New York: St. Martin's Press, 1988), 320.

she claimed she knew nothing [about Stalin's show-trials]. Paul Johnson, *Intellectuals*, 295.

"whatever Zilboorg did for Field's mental health . . . ." Wright, *Lillian Hellman: The Image, The Woman*, 171.

Zilboorg['s] coordinating role. Jack Alexander, "Do-Gooder," *Saturday Evening Post*, 6 Dec. 1941, 14–108.

Zilboorg as Hellman's guest. Rollyson, *Lillian Hellman*, 158.

Zilboorg censured by the APA. Ibid., 171.

George Gershwin as Zilboorg's patient. Ibid., 157.

Richard Rodgers . . . used psychoanalytic ideas. Sievers, *Freud on Broadway*, 363.

"shown an early affinity." Ibid., 364.

Sievers claimed. Ibid., 364–66.

"clearly indebted to psychoanalysis." Ibid., 369.

Inside U.S.A. Ibid.

"been replaced by . . . psychoanalysis." Krin Gabbard and Glen O. Gabbard, *Psychiatry and the Cinema* (Chicago: University of Chicago Press, 1987), 50.

"primarily . . . common sense." Ibid., 56.

referred to as a "quack": Ibid., 57.

Fenichel a Party member. Paul Roazen, *Freud and His Followers* (New York: New York University Press, 1984), 505.

"a dedicated Socialist." Otto

Friedrich, *City of Nets: A Portrait of Hollywood in the 1940s* (New York: Harper and Row, 1986), 222.

all his employees undergo psychoanalysis: Roazen, *Freud and His Followers*, 170.

"cherished . . . critique of modern society. Friedrich, *City of Nets*, 222.

"the greatest love specialist." *New York Times*, 24 Jan. 1925, quoted by Catherine Covert, "Freud on the Front Page," 168.

"to cooperate in . . . stories of history." Irving Schneider, "The Theory and Practice of Movie Psychiatry," *American Journal of Psychiatry* 144 (1987): 996–1002.

"Hollywood found that psychoanalysis was fun." Friedrich, *City of Nets*, 224.

"Hollywood was full of . . . explanations." Ibid., 222.

"doorbell . . . to be heard." Ibid., 224.

Darryl Zanuck's psychoanalyst. Goodman, *The Fifty-Year Decline*, 249.

"Psychoanalysts . . . let the matter drop." Ibid.

"in Los Angeles . . . psychoanalytic jargon." "Midyear Mood of America," *Newsweek*, 4 July 1955, 46.

"was not only . . . source." Irving Schneider, "Images of the Mind: Psychiatry in the Commercial Film," *American Journal of Psychiatry* 134 (1977): 613–20.

Ben Hecht. Friedrich, *City of Nets*, 224.

"Our story . . . and confusion disappear." Gabbard, *Psychiatry and the Cinema*, 64.

"contain . . . in the American cinema." Ibid., 62.

"authoritative voices of . . . well-being." Ibid., 84.

"the oracular psychiatrist . . . wounded people." Ibid., 76.

"omniscient psychiatrist . . . three interviews." Franklin Fearing, "Psychology and the Films," *Hollywood Quarterly* 2 (1947): 118–121.

6.
Freud in the Nursery

"Most of the damage . . . of the century." Martin L. Gross, *The Psychological Society* (New York: Random House, 1978), 247, quoting Ames.

"I'm still basically a Freudian." Dr. Benjamin Spock, interview with the author, 2 June 1989.

"the amateur mother of yesterday." Sarah Comstock, "Mothercraft: A New Profession for Women," *Good Housekeeping* June 1914, 672–78.

"unrestrained . . . civilized beings." Anna Freud, foreword to Edith Buxbaum, *Your Child Makes Sense* (London: George Allen and Unwin, 1951) vii.

"that children . . . aggressive strivings." Ibid., ix.

"new wisdom . . . young children." Max Eastman, "Exploring the Soul and Healing the Body," *Everybody's Magazine* June 1915, 741–50.

a Freudian-inspired account. Peter C. MacFarlane, "Diagnosis by Dreams," *Good Housekeeping* Feb. 1915, 125–33.

455 articles on child rearing: A. Michael Sulman, "The Freudianization of the American Child: The Impact of Psychoanalysis in Popular Periodical Literature in the United States, 1919–1939" (Ph.D. diss., University of Pittsburgh, 1972), 96.

"only 17 percent of these articles . . . ." David R. Miller and Guy E. Swanson, *The Changing American Parent: A Study in the Detroit Area* (New York: John Wiley, 1958), 186.

"appeared to be . . . Freudian theory." Geoffrey H. Steere, "Freudianism and Child-Rearing in the Twenties," *American Quarterly* 20 (1968): 759–67.

"dozen healthy infants," ". . . race of his ancestors." James B. Watson, *Behaviorism* (New York: W. W. Norton, 1930), 104.

"If you expected . . . your child." James B. Watson, *Psychological Care of Infant and Child* (New York: W. W. Norton, 1928), 82.

Watson . . . overcome such obstacles. Christina Hardyment, *Dream Babies: Three Centuries of Good Advice on Child Care* (New York: Harper & Row, 1983), 170–71.

"a godsend to parents." Ibid., 173.

"on every intelligent mother's shelf." Ibid.

"a quiet goodnight . . . let him howl." Ibid., 186, quoting Watson.

"There is a . . . difficult task." Watson, *Psychological Care*, 81–82.

"If you haven't a nurse . . . use a periscope." Ibid., 84–85.

"Somehow I can't . . . too." Ibid., 84.

"mother love . . . wound." Ibid., 87.

"there is no . . . characteristics." Watson, *Behaviorism*, 94.

"Isn't it just possible . . . I raise it." Hardyment, *Dream Babies*, 173, quoting Watson.

Spock's father. Benjamin Spock and Mary Morgan, *Spock on Spock: A Memoir of Growing Up With The Century* (New York: Pantheon, 1989), 14.

Spock's mother. Ibid., 18.

"My mother was a tyrant." Peter Castro, "Chatter," *People*, 5 Aug. 1991, 90, quoting Spock.

"a private . . . to keep warm." Henry Allen, "Bringing Up Benjamin Spock," *Washington Post*, 27 Nov. 1989, B:1.

"worried about lions. . . ." Benjamin Spock, "Where I Stand And Why," *Redbook*, July 1967, 20–33.

"My mother . . . my children." Spock and Morgan, *Spock on Spock*, 60.

"had never . . . had a date." Ibid., 65.

He had completed three years at Yale. Benjamin Spock, "How My Ideas Have Changed," *Redbook* Oct. 1963, 51–126.

Spock's introduction to . . . Freud. Lynn Z. Bloom, *Doctor Spock: Biography of a Conservative Radical* (Indianapolis: Bobbs-Merrill, 1972), 95.

the psychological aspects. Spock, "How My Ideas Have Changed," 51–126.

"Taking care of schizophrenic . . . myself." Spock and Morgan, *Spock on Spock*, 102–103.

"Over a three-year period . . . with me." Ibid., 109–10. Details on Spock's psychoanalytic training are from this book and from Bloom, *Doctor Spock*, 71–72, 95–96.

"all children . . . different temperaments." Benjamin Spock, "Do Parents Cause Children's Emotional Troubles?" *Redbook*, June 1966, 20–23.

"When I began . . . saw it come out." Spock and Morgan, *Spock on Spock*, 16.

eminent faculty members. Bloom, *Doctor Spock*, 205.

"with such skilled counselors . . . difficulties." Benjamin Spock, "A Redbook Dialogue." *Redbook*, April 1972, 80–141.

"the children in the study." Bloom, *Doctor Spock*, 209. See also Spock, "A Redbook Dialogue" in which he says of the children "the usual problems arose."

"was slower and . . . anticipated." Benjamin Spock, "Toilet Training After 18 Months," *Redbook*, July 1968, 22–23.

"at the first signs . . . a while." Ibid.

"inconsistent and vacillating . . . any resistance." Benjamin Spock and Molly Bergen, "Parents' Fear of Conflict in Toilet Training," *Pediatrics* 34 (1964): 112–16. Reproduced by permission of *Pediatrics* and the Williams and Wilkins Company.

"turned out quite well." Dr. Benjamin Spock, interview with the author, 2 June 1989.

"the whole Oedipal situation . . . again." Ibid.

"Trust yourself . . . you do." Benjamin Spock, *The Common Sense Book of Baby and Child Care* (New

York: Duell, Sloan and Pearce, 1946), 1.

"the theoretical . . . is Freudian." Spock interview with the author, 2 June 1989.

"Dr. Spock . . . on child rearing." Bloom, *Doctor Spock*, 126.

"trying to take . . . their babies." Spock and Morgan, *Spock on Spock*, 130.

"children and . . . principles." Bloom, *Doctor Spock*, 75.

"were disproportionately . . . themselves." Spock and Morgan, *Spock on Spock*, 131.

"find ways . . . I had experienced." Spock, "Where I Stand and Why," 20–33.

chapter in a medical text. Benjamin Spock and Mabel Huschka, "The Psychological Aspects of Pediatric Practice," in *Practitioner's Library of Medicine and Surgery*, vol. 13, (New York: Appleton-Century, 1938), cited by Bloom, 84.

"When a baby . . . finicky person." Spock, *The Common Sense Book*, 195–96.

"He's apt to say . . . me, too." Ibid., 299.

"We realize . . . normal development." Ibid., 301.

"He may develop . . . children." Ibid., 303.

"libido might be forced into collateral roads." Sigmund Freud, "Three Contributions to the Theories of Sex" in *The Basic Writings of Sigmund Freud*, ed. Abraham A. Brill (New York: Modern Library, 1938), 594.

"psychoanalysts . . . about masturbation." Benjamin Spock, "Teaching Children Good Attitudes Toward Good Health," *Redbook*, Aug. 1972, 10–23.

"Freud taught . . . one degree or other." Benjamin Spock, "Should Girls Be Raised Exactly Like Boys?" *Redbook*, Feb. 1972, 24–28.

"[It was] that crucial stage . . . of guilt." Benjamin Spock, "What I Think About Nudity in the Home," *Redbook*, July 1975, 29–33.

"excitability . . . parents disapprove." Ibid.

"psychoanalytic experience . . . the same sex." Benjamin Spock, "Kinds of Rebellion in Adolescence," *Redbook*, July 1966, 20–25.

"I think it's wise . . . in your 'bed.'" Spock, "What I Think About Nudity in the Home," 29–33.

"is apt to be upset . . . resentment." Benjamin Spock, "A Little Excitement Goes a Long Way," *Redbook*, June 1965, 24–28.

"parental tickling . . . for children." Benjamin Spock, "Teaching Lovingness to Children," *Redbook*, July 1971, 26–37.

"momentarily with the hand . . . for any baby." Spock, "A Little Excitement Goes a Long Way," 24–28

"tossing the baby in the air." Ibid.

"pretend to be lions . . . nervous symptoms." Benjamin Spock, "A Father's Companionship," *Redbook*, Oct. 1974, 25–29.

"in the boy's unconscious . . . such play." Bloom, *Doctor Spock*, 92.

"getting down on . . . lion." Ibid.

"nine years of psychoanalysis." Ibid., 91.

"who run into study . . . psychotherapy." Spock, "Kinds of
Rebellion in Adolescence," *Redbook*, July 1966, 20–25.

"Parents should ask . . . great or
small." Benjamin Spock, "When a
Child Needs Psychiatric Help,"
Redbook, May 1966, 19–22.

"brought up in a Republican family
. . . ." Spock, "Where I Stand and
Why," 20–33.

"had been a Socialist in college."
Bloom, *Doctor Spock*, 59.

"to alleviate the plight . . . ." Ibid.,
83.

"who taught us all . . . 'OMM-
MMM.'" Spock and Morgan,
Spock on Spock, 190.

"for conspiracy to counsel . . .
draft." Ibid., 198.

"that a new . . . Democratic Party."
Bloom, *Doctor Spock*, 337.

"was raised on a book . . . they
espouse." Ibid., 132.

"Feed 'em whatever . . . teaching."
Ibid. See also Allen, "Bringing Up
Benjamin Spock."

"I've never considered myself . . . a
permissivist": Benjamin Spock,
"What I Said In February About
Raising Children—And What I
Did Not Say," *Redbook*, June 1974,
22–31.

"that they would arouse . . . their
children." Spock and Bergen,
"Parents' Fear of Conflict in Toilet
Training."

"The parents were afraid . . . their
children." Spock, "A Redbook
Dialogue," 80–141.

"All the mothers ignored . . . the
bathroom." Spock and Bergen,
"Parents' Fear of Conflict in Toilet
Training," 112–16.

these four were *more* successful.
Ibid.

"in which the . . . working class."
Ibid.

"few of the families . . . in psychology." Spock, "Toilet Training,"
Redbook, Nov. 1963, 38–46.

"without severity . . . to the personality." Ibid.

"The fear of arousing . . . has
helped." Spock and Bergen, "Parents' Fear of Conflict in Toilet
Training," 112–16.

"It's professional people . . .
opened." Allen, "Bringing up
Benjamin Spock."

"was sure that . . . certain parents."
Spock, "How My Ideas Have
Changed," 51–126.

"throw food on the floor." Ibid.

"there now were . . . parental hesitancy." Ibid.

"took care to tone . . . about
them." Spock, "Toilet Training,"
38–46.

"occur mainly in . . . child psychology." Benjamin Spock, *Baby and
Child Care* (1968 edition), 259.

"the first medium . . . is valuable."
Benjamin Spock, "Helping Your
Children to Learn About Money,"
Redbook, Dec. 1967, 20–22.

"a generous lacing . . . in child-care
history." Hardyment, *Dream
Babies*, 233.

"This vitreus monster . . . just
anyone." Selma H. Fraiberg,
*The Magic Years: Understanding
and Handling the Problems of
Early Childhood* (New York:
Charles Scribners' Sons, 1959),
94.

"Spoil That Baby." Gross, *The Psychological Society*, 266.

"Instead of getting . . . character."
Ibid.

"by my child development friends."
Margaret Mead, *Blackberry Winter: My Earlier Years* (New York: Simon and Schuster, 1972), 248.

"unusual to find . . . all authority."
Margaret Mead, "Margaret Mead Answers Questions," *Redbook*, Feb. 1964, 12.

"the air of a Nativity pageant." Jane Howard, *Margaret Mead: A Life* (New York: Simon and Schuster, 1984), 217.

Mead wanted a record. Mead, *Blackberry Winter*, 261.

"at the slightest whimper." Howard, *Margaret Mead: A Life*, 218.

"She never insisted . . . as a child." Mary C. Bateson, *With A Daughter's Eye: A Memoir of Margaret Mead and Gregory Bateson* (New York: William Morrow, 1984), 72–73.

"a training institute . . . child care." Bloom, *Doctor Spock*, 85. See also Spock and Morgan, 110–11 and Margaret M. Caffrey, *Ruth Benedict: Stranger in This Land* (Austin: University of Texas Press, 1989), 244.

Ernest Jones had published. John C. Burnham, "Psychoanalysis in American Civilization Before 1918," (Ph.D. diss., Stanford University, 1958), 344.

his book . . . was published. Wilfred Lay, *The Child's Unconscious Mind, The Relations of Psychoanalysis to Education, A Book For Teachers and Parents* (New York: Dodd, Mead, and Co., 1919).

"a kind of mental . . . emotional rapport." Roy Lubove, *Professional Altruist: The Emergence of Social Work as a Career 1880–1930* (New York: Cambridge University Press, 1965), 100–101.

"that academic credit . . . educational curriculum." Reuben Fine, *A History of Psychoanalysis* (New York: Columbia University Press, 1979), 564.

sixty thousand school guidance counselors. Gross, *The Psychological Society*, 272.

"schools are our community mental health centers." Ibid., 271.

"a dubious character . . . quasi-pornographic." Paul Roazen, *Freud and His Followers* (New York: New York University Press, 1984), 211.

"Vegeto therapy." Alexander S. Neill, *Neill! Neill! Orange Peel!* (New York: Hart Publishing Company, 1972), 511. The information on Neill's three psychoanalyses is also from this book.

"carries on the . . . to be a hypocrite." Ibid., 510.

"A Frenchwoman . . . make him behave." Spock, "Do Parents Cause Children's Emotional Troubles?", 20–23.

"Spock and Gesell . . . household." Isaac Rosenfeld, "Life In Chicago," *Commentary* 23 (1957): 530–31.

7.
Freud in <u>Jails and Prisons</u>

"I believe . . . a metaphysical one." William A. White, *Forty Years of Psychiatry* (New York: Nervous and Mental Disease Publishing Company, 1933), 78.

theories of Cesare Lombroso. For a summary of his work, see Stephen J. Gould, *The Mismeasure of Man* (New York: W. W. Norton, 1981), 132–42.

The Jukes. Hamilton Cravens, *The Triumph of Evolution: American Scientists and the Heredity-Environmental Controversy, 1900–1941* (Philadelphia: University of Pennsylvania Press, 1978), 3–4.

Healy . . . discovered Freud's work. John C. Burnham, "Psychoanalysis in American Civilization Before 1918," (Ph.D. unpublished doctoral diss., Stanford University, 1958), 257.

"mental conflicts and repressions." William Healy, *The Individual Delinquent* (Montclair, N.J.: Patterson Smith, 1969), 352. Originally published in 1915.

"most cases . . . sex experiences." Ibid., 353.

"the complex . . . and action." William Healy, *Mental Conflicts and Misconduct* (Boston: Little Brown and Company, 1917), 23.

"We ourselves have . . . impulses." Ibid., 29.

"peered anxiously . . . psychoanalyzed." Roy Lubove, *Professional Altruist: The Emergence of Social Work as a Career* (New York: Cambridge University Press, 1965), 88.

"that all . . . is mental hygiene." Ibid., 113.

"American social workers . . . ." *The Freudian Fallacy: An Alternative View of Freudian Theory* (Garden City, N.Y.: Dial Press, 1984), 246.

"impact on criminology . . . profound": Burnham, "Psychoanalysis in American Civilization Before 1918," 261.

"not less than two-thirds . . . confinement." Bernard Glueck, "Concerning Prisoners," *Mental Hygiene* 2 (1918): 177–218.

"a new type . . . criminal himself." Ibid.

"59 per cent." Ibid.

"an offending pathogenic . . . subject." Bernard Glueck, *Studies in Forensic Psychiatry* (Boston: Little Brown, 1916), 244–45.

"wide-spread superstition." Glueck, "Concerning Prisoners."

"the penal problem . . . a psychiatric one." Margo Horn, *Before It's Too Late: The Child Guidance Movement in the United States, 1922–1945* (Philadelphia: Temple University Press, 1989), 22, quoting Glueck.

"a system of psychopathic . . . offenders." Glueck, "Concerning Prisoners," 177–218.

"Criminology . . . at the psychological level." Glueck, *Studies in Forensic Psychiatry*, vii.

"Indeed . . . correction and reformation." Ibid., v.

William A. White. Information on White is from Arcangelo R. T. D'Amore, ed., *William Alanson White: The Washington Years 1903–1937.* DHEW publication no. (ADM) 76–298. (Washington: Government Printing Office, 1976).

"It has been demonstrated . . . past." William A. White, *Insanity and the Criminal Law* (New York: Macmillan, 1923), 37–38.

"that the murderer . . . most surprised." Ibid.

"the discarding of . . . responsibility." White, *Forty Years of Psychiatry*, 78.

"that prisons . . . be abolished." William A. White, *Crime and Criminals* (New York: Farrar and Rinehart, 1933), 231.

"gradual transformation of . . . human conduct." D'Amore, *William White*, 51, quoting White.

"making the return . . . the past." Ibid.

"Save our boys . . . hanged." Arthur Weinberg and Lila Weinberg, *Clarence Darrow: A Sentimental Rebel* (New York: G. P. Putnam's Sons, 1980), 298.

"a milestone defense." Ibid., 297.

"For the first time . . . law." D'Amore, *William White*, 110, quoting Darrow.

"a short story he had published." Clarence Darrow, "A Skeleton in the Closet" in *A Persian Pearl and Other Essays* (New York: Haskell House, 1974), 235–45. Originally published in 1899.

"they were . . . responsible." Weinberg, *Clarence Darrow*, 312.

"It seems to me . . . used." Clarence Darrow, *Crime: Its Causes and Treatment* (New York: Thomas Y. Crowell, 1922), 274–75.

"all prisons . . . in a hospital." Ibid., 278.

"any sum he cared to name." Catherine L. Covert, "Freud on the Front Page: Transmission of Freudian Ideas in the American Newspaper of the 1920's" (Ph.D. diss., Syracuse University, 1975), 124.

"to charter . . . other passengers." Weinberg, *Clarence Darrow*, 301.

"The behavior . . . yesterday's impressions." *Boston Herald*, 3 Aug. 1924, cited by Covert, 27.

"mentally diseased . . . "; ". . . pie for supper." Hal Higdon, *The Crime of the Century* (New York: G. P. Putnam's Sons, 1975), 216.

"To my mind . . . split personality." Ibid., 217.

"governed . . . with this crime." Ibid., 218.

"is suffering . . . the situation." Ibid.

"the inevitable . . . personalities." Ibid.

"no one ever knew . . . ." *Chicago Tribune*, 5 June 1924, cited by Covert, 67.

"was permitted to testify . . . the two boys." D'Amore, *William White*, 110.

"Dickie . . . perfectly good." Higdon, *The Crime of the Century*, 206.

"prudish and austere." *New York Times*, 3 Aug. 1924, cited by Covert, 7.

"pushed . . . his schoolwork." Higdon, *The Crime of the Century*, 207.

"Miss Struthers and Mathilda Wantz . . . ." Ibid., 226.

"Dickie . . . antisocial tendencies." *New York Times*, 3 Aug. 1924, cited by Covert, 6.

"still . . . talking to his teddy bear." Higdon, *The Crime of the Century*, 212.

"There is a tendency . . . in reality." Ibid., 208.

"I am . . . *Who's Who*." Ibid., 212.

"were perfectly charming." Ibid.

"If the fathers . . . law-abiding." Ibid., 208–209.

"Loeb . . . Giant Ego." *New York*

Journal, 2 Aug. 1924, cited by Covert, 1.

"putting the boys . . . in a criminal court." *New York Herald-Tribune*, 2 Aug. 1924, cited by Covert, 6.

"the clarity . . . of the body." *New York Journal*, 2 Aug. 1924.

"patients . . . examination." *New York Herald-Tribune*, 2 Aug. 1924.

"finally forced . . . flushed and angered." Covert, "Freud on the Front Page," 46.

"the careful analysis . . . to criminology." Weinberg, *Clarence Darrow*, 312.

"that similar . . . abnormalities." Ibid.

"the real responsibility." *New York Times*, 4 Aug. 1924, cited by Covert, 67.

White was subsequently investigated. D'Amore, *William White*, 4, 140.

Darrow . . . one of the principal speakers. Peter Gay, *Freud: A Life for Our Time* (New York: W. W. Norton, 1988), 574.

"a sort of crash course . . . Freudian thought." Covert, "Freud on the Front Page," 2, vi.

Nathan Freudenthal Leopold: So listed in the Library of Congress card catalogue.

"Maybe . . . my subconscious mind." *New York Graphic*, 15 Sept. 1924, cited by Covert, 269.

"disciplinary measure for children." Associated Press, 1926, cited by Covert, 270.

"I remember thinking how absurd . . . own conduct." Justin Kaplan, *Lincoln Steffens: A Biography* (New York: Simon and Schuster, 1974), 201.

"Please mamma . . . than I." Lawrence J. Friedman, *Menninger: The Family and the Clinic* (New York: Alfred A. Knopf, 1990), 13.

"I guess . . . have a baby now." Ibid., 145.

"feebleminded." Ibid., 29.

papers on the post-influenzal schizophrenia syndrome. Karl A. Menninger, "Reversible Schizophrenia," *American Journal of Psychiatry* 1 (1922): 573–88; Karl A. Menninger, "Influenza and Schizophrenia," *American Journal of Psychiatry* 5 (1926): 469–529.

brief trial of psychoanalysis. Friedman, *Menninger*, 47.

"I don't think . . . that long." Ibid., 145.

"among the very best." Paul Roazen, *Freud and His Followers* (New York: New York University Press, 1984), 510.

"encouraged him to have a mistress." Friedman, *Menninger*, 82.

termination of this "foursome." Ibid., 141.

"he . . . to take a lover." Ibid., 83.

"told Freud . . . very much." Ibid., 108.

"utterly impersonal." Ibid.

"never had . . . ill patients." Ibid., 109.

"my narcissism received . . . a terrific blow." Ibid., 108.

He subsequently wrote to Freud. Ibid., 109–10.

"Freud did not treat . . . my life." Ibid., 110.

"launched him on . . . modern times." Ibid.

Menninger compared Freud to Plato and Galileo. Karl A. Menninger, "Death of a Prophet," *New*

Republic, Aug. 9, 1939, 23–25. Despite the title of this article it was published six weeks prior to Freud's death and was not an obituary.

"genius . . . character." Karl Menninger, "Sigmund Freud," *Nation* Oct. 7, 1939, 373–74.

"more Freudian than Freud." Obituary of Karl Menninger, *New York Times*, 19 July 1990, B:6.

"stressed the role of . . . development." Roazen, *Freud and His Followers*, 144.

"was unquestionably Freud's favorite in Vienna." Roazen, *Freud and His Followers*, 421–22.

"closer . . . daughter Anna." Ibid.

"one of Freud's most brilliant pupils." Ibid., 156.

Brunswick had divorced. Roazen, *Freud and His Followers*, 423–26.

Brunswick had become addicted. Ibid., 427 ff. See also Friedman, 86.

"tendency to fall asleep . . . analytic hour." Friedman, *Menninger*, 86.

"occasionally ordering things . . . stores." Steven Marcus, *Freud and the Culture of Psychoanalysis* (Boston: G. Allen and Unwin, 1984), 214. This account is of a woman who was in psychoanalysis with Dr. Brunswick at the same time as Menninger.

"chronically insecure." Friedman, *Menninger*, 125.

"moody and unpredictable." Ibid., 190.

strike by psychiatrists. Ibid., 134–35.

revolt of the staff. Ibid., 304–33.

misogynistic articles. K. A. Menninger, "Men, Women and Hate,"

Atlantic Monthly Feb. 1939, 158–68 and "Parents Against Children," *Atlantic Monthly* Aug. 1939, 163–75.

"the childhood experience . . . too rapidly." Menninger, "Parents Against Children." All other quotes are from this article.

Menninger knew Healy. Friedman, *Menninger*, 92.

"medieval stupidities." Karl A. Menninger, "Psychiatry and the Prisoner," *Proceedings of the National Conference of Social Work*, (1925): 552–55. All other quotes are from this article.

"Alexander, Healy . . . effective control." Karl A. Menninger, "Combatting Man's Destructive Urge," *Survey Graphic* Oct. 1937, 520–23.

"the deductive genius . . . of destructiveness." Menninger, "Parents Against Children."

"it is . . . of war." Ibid.

question-and-answer column. Karl Menninger, "Mental Hygiene in the Home," *Ladies Home Journal* Oct. 1930, 109, Nov. 1930, 101, and Dec. 1903, 75. In these columns Menninger frequently blamed the mothers for causing their childrens' problems.

Time cover story. "Are You Always Worrying?" 25 Oct. 1948, 64–72.

"the world's best-known psychiatric center." "Menninger Appeal," *Newsweek* 17 April 1950), 50–51.

"a thunderous . . . law enforcement." *New York Times* review quoted on cover of Karl Menninger, *The Crime of Punishment* (New York: Viking Press, 1969), paperback edition.

"a model of rationalism." Ibid.

"the beginning of . . . the criminal's mind." "A Psychiatrist Views Crime," *Time*, Dec. 6, 1968, 117–18.

articles by Menninger. Karl Menninger, "The Crime of Punishment," *Saturday Review of Literature* Sept. 7, 1968, 21–25; Karl Menninger, "Punishment as Crime," *Saturday Evening Post* Oct. 5, 1968, 16–22; Karl A. Menninger, "A Psychiatrist Looks at Violence," *Catholic World* Sept. 1969, 262–64.

"The offenders . . . from *our* standpoint." Karl Menninger, "Verdict Guilty—Now What?" *Harper's* Aug. 1959, 60–64.

"I suspect that . . . against them." Menninger, *The Crime of Punishment*, 28.

"the great majority of . . . them." Ibid., 265.

"before . . . a modern therapeutic one." Menninger, "Punishment as Crime," 16–22.

"This would . . . present form and function." Menninger, "The Crime of Punishment," *Saturday Review of Literature*.

"the existence of . . . criminal responsibility." Winfred Overholser, "Major Principles of Forensic Psychiatry" in Silvano Arieti, ed., *American Handbook of Psychiatry*, vol. 2 (New York: Basic Books, 1959), 1897.

"the present commissioner . . . of the state." Stanley P. Davies, "Mental Hygiene and Social Progress," *Mental Hygiene* 13 (1929): 226–49.

"fifty . . . to be examined." Robert N. Poctor, *Racial Hygiene: Medicine Under the Nazis* (Cambridge: Harvard University Press, 1988), 203.

"Predisposition . . . and parental control." Seymour L. Halleck, *Psychiatry and the Dilemmas of Crime* (Berkeley: University of California Press, 1967), 88.

"They were never . . . their parents." Benjamin Spock, "Love and Good Behavior," *Redbook*, Aug. 1976, 23–26.

"We can accept . . . way of living." Margaret Mead, "A Life for a Life: What That Means Today," *Redbook*, June 1978, 56–60.

"the discarding of the concept of responsibility." White, *Forty Years of Psychiatry*, 78.

"Our traditions . . . criminal responsibility." Richard Arens, *Make Mad the Guilty* (Springfield, IL.: Charles C. Thomas, 1969), vii.

"more revolutionary . . . segregation." Ibid., quoting Menninger.

"a psychiatric . . . clown act." G. Wright, "Sirhan's Psyche Show," *San Francisco Examiner and Chronicle*, 20 April 1969, 8.

"read everything . . . to be a psychoanalyst." Covert, "Freud on the Front Page," 72.

insanity increased fivefold between 1965 and 1976. Peter Meyer, *The Yale Murder* (New York: Empire Books, 1982), 200.

admitted purchasing a book. Willard Gaylin, *The Killing of Bonnie Garland* (New York: Simon and Schuster, 1982), 83–84. Gaylin, who obtained the information from Herrin in an interview, said that the book was "about a police officer who, having committed a murder, is defended by psychiatrists."

"transient situational reaction." Meyer, *The Yale Murder*, 215, 219.

"the Oedipal situation in life." Ibid., 217.

"an intense fear of . . . loved ones." Ibid., 179.

"Psychiatrically speaking . . . guilty of an abscess." Gaylin, *The Killing of Bonnie Garland*, 253.

"If you . . . are breaking up." Ibid., 270.

"The Freudian faith . . . works both ways." Vladimir Nabokov, *Strong Opinions* (New York: Vintage Books, 1973), 116.

"If Menninger was . . . one of its victims." Meyer, *The Yale Murder*, 266.

"ought to be . . . that disease." Margo Horn, *Before It's Too Late: The Child Guidance Movement in the United States, 1922–1945* (Philadelphia: Temple University Press, 1989), 28.

Cambridge-Sommerville Delinquency Project. This project was nicely summarized by Dr. Joan McCord in "Crime in Moral and Social Contexts—the American Society of Criminology, 1989 Presidential Address," *Criminology* 28 (1990): 1–26 and "Consideration of Some Effects of a Counseling Program" in Susan E. Martin, Lee B. Sechrest, and Robin Redner, eds., *New Directions in the Rehabilitation of Criminal Offenders* (Washington, D.C.: National Academy Press, 1981), 394–405.

"the boys who had . . . fewer crimes." McCord, "Crime in

Moral and Social Contexts," 1–26.

"The evaluation . . . from the program." Ibid.

"as adults . . . committed more than one crime." Joan McCord, "A Thirty-Year Follow-Up of Treatment Effects," *American Psychologist* 33 (1978): 284–89.

" 'More' was 'worse.' " McCord, "Consideration of Some Effects," 394–405.

"the supportive attitudes . . . they could receive." Ibid.

"prisons and punishment should both be abolished." White, *Careers and Criminals*, 231.

"former Patuxent inmates are . . . in the study." Howard Schneider, "Patuxent's Counseling Said to be Ineffective," *Washington Post*, 6 Feb. 6 1991, D:6.

"have no . . . recidivism." Ibid.

Charles Wantland. Paul Duggan and Debbie M. Price, "Patuxent Twice Freed P. G. Slayer," *Washington Post*, 10 Dec. 1988, A:1. The information on Billy Ray Prevatte, James Stavarakas and Robert Angell was also taken from this article.

William Snowden. "Patuxent Had Twice Paroled Man Charged in Md. Slaying," *Washington Post*, 3 Jan. 1991, D:3.

"nothing but a psychiatric sandbox." Paul Duggan, "Patuxent Board Faces Unhappy Choices in Deciding Holiday Furloughs," *Washington Post*, 27 Nov. 1988, A:1.

"the effect of individual . . . recidivism." Douglas Lipton, Robert Martinsen, and Judith Wilks, *The Effectiveness of Correctional Treatment*

(New York: Praeger, 1975), 210.

"no clearly . . . made." Ibid., 213.

"pragmatically . . . recidivism." Ibid., 210.

"treatment . . . orientation." Ibid., 213.

"nearly 63 percent . . . within three years." "Most Ex-Inmates Rearrested Within 3 Years, Study Finds," *Washington Post*, 3 April 1989, A:5.

Willie Horton. Information on the Horton case was taken from the Pulitzer Prize-winning series in the Lawrence [Massachusetts] *Eagle-Tribune*, 17 Apr., 7, 13, 27 May, 7 June, 21 July, 16 Aug., 6, 27 Dec. 1987. It was this series by a local newspaper which originally exposed the Horton story.

"revolving-door . . . for parole." Jack W. Germond and Jules Witcover, *Whose Broad Stripes and Bright Stars?* (New York: Warner Books, 1989), 11.

"Is this your pro-family team for 1988." Ibid., 423.

"many Democrats . . . of public policy." Meg Greenfield, "No Furlough From Crime," *Washington Post*, 6 Dec. 1988, A:21.

8.
Philosopher Queen and Psychiatrist Kings: The Freudianization of America

"Above all . . . diffused throughout our culture." Morris Dickstein, *Gates of Eden: American Culture in the Sixties* (New York: Penguin Books, 1977), v.

"The radicalism of the 1960s . . . was economic." Richard H. Pells, *The Liberal Mind in a Conservative Age: American Intellectuals in the 1940s and 1950s* (New York: Harper and Row, 1985), 403.

"everybody . . . is his chief opposition." Brock Brower, "Who's In Among the Analysts," *Esquire* July 1961, 78–84.

"revolution has been . . . organized religion." Charles J. Rolo, "The Freudian Revolution," *Atlantic Monthly* July 1961, 62.

"Mother of the World." Robert Cassidy, *Margaret Mead: A Voice for the Century*, (New York: Universe Books, 1982), 10.

"gave more than 100 speeches . . . and articles." Margaret Mead, "September to June: An Informal Report," *Redbook*, June 1963, 32.

"a woman of enormous self-confidence." Irene Kubota, "An Interview With Margaret Mead," *Redbook*, Aug. 1974, 31.

"liberally endowed . . . convictions." Marvin Harris, *The Rise of Anthropological Theory* (New York: Crowell, 1968), 412.

"her egotism was ungovernable." Jane Howard, *Margaret Mead: A Life* (New York: Simon and Schuster, 1984), 406.

"Mother-Goddess Mead." David Cort, "Margaret Mead for President," *Monocle*, Summer-Fall, 1963, 27–30.

"Hello, isn't . . . to know?" Howard, *Margaret Mead: A Life*, 386.

investigations carried out by Freud. Margaret Mead, "A New Understanding of Childhood," *Redbook*, Jan. 1972, 54.

"the source of . . . his relations to

people." Margaret Mead, "Margaret Mead Answers," *Redbook*, Oct. 1965, 22.

"there are also . . . mishaps in upbringing." Margaret Mead, "Margaret Mead Answers," *Redbook*, Dec. 1964, 6.

"Comparative studies suggest . . . routine." Margaret Mead, "Margaret Mead Answers," *Redbook*, Jan. 1965, 6.

"modern scientific treatment." Margaret Mead, "Margaret Mead Answers," *Redbook*, June 1966, 30.

"Freud's belief was . . . more productive." Margaret Mead, "A New Understanding of Childhood," *Redbook*, Jan. 1972, 54.

"A willingness to look . . . as a person." Margaret Mead, "Margaret Mead Answers," *Redbook*, Nov. 1978, 37–39.

Mead . . . had not undergone psychoanalysis." Margaret Mead, "Margaret Mead Answers," *Redbook*, Dec. 1964, 7.

never "spent three uninterrupted days with her daughter." Howard, *Margaret Mead: A Life*, 243.

"for almost a year . . . or not at all." Mary C. Bateson, *With a Daughter's Eye: A Memoir of Margaret Mead and Gregory Bateson* (New York: William Morrow, 1984), 99.

Mead as a lifelong liberal. S. Toulmin, "The Evolution of Margaret Mead, *New York Review of Books*, 6 Dec. 1984, 3–9. See also Margaret Mead, "Margaret Mead Answers," *Redbook*, June 1965, 10.

"There are no elders . . . unknown." Sheila Johnson, "A Look at Margaret Mead," *Commentary* 55 (1973): 70–72.

the legalization of marijuana. Howard, *Margaret Mead: A Life*, 390.

"aids to therapy." Margaret Mead, *New Lives for Old* (New York: William Morrow, 1956), 524. See also Margaret Mead, "Margaret Mead Answers," *Redbook*, Jan. 1968, 32.

she gave as many as eighty lectures. Winthrop Sargeant, "It's All Anthropology," *New Yorker*, 30 Dec. 1961, 31–44.

"with the same range of potentialities." Margaret Mead, *Blackberry Winter: My Earlier Years* (New York: Simon and Schuster, 1972), 224.

"the most powerful influence on . . . his culture." Margaret Mead, "Margaret Mead Answers," *Redbook*, Feb. 1963, 21.

"measurable differences in their capacity . . . civilization." Margaret Mead, *New Lives for Old* (New York: William Morrow, 1960), 436.

"a reduction . . . of crime." Margaret Mead, "The Nudist Idea," *Redbook*, July 1968, 43.

criticized laws against homosexuality. See, for example, her columns in *Redbook* Dec. 1964, March 1968, and April 1968.

"bisexual potentialities are normal." Margaret Mead, "Margaret Mead Answers," *Redbook*, July 1963, 29.

"probably because . . . of the opposite sex." Margaret Mead, "Bisexuality: What's It All About?" *Redbook*, Jan. 1975, 29.

"If the term *natural* be taken to . . . potentiality." Margaret Mead, "Cultural Determinants of Sexual

Behavior," in *Sex and Internal Secretions*, ed. William C. Young (Baltimore: Williams and Wilkins, 1961), 1471.

"running around wearing a . . . human being." Kubota, "An Interview with Margaret Mead," 31.

"seemed to me . . . progressivism." David Riesman, *Individualism Reconsidered and Other Essays* (Glencoe: The Free Press, 1954), 306.

"the theorists . . . of the sixties." Dickstein, *Gates of Eden*, 70.

"a typical thirties radical." Thomas B. Morgan, "How Hieronymus Bosch (XVth Century) and Norman O. Brown (XXth) Would Change the World," *Esquire* March 1963, 100–135. Used with permission of *Esquire* and the Hearst Corporation.

Marcuse. For his background see Richard Goodwin, "The Social Theory of Herbert Marcuse," *Atlantic Monthly* June 1971, 68–85.

"The most exciting works available." Greg Calvert and Carl Neiman, *A Disrupted History: The New Left and the New Capitalism* (New York: Random House, 1971), 37.

conference on Marcuse's works. Edward J. Bacciocco, *The New Left in America: Reform to Revolution 1956 to 1970* (Stanford: Hoover Institution Press, 1974), 186.

Marcuse's stepson. Allen J. Matusow, *The Unraveling of America: A History of Liberalism in the 1960s* (New York: Harper and Row, 1984), 332.

"The riot is the social extension of the orgasm." Thornton, *The Freudian Fallacy: An Alternative View of Freudian Theory* (Garden City, N.Y.: Dial Press, 1984), 250.

Marcuse . . . went into hiding. Paul Robinson, *The Freudian Left: Wilhelm Reich, Géza Róheim, Herbert Marcuse* (New York: Harper and Row, 1969), xii.

"The ideological leader of the New Left." Massimo Teodori, ed., *The New Left: A Documentary History* (Indianapolis: The Bobbs-Merrill Company, 1969), 469.

"a prophet of . . . irrational form." Leszek Kolakowski, *Main Currents of Marxism: Its Origin, Growth and Dissolution* (Oxford: Clarendon Press, 1978), 415.

"a moral imperative." Herbert Marcuse, "Letter to Angela Davis," *Ramparts* Feb. 1971, 22.

"as the true proletariat." Robinson, *The Freudian Left*, 243.

"Our civilization is . . . the sense of guilt." Sigmund Freud, *Civilization and Its Discontents*, in *The Standard Edition of the Complete Psychological Works of Sigmund Freud*, vol. 21, ed. James Strachey (London: Hogarth Press, 1961), 86, 134.

"The sickness of . . . his civilization." Herbert Marcuse, *Eros and Civilization: A Philosophical Inquiry Into Freud* (New York: Vintage Books, 1955), 224.

"a new stage of civilization." Ibid., viii.

"being is essentially the striving for pleasure." Ibid., 113.

"a new erotic pastoral." Richard King, *The Party of Eros: Radical Social Thought and the Realm of*

Freedom (Chapel Hill: University of North Carolina Press, 1972), 138.

"play . . . the desired human activities." Ibid., 136.

"smitten with . . . *Eros and Civilization*." Philip Gold, "The 19th Century on the Couch," *Insight*, 26 Nov. 1990, 56–57.

"the most significant . . . publication." Clyde Kluckhohn, from a book review of *Eros and Civilization* in the *New York Times*, no date given, on the cover of the book.

"The sexual deviant . . . hero of *Eros and Civilization*." Robinson, *The Freudian Left*, 241.

"the chief spokesman . . . for radicalism." King, *The Party of Eros*, 78.

"the repression of infantile . . . kind." Ibid., 84.

"gonad theory of revolution." Ibid., quoting C. Wright Mills and P. J. Salter.

Nineteen publishers had rejected the manuscript. Pells, *The Liberal Mind in a Conservative Age*, 208.

pretext of the book. Paul Goodman, *Growing Up Absurd: Problems of Youth in the Organized Society* (New York: Vintage Books, 1960), xvi.

"organic integration of work, living and play." Pells, *The Liberal Mind in a Conservative Age*, 213, quoting Goodman.

"in Goodman's terms . . . regeneration." King, *The Party of Eros*, 111.

one of the campus "bibles." Pells, *The Liberal Mind in a Conservative Age*, 208.

survey of SDS leaders. Kirkpatrick

Sale, *SDS* (New York: Random House, 1973), 205.

"My homosexual acts . . . as a right." Dickstein, *Gates of Eden*, 77, quoting Goodman.

Goodman's autobiographical notes. Paul Goodman, *Five Years* (New York: Brussel and Brussel, 1966).

"deeply stirred by . . . the Left." Morgan, "How Hieronymus Bosch," 100–135.

"I have never had . . . effect on me." Ibid.

"There is . . . repression of himself." Norman O. Brown, *Life Against Death: The Psychoanalytical Meaning of History* (New York: Vintage Books, 1959), 3.

"Assuming . . . toilet-training patterns." Morgan, "How Hieronymus Bosch," 100–135.

"What the great world needs . . . strife." Brown, *Life Against Death*, 322.

"We, however . . . to know himself." Ibid., xiii.

"delighting in . . . sensuous life." Ibid., 308.

"Here again . . . bodily organs." Ibid., 308.

"Freud and Marx . . . together." "Freud's Disciple," *Time* 15 July 1966, 82, quoting Brown.

"overwhelmed . . . by a major thinker." Norman Podhoretz, *Breaking Ranks: A Political Memoir* (New York: Harper and Row, 1979), 48.

"One of the most interesting . . . I know." Lionel Trilling, review of *Life Against Death*, by Norman O. Brown, in *Mid Century* (quoted on back cover of the book).

"one . . . to be with it." "Freud's

Disciple," *Time*, 82.

"one of the . . . counter culture." Theodore Roszak, *The Making of a Counter Culture* (Garden City: Anchor Books, 1969), 84.

"a . . . Dionysus with footnotes." Ibid., 115.

"for the young men . . . American illness." Frederick J. Hoffman, "Philistine and Puritan in the 1920s," *American Quarterly* 1 (1949): 247–63.

"I can recall . . . the corner." Dickstein, *Gates of Eden*, 82.

third of the intellectuals. Charles Kadushin, *The American Intellectual Elite* (Boston: Little Brown, 1974), 22, 34.

approximately 200 leading American intellectuals. Ibid., 19.

"to name . . . in the intellectual community." Ibid., 30–31.

"Freud has been marvelous": Alfred Kazin, "The Lessons of the Master," *Reporter* 16 Apr. 1959, 39–41.

"nearly half . . . were Jews." Kadushin, *American Intellectual Elite*, 23; see also the detailed breakdown, 35.

"Nothing I . . . still crying." Susan Sontag, *On Photography* (New York: Farrar, Straus and Giroux, 1977), 20.

"Arendt's book provoked . . . discussion." Alexander Bloom, *Prodigal Sons: The New York Intellectuals and Their World* (New York: Oxford University Press, 1986), 329.

bitterly argued debate. Ibid., 330.

"Next . . . Babbit began to look good." Arthur Schlesinger, Jr., "Our Country and Our Culture," *Partisan Review* 19 (1952): 591.

intellectuals without Ph.D.'s. Bloom, *Prodigal Sons*, 311.

others had conferred upon them Ph.D.'s. Ibid.

Philip Rahv was given a professorship. Ibid.

40 percent . . . were professors. Kadushin, *American Intellectual Elite*, 30.

Kristol and Podhoretz . . . supported Ronald Reagan. Bloom, *Prodigal Sons*, 374.

"Being an intellectual . . . way of life." Ibid., 315.

"had had to defend a . . . sofa." Ibid.

"*Partisan Review* was . . . the *New Yorker*." Ibid., 311, quoting an interview with Midge Decter.

"What does . . . in the United States." "Parnassus—Coast to Coast," *Time*, 11 June 1956, quoted in Bloom, 207.

"he began to flatter . . . for supper." Bloom, *Prodigal Sons*, 324, quoting Midge Decter.

"We became . . . royalty." Ibid., 324.

1966 study of psychoanalysts. Arnold A. Rogow, *The Psychiatrists* (New York: G. P. Putnam's Sons, 1970), 126.

62 percent of the psychoanalysts. Ibid., 124.

Goldwater 10 percent. Everett C. Ladd and Seymour M. Lipset, "Politics of Academic Natural Scientists and Engineers," *Science* 176 (1972): 1091–1100.

voting preference. Rogow, *The Psychiatrists*, 72.

"a permanent minority." Robert A. Rutland, *The Democrats: From Jefferson to Carter* (Baton Rouge:

Louisiana State University Press, 1979), 190.

"concerned chiefly with human rights." Ibid., 184.

"social values more . . . profit." Arthur Schlesinger, *The New Deal in Action* (New York: Macmillan, 1940), 24.

Harry Hopkins. See George McJimsey, *Harry Hopkins: Ally of the Poor and Defender of Democracy* (Cambridge: Harvard University Press, 1987).

comprehensive civil rights program. Alonzo L. Hamby, *Beyond the New Deal: Harry S. Truman and American Liberalism* (New York: Columbia University Press, 1973), 243.

a proposal . . . for the Mental Health Act. Arthur M. Schlesinger, *History of United States Political Parties* vol. 4 (New York: Chelsea House, 1973), 2716.

the ostensible reason: For a review of this, see E. Fuller Torrey, *Nowhere to Go: The Tragic Odyssey of the Homeless Mentally Ill* (New York: Harper and Row, 1988), chapter 3.

"have some . . . dynamics." William C. Menninger, "Presidential Address," *American Journal of Psychiatry* 106 (1949): 1–12.

"Modern psychiatry . . . of his relationships." Francis J. Braceland, "Psychiatry and the Science of Man," *American Journal of Psychiatry* 114 (1957): 1–9.

"would require the . . . patient's environment." Robert H. Felix, "The Relation of the National Mental Health Act to State Health Authorities," *Public Health Reports*,

10 Jan. 1947, 41–49.

"a plot . . . a concentration camp." Donald Robinson, "Conspiracy USA: The Far Right's Fight Against Mental Health," *Look*, 26 Jan. 1965, 30–32.

"psychiatry is . . . American thinking." Judd Marmor, Viola W. Bernard, Perry Ottenberg, "Psychodynamics of Group Opposition to Health Programs," *American Journal of Orthopsychiatry* 30 (1960): 330–45.

"mental health is a Marxist weapon." Robinson, "Conspiracy USA," 30–32.

"Mental health is . . . the Marxist ideology." Ibid.

Rosemary . . . mentally ill as well. The evidence is strong that Rosemary Kennedy developed a severe mental illness, either schizophrenia or manic-depressive illness, in her late teens and that is why a lobotomy was carried out on her. See Torrey, *Nowhere to Go*, 103–106 for a discussion of this.

"serves . . . a valid mental hygiene." William C. Menninger, "Analysis of Psychoanalysis," *New York Times Magazine*, 18 May 1947, 12–50.

wars were "mental health problems." Rogow, *The Psychiatrists*, 147, quoting Stevenson.

"if the race is . . . responsibility." G. Brock Chisholm, "The Reestablishment of Peacetime Society: The Responsibility of Psychiatry," *Psychiatry* 9 (1946): 3–11.

"education, social work . . . the total social environment." Robert H. Felix, *Mental Health and Social Welfare* (New York: Columbia

University Press, 1961), 21.

"to improve . . . environmental conditions." Stanley F. Yolles, "Social Policy and the Mentally Ill," *Hospital and Community Psychiatry* 20 (1969): 21–42.

"in addition to . . . reading difficulties." Stanley F. Yolles, "The Role of the Psychologist in Comprehensive Community Mental Health Centers: The National Institute of Mental Health View," *American Psychologist* 21 (1966): 37–41.

"The conditions of . . . the modern psychiatrist." Stanley F. Yolles, "Intervention Against Poverty: A Fielder's Choice for the Psychiatrist," *American Journal of Psychiatry* 122 (1965): 324–25.

"a socially defined . . . problem." Leonard J. Duhl and Robert J. Leopold, *Mental Health and Urban Social Policy* (San Francisco: Jossey-Bass, 1968), 3.

"construct . . . mentally healthy individuals." Ibid.

"The totality . . . is conducive to mental health." Leonard J. Duhl, "The Shame of the Cities," *American Journal of Psychiatry* 124 (1968): 1184–89.

Never before had federal funds. Torrey, *Nowhere to Go*, 129–30.

"to resolve . . . technical progress." Anthony F. Panzetta, *Community Mental Health: Myth and Reality* (Philadelphia: Lea and Febiger, 1971), 111.

Lincoln Hospital Mental Health Services. Torrey, *Nowhere to Go*, 133–37.

"the Soviet Union has . . . all human behavior." Harold G.

Whittington, "The Third Psychiatric Revolution—Really?" *Community Mental Health Journal* 1 (1965): 73–80.

"the Russian fantasy . . . of our social order." Lawrence S. Kubie, "Pitfalls of Community Psychiatry," *Archives of General Psychiatry* 18 (1968): 257–66.

"the poor tend to . . . the body." Michael Harrington, *Fragments of the Century* (New York: Touchstone Books, 1972), 184.

Harrington's own four-year psychoanalysis. Ibid., 169.

"I had read my Freud and . . . psychoanalyzed." Ibid., 166.

"two frantic weeks of . . . work days." Ibid., 174.

"grew up . . . concepts were prominent." Frank Mankiewicz, telephone interview with author 15 Sept. 1990. Mankiewicz confirmed Harrington's account of the planning for the war on poverty.

"an ex-Trotskyist and union organizer." Harrington, *Fragments of the Century*, 174.

Leonard Tennenhouse ed., *The Practice of Psychoanalytic Criticism* (Detroit: Wayne State University Press, 1976).

Morton Kaplan and Robert Kloss, *The Unspoken Motive: A Guide to Psychoanalytic Literary Criticism* (New York: The Free Press, 1973).

Frederick Crews, *Out of My System: Psychoanalysis, Ideology and Critical Method* (New York: Oxford University Press, 1975).

"the only . . . theory . . . mankind has devised." Ibid., 4.

the Italian word for vulture. Robert Coles, "Shrinking History—Part One," *New York Review of Books*, 22 Feb. 1973, 15.

"The authors have . . . false conclusions." Martin L. Gross, *The Psychological Society* (New York: Random House, 1978), 73, quoting Tuchman.

Erikson . . . had had no education. H. Stuart Hughes, *The Sea Change: The Migration of Social Thought, 1930–1965* (New York: Harper and Row, 1975), 219.

characteristics in Luther's personality. Erik H. Erikson, *Young Man Luther: A Study in Psychoanalysis and History* (New York: W. W. Norton and Company, 1958), 245.

"We must conclude . . . anal defiance." Ibid., 247.

"originology . . . to be its 'origin.' " Robert Coles, *Erik H. Erikson: The Growth of His Work* (Boston: Little Brown, 1970), 63, quoting Erikson. The Coles biography is an excellent source for understanding this complex man.

1977 survey. George M. Kren, "Psychohistory in the University," *Journal of Psychohistory* 4 (1977): 339–50.

"Psycho-history derives . . . from psychoanalysis." Gertrude Himmelfarb, "The New History," *Commentary*, Jan. 1975, 72–78.

"in this psychoanalytically based . . . history." Gross, *The Psychological Society*, 66.

"Richard felt . . . wishes toward him." James W. Hamilton, "Some Reflections on Richard Nixon in the Light of His Resignation and Farewell Speeches," *Journal of Psychohistory* 4 (1977): 491–511.

book or literary journal editors. A survey was sent by the author to random members of the American Psychiatric Association practicing in New York City in 1990 and who had been in practice for at least thirty years. A majority of respondents had little or no knowledge of the subject, while a minority were clearly familiar with the issue. Although this was not intended to be a scientific survey, it corroborated anecdotal information from individuals in the publishing industry that they tend to use a small circle of psychiatrists and psychoanalysts. The respondents were promised anonymity.

"Just as . . . inspired fiction followed." Alfred Kazin, "The Language of Pundits," *Atlantic Monthly* July 1961, 73–78.

Lessing later acknowledged. Jeffery Berman, *The Talking Cure*, 179.

Dr. Nolan . . . Plath's own psychoanalyst. Ibid., 25.

sold over five million copies. Ibid., 155.

"characters are . . . in literature." Ibid., 239.

"for many years." Ibid., 25.

"no novelist has . . . Roth does." Ibid., 253.

"the most virulently anti-Freudian artist of the century." Ibid., 229.

"no novelist has waged a more relentless campaign." Ibid., 211.

"Viennese quack." Vladimir Nabokov, *Speak, Memory: An Autobiography Revisited* (New York: Vintage Books, 1989), 300. First published in 1951.

"one of the vilest deceits . . . on others." Ibid., 215.

"Let the credulous . . . private parts." Ibid.

"in the course of evening walks in Vienna." Paul Roazen, *Freud and His Followers* (New York: New York University Press, 1984), 329.

"the witchdoctor Freud." Brian Boyd, *Vladimir Nabokov: The Russian Years* (Princeton: Princeton University Press, 1990), 260.

"the vulgar . . . life of their parents." Nabokov, *Speak, Memory,* 20.

"as a kind of internal Marxism." Berman, *The Talking Cure,* 214.

"nothing but a kind of microcosmos of communism." Ibid., 217.

"the difference . . . of spacing." Ibid., 222.

"to analyze or anal-ize his works." Ibid., 212.

"jog on . . . sexual myth." Nabokov, *Speak, Memory,* 300.

"in the fifties . . . to be in treatment." A survey was sent by the author to random members of the American Psychiatric Association practicing in Los Angeles in 1990 and who had been in practice there for at least thirty years. The respondents were promised anonymity.

According to Dr. Irving Schneider. Irving Schneider, "Images of the Mind," 613–20.

Sartre "knew Freud's work . . . about the human mind." John Huston, *An Open Book* (New York: Alfred A. Knopf, 1980), 294.

"Monroe's own analyst objected."

Krin Gabbard and Glen O. Gabbard, *Psychiatry and the Cinema* (Chicago: University of Chicago Press, 1987), 109.

"Freud's descent into . . . the ligh Ibid., 108.

"revealing . . . Cecily's father." Stuart M. Kaminsky, *John Huston: Maker of Magic* (Boston: Houghton Mifflin, 1978), 141.

"Know thyself . . . Let us hope." Gabbard, *Psychiatry and the Cinema,* 110.

"more than . . . depictions of psychiatric treatment." Ibid., 96.

"but I . . . pay for the sessions you miss." Ibid., 132.

one-third of the psychiatrists. Rogow, *The Psychiatrists,* 62.

a survey of 30 social workers. J. Peek and C. Plotkin, "Social Caseworkers in Private Practice," *Smith College Studies in Social Work* 21 (1951): 165–97.

"psychoanalytic syndrome . . . of interpersonal relations." Charles Kadushin, "The Friends and Supporters of Psychotherapy on Social Circles in Urban Life," *American Sociological Review* 31 (1966): 786–802.

"They all make money." Fritz S. Perls, *In and Out the Garbage Pail* (Lafayette, CA: Real People Press, 1969), pages unnumbered. Perls identifies this man as "Hirschman" in his autobiography but he almost certainly was referring to Edward Hitschmann who was working with Federn et al. at that time.

"the Edison of Psychiatry . . . bearers of light." Ibid.

"I came from South Africa . . . ." Ibid.

"the mistakes . . . of my life." Ibid.

"I am really beginning . . . my life." Ibid.

"Freud took the first step . . . of psychiatry." Ibid.

"a playground for . . . pudgy egos." Art Harris, "Esalen: From '60s Outpost to the Me Generation," *Washington Post*, 24 Sept. 1978, C:1–4.

primal scream therapy. E. Fuller Torrey, "The Primal Therapy Trip: Medicine or Religion?" *Psychology Today*, Dec. 1976, 62–68.

"all addictions . . . nonorganic psychosis." E. Michael Holden, "Primal Therapy" in *The Psychotherapy Handbook*, ed. Richie Herink, (New York: New American Library, 1980), 495.

study of primal therapy. Tomas Videgard, *The Success and Failure of Primal Therapy* (Stockholm: Almquist and Wiksell, 1984).

"As little boys . . . will be handled." Margaret Mead, *Male and Female: A Study of Sexes in a Changing World* (New York: Penguin Books, 1962), 114, 117. Originally published in 1949.

estimated ten million. The figure from 1980 was estimated to be 9.6 million so today it would be at least 10 million. See Council Report, "The Future of Psychiatry," *Journal of the American Medical Association* 264 (1990): 2542–48.

31 percent. Gross, *The Psychological Society*, 318, citing Dr. George Vaillant's study.

Several studies have shown. For example, see David J. Knesper, John R. C. Wheeler, and David J. Pagnucco, "Mental Health Services Providers' Distribution Across Counties in the United States," *American Psychologist* 39 (1984): 1424–34.

"It was . . . their Ph.D. dissertations." Roazen, *Freud and His Followers*, 141.

"The optimum conditions . . . among the healthy." Ibid., 160.

"rascally, yea—forsooth knaves." William Shakespeare, *King Henry IV*, Part II, Introduction, ii.

"dysfunctional parenting." Pia Melody, *Facing Codependence* (New York: Harper and Row, 1989), 117.

"child abuse . . . or inadequate nurturance." Wendy Kaminer, "Chances Are You're Codependent Too," *New York Times Book Review*, 11 Feb. 1990, 3–27.

"because of . . . life." Mellody, *Facing Codependence*, 3.

"Recovery involves . . . or abusive." Ibid., 117.

Bradshaw profiled: Emily Mitchell, "Father of the Child Within," *Time*, 25 Nov. 1991, 82–83.

"A lot of what we consider . . . abusive." David Gelman, "Making It All Feel Better," *Newsweek*, 26 Nov. 1990, 66–68.

"neglected . . . by parents." Ibid.

"dogma-eat-dogma world." Dava Sobel, "Freud's Fragmented Legacy," *New York Times Magazine*, 26 Oct. 1980, 28–108.

the nuclear in nuclear family. Jamie Diamond, "How Not to Get the DT's When Happy Or, Why It's Easier to Blame the Potato Chip Instead of Yourself," *Lears*, Sept. 1990, 81–82. The exact quote used by Diamond in this excellent article is: "And those poor people with toxic parents must think that the nuclear in nuclear reactor is the same nuclear as in nuclear family."

"over 175,000 copies in print." Donna Ewy and Roger Ewy, *Preparation for Breastfeeding* (Garden City, N.Y.: Doubleday and Company, 1975).

"the average . . . on its shelves." Nancy McGrath, "By the Book," *New York Times Magazine*, 27 June 1976, 26–27.

"Children do . . . with the mother." Ellen Galinsky, *Between Generations: The Six Stages of Parenthood* (New York: New York Times Books, 1981), 171.

"A word about . . . replacing the parent." William Sears, *Creative Parenting* (New York: Everest House, 1982), 327.

"the decisive significance of early childhood." Alice Miller, *Thou Shalt Not Be Aware: Society's Betrayal of the Child* (New York: Meridian Books, 1986), 52.

"When Young Children Need Therapy." Julius Segal and Zelda Segal, "When Young Children Need Therapy," *Parents*, Feb. 1990, 184.

"Rebecca Shahmoon Shanok . . . in private practice." Rebecca S. Shanok, "When You Share Family Stories," *Parents*, Feb. 1990, 187.

"in clinics . . . of cribside therapy." David Gelman, "A Is For Apple, P Is For Shrink," *Newsweek*, 24 Dec. 1990, 64–66.

"The legacy of . . . the current work." Ann Crittenden, "New Insights Into Infancy," *New York Times Magazine*, 13 Nov. 1983, 84–96.

"inadequate parenting . . . psychological deprivation." Judy Mann, "Son Suing His Parents Opens a Pandora's Box," *Washington Post*, 10 Jan. 1979, B:1.

two teenage sisters sued their parents. Jacqueline Trescott, "Children v. Parents," *Washington Post*, 4 April 1979, B:1.

"he has done . . . for his own life." Ellen Goodman, "Psychological Malparenting: Excuses, Excuses," *Washington Post*, 19 June 1978, A:23.

Freudiana. Michael Z. Wise, "In Vienna, Id's Show Time!," *Washington Post*, 19 Dec. 1990, C:1.

oratorio Oedipus Tex. James R. Oestreich, "Works of P. D. Q. Explore Rap and Freudian Subtleties," *New York Times*, 29 Dec. 1990, C:18.

"A Chaste Lounge." Anna Kisselgoff, "Freudian Fears and Disney From a Comic Paul Taylor," *New York Times*, 20 April 1990, B:1.

the portrayal of psychoanalysis in the movies. Alessandra Stanley, "Mental Images: Psychoanalysis on the Screen," *New York Times*, 16 Nov. 1990, C:1.

a book about Joel Steinberg. Joyce Johnson, *What Lisa Knew* (New York: G. P. Putnam's Sons, 1990).

"an only child . . . more personal issues." Ed Bruske, "Lawyer Sentenced for Cheating Client," *Washington Post*, 6 April 1990, A:12.

"to show that Mr. List . . . their souls." Joseph F. Sullivan, "Judge Narrows Verdict in Jersey Family

Murder," *New York Times*, 10 April 1990, B:2.

"It is commonplace . . . or not." Peter Gay, *Freud: A Life for Our Time* (New York: W. W. Norton, 1988), xvii.

"To us he is . . . opinion." W. H. Auden, *Selected Poetry* (New York: Random House, 1971), 57.

9.
The Scientific Basis of Freudian Theory

"For no other system . . . of human behavior." Alfred Kazin, "The Freudian Revolution Analyzed," *New York Times Magazine*, 6 May 1956, 22–40.

"now found . . . culture substance." Seymour Fisher and Roger P. Greenberg, *The Scientific Credibility of Freud's Theories and Therapy* (New York: Basic Books, 1977), viii.

"have begun to merge . . . in common." Harold Bloom, "Freud, the Greatest Modern Writer," *New York Times Book Review*, 23 Feb. 1986, 26.

"in the end . . . the Freudian Century?" Benjamin Nelson, ed., *Freud and the 20th Century* (New York: Meridian Books, 1987), 9.

"psychoanalysis . . . is occurring now." Marshall Edelson, *Psychoanalysis: A Theory in Crisis* (Chicago: University of Chicago Press, 1988), xi, xii.

"universally recognized scientific achievements." Thomas Kuhn, *The Structure of Scientific Revolutions* (Chicago: University of Chicago Press, 1962), x.

"ceased to function . . . led the way." Ibid., 91.

Brill . . . spoke of the "laws." John C. Burnham, "Psychoanalysis in American Civilization Before 1918" (Ph.D. diss., Stanford University, 1958), 146.

"psychoanalysis is . . . study of the mind." Abraham A. Brill, "A Psychoanalyst Scans His Past," *Journal of Nervous and Mental Disease* 95 (1942): 537–49.

"this new therapy is . . . a science." Peter C. MacFarlane, "Diagnosis by Dreams," *Good Housekeeping* Feb. 1915, 125–33.

"a hardheaded man of science." Peter Gay, *Freud: A Life For Our Time*, 56.

"I am not really . . . that type of being." Ernest Jones, *The Life and Work of Sigmund Freud*, vol. 1 (New York: Basic Books, 1953), 348 quoting a Freud letter of 1 Feb. 1900.

"those critics . . . seeing with them." Russell Jacoby, *The Repression of Psychoanalysis: Otto Fenichel and the Political Freudians* (New York: Basic Books, 1983), 138.

Freud responded testily. Fisher and Greenberg, *Freud's Theories and Therapy*, ix.

"the basic concepts of science . . . account for experience." Heinz Hartmann, "Psychoanalysis as a Scientific Theory," in *Psychoanalysis, Scientific Method and Philosophy*, ed. Sidney Hook, (New York: Grove Press, 1959), 29.

"We possess the truth, I am sure of it." Ibid., 12–13.

"a scientific fairy tale." Gay, *Freud: A Life for Our Time*, 93.

"If the patient loved . . . 50 Kronen each." Peter F. Drucker, *Adventures of a Bystander* (New York: Harper and Row, 1978), 89.

"this book is indicative of . . . advance." Unsigned book review of Sigmund Freud, *The Interpretation of Dreams*, *Nation* May 1913, 503–05.

"well founded . . . unscientific method." C. Ladd Franklin, "Freudian Doctrines," *Nation* 19 Oct. 1916, 373–74.

"upon the same ground as . . . the moon." Anonymous, "An American Expert's Indictment of American Dream Analysis as a Psychological Humbug," *Current Opinion* Sept. 1916, 34–35.

"Astrology . . . to Medieval Symbolism." Burnham, "Psychoanalysis in American Civilization Before 1918," 100.

"the premature crystallization of spurious orthodoxy." Hans J. Eysenck and Glenn D. Wilson, *The Experimental Study of Freudian Theories* (London: Methuen, 1973), 395.

"Although we have . . . the problem." Gardner Murphy, Lois B. Murphy and Theodore Newcomb, *Experimental Social Psychology* (New York: Harper, 1937), 575.

"empirical data bearing on . . . personality." Harold Orlansky, "Infant Care and Personality," *Psychological Bulletin* 46 (1949): 1–48.

"anyone who tries . . . publications at all." Ernest R. Hilgard, Lawrence S. Kubie, and E. Pumpian-Mindlin, *Psychoanalysis as Science* (New York: Basic Books, 1952), 44.

"only two studies give . . . procedures." Paul Kline, *Fact and Fantasy in Freudian Theory* (London: Methuen, 1972), 93.

"Freudian theory . . . riddled with subjective interpretation." Ibid., ix.

"It is true . . . stages of development." Fisher and Greenberg, *Freud's Theories and Therapies*, 393.

"not one study . . . of the population." Eysenck and Wilson, *Freudian Theories*, 392.

"a medieval morality play . . . to deserve scientific status." Hans J. Eysenck, *Decline and Fall of the Freudian Empire* (London: Penguin Books, 1985), 35.

"without doubt . . . tellers of fairy tales." Ibid., 208.

interview with Dr. Sontag. Dr. Lester Sontag, telephone interview with author, 25 Sept. 1989.

"people's childhoods . . . would expect." David Goleman, "Traumatic Beginnings: Most Children Seem Able to Recover," *New York Times*, 13 March 1984, C:1, quoting Dr. Vaillant.

"too early . . . or too libidinous." Fisher and Greenberg, *Freud's Theories and Therapy*, 145.

"can go over . . . easily joined." Sigmund Freud, "Character and Anal Eroticism" (1908), in *The Complete Psychological Works of Sigmund Freud*, vol. 9, ed. James Strachey, (London: Hogarth Press, 1966), 169–75.

"all collectors are anal erotic." Kline, *Fact and Fantasy*, 10, quoting Jones.

"persons who fit the anal . . . phase." Fisher and Greenberg,

Freud's Theories and Therapies, 141.
the length of breast-feeding. Ibid.,
110.

As has been pointed out. Sibylle
Escalona, "Problems in Psycho-
Analytic Research, *International
Journal of Psycho-Analysis* 33
(1952): 11–21.

"It can be . . . the average man."
Fisher and Greenberg, *Freud's Theories and Therapy*, 199.

"facilitated by . . . the part of the
father." Ibid., 395.

"bound securely to a wooden cradle
. . . is visible." Orlansky, "Infant
Care and Personality," 1–48.

"the child is expected to . . . take
care of his defecation needs
alone." John W. Whiting and Irvin
L. Child, *Child Training and Personality: A Cross-Cultural Study*
(New Haven: Yale University
Press, 1953), 73–74.

Anecdotal accounts. Thomas
Maeder, *Children of Psychiatrists
and Psychotherapists* (New York:
Harper and Row, 1989).

suicide rate. Bureau of the Census,
*Statistical Abstract of the United
States*, 1985–1989 (Washington:
U. S. Government Printing
Office), tables on death rates from
selected causes.

divorce and crime rates. "Divorces
and Annulments and Rates:
United States, 1940–87," *Monthly
Vital Statistics Report* 38 (May 15,
1990): 7; "Crime Index Rate,
1960–89," Federal Bureau of
Investigation, mimeo.

study of breast-size preference.
Kline, *Fact and Fantasy*, 91. See
also Eysenck and Wilson, 387.

"A theory must not be . . . its

opposite." Ernest Nagel,
"Methodological Issues in Psychoanalytic Theory," 40.

"Investigation into . . . these experiences." Sigmund Freud, "My
Views on the Part Played by Sexuality in the Etiology of the Neuroses" (1905), in *Collected Papers*,
vol. 1, ed. Ernest Jones, (New
York: International Psychoanalytic
Press, 1924), 272.

"came to view the . . . actual."
Philip Rieff, *Freud: The Mind of
the Moralist* (Chicago: University
of Chicago Press, 1959), 50.

"No other discipline . . . empirically testable." Eysenck, *Decline
and Fall*, 150.

Studies of 20 pairs in America.
Horatio H. Newman, Frank N.
Freeman, and Karl J. Holzinger,
*Twins: A Study of Heredity and
Environment* (Chicago: University
of Chicago Press, 1937).

12 pairs in Denmark. Niels Juel-
Nielsen, "Individual and Environment," *Acta Psychiatrica Scan-
dinavica Supplementum* 183 (1964):
11–144.

44 pairs in England. James Shields,
*Monozygotic Twins Brought Up
Apart and Brought Up Together*
(London: Oxford University Press,
1962).

"who, though they knew . . .
saleswomen." Ibid., 153.

study of 850 high school twin pairs.
John Loehlin and Robert Nichols,
*Heredity, Environment and Personal-
ity: A Study of 850 Sets of Twins*
(Austin: University of Texas Press,
1976).

study of newborn twins. D. G.
Freedman and Barbara Keller,

"Inheritance of Behavior in Infants," *Science* 140 (1963): 196–98.

Bridget and Dorothy. Constance Holden, "Identical Twins Reared Apart," *Science* 207 (1980): 1323–28.

"genetic factors exert . . . variability." Thomas J. Bouchard, David T. Lykken, Matt McGue, Nancy L. Segal, and Auke Tellegen, "Sources of Human Psychological Differences: The Minnesota Study of Twins Reared Apart," *Science* 250 (1990): 223–28.

"that . . . attributed to genetic diversity." Auke Tellegen, David T. Lykken, Thomas J. Bouchard, Kimberly J. Wilcox, Nancy L. Segal, and Stephen Rich, "Personality Similarity in Twins Reared Apart and Together," *Journal of Personality and Social Psychology* 54 (1988): 1031–39.

religiosity and traditionalism. Niels G. Waller, Brian A. Kojetin, Thomas J. Bouchard, David T. Lykken, and Auke Tellegen, "Genetic and Environmental Influences on Religious Interests, Attitudes, and Values: A Study of Twins Reared Apart and Together," *Psychological Science* 1 (1990): 1–5.

Swedish study. Nancy L. Pedersen, Robert Plomin, Gerald E. McClearn, and L. Friberg, "Neuroticism, Extraversion and Related Traits in Adult Twins Reared Apart and Reared Together," *Journal of Personality and Social Psychology* 55 (1988): 950–57.

one-third of their attitudes about responsibility. Nancy L. Pedersen,

Margaret Gatz, Robert Plomin, John R. Nesselroade, and Gerald E. McClearn, "Individual Differences in Locus of Control During the Second Half of the Life Span for Identical and Fraternal Twins Reared Apart and Reared Together," *Journal of Gerontology* 44 (1989): 100–105.

"that twins reared . . . together." H. Langinvainio, J. Kaprio, M. Koskenvuo, and J. Lonnqvist, "Finnish Twins Reared Apart: III Personality Factors," *Acta Geneticae Medicae et Gemellologiae* 33 (1984): 259–64.

"rough estimate of broad heritability." J. P. Rushton, D. W. Fulker, M. C. Neale, R. A. Blizard, and H. J. Eysenck, "Altruism and Genetics," *Acta Geneticae Medica et Gemellologiae* 33 (1984): 265–71.

shyness is another personality trait. Denise Daniels and Robert Plomin, "Origins of Individual Differences in Infant Shyness," *Developmental Psychology* 21 (1985): 118–21; Jerome Kagan, J. Steven Reznick, Nancy Snidman, "Biological Basis of Childhood Shyness," *Science* 240 (1988): 167–71.

twin study based on Denmark's national twin registry. C. Robert Cloninger and Irving I. Gottesman, "Genetic and Environmental Factors in Antisocial Behavior Disorders. In Sarnoff A. Mednick, Terrie Moffitt, and Susan A. Stack, eds., *The Causes of Crime* (Cambridge: Cambridge University Press, 1987), 92–109.

study in Norway. O. S. Dalgaard and Einar Kringlen, "A Norwegian Study of Criminality," *British Journal of Criminality* 16 (1976): 213–32.

extensive adoption study. Sarnoff A. Mednick, William F. Gabrielli, and Barry Hutchings, "Genetic Factors in the Etiology of Criminal Behavior," in Mednick et al., *The Causes of Crime*, 74–91.

"some factor transmitted by criminal . . . activity." Sarnoff A. Mednick, William F. Gabrielli, Barry Hutchings, "Genetic Influences in Criminal Convictions: Evidence From an Adoption Cohort," *Science* 224 (1984): 891–94.

"While the adoption studies . . . matter very much." Christopher Jencks, "Genes and Crime," *New York Review of Books*, 12 Feb. 1987, 33–41.

"after 30 generations . . . the low lines." Robert Plomin, "The Role of Inheritance in Behavior," *Science* 248 (1990): 183–88.

"heredity is . . . physical size." John P. Scott and John L. Fuller, *Genetics and the Social Behavior of the Dog* (Chicago: University of Chicago Press, 1965), 378.

"pronounced individual . . . personality." Jane Goodall, *The Chimpanzees of Gombe: Patterns of Behavior* (Cambridge: Harvard University Press, 1986), 172.

"Continuity over . . . our subjects." Alexander Thomas and Stella Chess, "Genesis and Evolution of Behavioral Disorders: From Infancy to Early Adult Life," *American Journal of Psychiatry* 141 (1984): 1–9.

"increasing acceptance of . . . psychology." Robert Plomin, "Environment and Genes: Determinants of Behavior," *American Psychologist* 44 (1989): 105–11.

"pairs of unrelated children . . . chosen at random." Norman D. Henderson, "Human Behavior Genetics," *Annual Review of Psychology* 33 (1982): 403–40.

"could account for no more than 5% of the variance." Ibid.

"the effect of . . . psychological traits." Bouchard et al.

"the influence of . . . phenotypic variance." Pederson et al.

"much more influenced by environment." Deborah Franklin, "What a Child is Given," *New York Times Magazine*, 3 Sept. 1989, 36–49, quoting Dr. Plomin.

six-year-old girl was attacked. Lenore C. Terr, "Childhood Traumas: An Outline and Overview," *American Journal of Psychiatry* 148 (1991): 10–20.

"previously . . . neighborhood." Ibid.

"the Oedipal period . . . is the source of all subsequent adult behaviors." Janet Malcolm, *Psychoanalysis: The Impossible Profession* (London: Picador Books, 1982), 158–59.

"how the child reacted to . . . his ego structure." Reuben Fine, *A History of Psychoanalysis* (New York: Columbia University Press, 1979), 161.

"the formerly institutionalized . . . problems." Wagner H. Bridger, "Early Childhood and Its Effects," *Harvard Mental Health Letter* Aug. 1991, 4–6.

"Short-term events . . . later improves." Ibid.

"the mothers . . . of siblings." Robert Plomin and Denise Daniels, "Why Are Children in the Same Family So Different From One Another?," *Behavioral and Brain Sciences* 10 (1987): 1–60, citing a study by Dunn et al., 1985.

"much difference." Ibid.

Two studies of twins: David C. Rowe and Robert Plomin, "The Importance of Nonshared (E1) Environmental Influences in Behavioral Development," *Developmental Psychology* 17 (1981): 517–31.

"in families . . . developments of children." R. Darrell Bock and Michele F. Zimowski, "Contributions of the Biometrical Approach to the Individual Differences in Personality Measures," *Behavioral and Brain Sciences* 10 (1987): 17–18. Quoted with the kind permission of *Behavioral and Brain Sciences* and the Cambridge University Press.

"each ego is endowed . . . for it." Peter B. Neubauer and Alexander Neubauer, *Nature's Thumbprint: The New Genetics of Personality* (Reading, Mass.: Addison-Wesley, 1990), 177, quoting Freud.

"One of the most fascinating . . . it will have." Margaret Mead, *Blackberry Winter: My Earlier Years* (New York: Simon and Schuster, 1972), 243, 260.

twin studies of Dr. Ronald S. Wilson: Ronald S. Wilson, "Twins: Early Mental Development," *Science* 175 (1972): 914–17; Ronald S. Wilson, "Synchronies in Mental Development: An Epigenetic Perspective," *Science* 202 (1978): 939–48.

"It is most important . . . at different times." Wilson, "Synchronies in Mental Development: An Epigenetic Perspective," quoting Gerald E. McClearn in *Carmichael's Manual of Child Psychology*, Paul H. Mussen, ed. (New York: Wiley, 1970), 61.

"no matter how well-fed . . . than average." Elizabeth Hall, "PT Conversation with Sandra Scarr: What's a Parent to Do?," *Psychology Today*, May 1984, 58–63.

"Parents may make the . . . nonshared environment." Stella Chess, "Let Us Consider the Roles of Temperament and of Fortuitous Events," *Behavioral and Brain Sciences* 10 (1987): 21–22.

"due to genotype-environment . . . of variance." Marvin Zuckerman, "All Parents Are Environmentalists Until They Have Their Second Child," *Behavioral Brain Sciences* 10 (1987): 42–44.

Dr. Irving Gottesman. Personal communication, March, 1991.

Dr. Edward O. Wilson. Daniel J. Kevles, *In the Name of Eugenics: Genetics and the Uses of Human Heredity* (New York: Alfred A. Knopf, 1985), 280.

"more than fifty students." Paul Selvin, "The Raging Bull of Berkeley," *Science* 251 (1991): 368–71.

"as outspoken and idiosyncratic Marxist." Kevles, *In the Name of Eugenics*, 281.

"biological determinism . . . in their own image." Richard C. Lewontin, Steven Rose, and Leon

J. Kamin, *Not In Our Genes* (New York: Pantheon, 1984), 15.

"share a commitment to . . . society." Ibid., ix.

"equal protection of the . . . rights of others." Irving I. Gottesman, "Genetic Aspects of Human Behavior: State of the Art," in Walter T. Reich, ed., *Encyclopedia of Bioethics* (New York: The Free Press, 1978), 529.

black athletes. "Minorities in Sports," *USA Today*, 19 Feb. 1991, C:10; Ira Berkow, "The Kangaroo Kid and Some Related Matters," *New York Times*, 1 May 1989, B:12.

"no group is . . . from the Caucasus." Lewontin et al., 126.

"In practically every . . . Spanish race." Franz Boas, "Are the Jews a Race?", *The World Tomorrow*, Jan. 1923, 5–6.

"within the same local . . . races." Lewontin et al., 126.

"The determinists would have . . . our genes." Ibid., 6.

"People can be made . . . desirable." Theodosius Dobzhansky, *Genetic Diversity and Human Equality* (New York: Basic Books, 1973), 87.

Phrenology evolved from . . . Franz Joseph Gall. Most of the data on phrenology is taken from John D. Davies, *Phrenology: Fad and Science* (New Haven: Yale University Press, 1955); the book provides an excellent history of the movement.

"regarded phrenology as . . . it had furnished him." Ibid., 85.

"a practical knowledge of . . . phrenology." Ibid., 163.

"that railroad trainmen . . . their heads." Ibid., 38.

"has assumed the majesty . . . of thinking beings." Ibid., 120.

"health, temperance . . . and religion." Ibid., 33.

"in some ways . . . of liberalism." Ibid., xi.

"phrenologists came to . . . retribution." Ibid., 99.

"prisons should rather be rehabilitation centers." Ibid.

"a quackery which succeeds by boldness." Ibid., 67.

"atheism, materialism . . . and free will." Ibid., 150.

"Avoid phrenologists . . . French infidels." Arthur Wrobel, introduction, in Arthur Wrobel, ed., *Pseudo-Science and Society in Nineteenth Century America* (Lexington: University of Kentucky Press, 1987), 12.

"the facial signs . . . insufficient evidence." Madeleine B. Stern, *A Phrenological Dictionary of Nineteenth-Century Americans* (Westport: Greenwood Press, 1982), 209–211.

10.
An Audit of Freud's American Account

"Does not every science come . . . of mythology?" Philip Rieff, *Freud: The Mind of the Moralist* (Chicago: University of Chicago Press, 1959), 204.

"America as a gigantic mistake." Peter Gay, *Freud: A Life for Our Time* (New York: W. W. Norton, 1988), 563.

"Hate America?" . . . more laughter. Max Eastman, "Significant Memory of Freud," *New Republic* 19 May 1941, 693–95. Reprinted with the kind permission of the New Republic, Inc.

"The only excuse . . . misdeed." Paul Roazan, *Freud and His Followers* (New York: New York University Press, 1984), 385.

"Americans as savages." Gay, *Freud: A Life for Our Time*, 563.

"that [psycho] analysis suits . . . a raven." Ibid.

In one letter. Freud to Smith E. Jelliffe, 9 Feb. 1939, Jelliffe Collection, Library of Congress.

handwriting to deteriorate. Ibid., 211.

American and English patients. Ibid., 388.

poet Hilda Doolittle's concern. Janice R. Robinson, *H. D.: The Life and Work of an American Poet* (Boston: Houghton Mifflin Co., 1982), 273–302.

Loeb "built a large house . . . as a recluse." Stephen Birmingham, *"Our Crowd": The Great Jewish Families of New York* (New York: Harper and Row, 1967), 255.

"What . . . if they bring no money?" Gay, *Freud: A Life for Our Time*, 563.

"America is useful." Ibid.

"Is it not sad that we are . . . human beings?" Ibid., 563–64.

"Freud disliked . . . egalitarianism between the sexes." Roazen, *Freud and His Followers*, 385.

American "petticoat government." Ibid.

America "is already threatened by the black race." Brock Brower, "Who's In Among the Analysts?"

Esquire July 1961, 78–84.

"continued to identify . . . as his own." Gay, *Freud*, 353.

Freud was uncertain [about Martin]. Freud to Oskar Pfister, 2 Jan. 1919, in Heinrich Meng and Ernst Freud, eds., *The Letters of Sigmund Freud and Oskar Pfister* (New York: Basic Books, 1963), 64–65.

son of Freud's favorite sister. Lucy Freeman and Herbert S. Strean, *Freud and Women* (New York: Continuum, 1987), 25. For several years Rosa lived in an apartment adjacent to Freud's own and thus Freud must have known this nephew very well.

[Freud] rarely voted. Paul Roazen, *Freud: Political and Social Thought* (New York: Alfred A. Knopf, 1968), 242.

"Politically I am just nothing." Ibid., 243.

[Freud's] skepticism about the Russian Revolution. Ernst Pfeiffer, ed., *Sigmund Freud and Lou Andreas-Salomé Letters* (New York: Harcourt Brace Jovanovich, 1966), 75.

"I have no hope . . . to improvement." Ernst L. Freud, ed., *The Letters of Sigmund Freud and Arnold Zweig* (New York: Harcourt Brace and World, 1970), 21.

"the psychological premises . . . an untenable illusion." Sigmund Freud, *Civilization and Its Discontents* in *The Standard Edition of the Complete Psychological Works of Sigmund Freud*, vol. 21, ed. James Strachey (London: Hogarth Press, 1961), 113.

"discuss . . . with an understanding

smile." Roazen, *Freud and His Followers*, 533.

Freud supported [Dollfus regime]. Ibid., 426, 534. Roazen claims that Dr. Ruth Brunswick and her husband Mark, both in psychoanalysis with Freud, were very disappointed at Freud's support for the Dollfuss regime.

"Benito Mussolini . . . the cultural hero." Ibid., 534.

"a believer . . . even mean." David Riesman, *Individualism Reconsidered and Other Essays* (Glencoe, Ill.: Free Press, 1954), 351, 354–55.

"In the depths of my heart . . . worthless." Freud to Lou Andreas-Salomé, 29 July 1929, in *The Letters of Sigmund Freud*, ed. Ernst L. Freud (New York: McGraw-Hill, 1964), 390.

"I have found little . . . at all." Freud to Oskar Pfister, 10 Sept. 1918, in Meng and Freud, 61–62.

"The unworthiness of human beings . . . on me." Ernest Jones, *The Life and Work of Sigmund Freud*, vol. 2 (New York: Basic Books, 1955), 182, quoting Freud.

only 3 percent of Freud's patients were poor. Isidor Wassermann, *American Journal of Psychotherapy* 12 (1958): 623–27.

"Freud did not accept . . . the ethics of healer." Peter F. Drucker, *Adventures of a Bystander* (New York: Harper and Row, 1978), 84, 87.

Max Eastman . . . extolled the merits of psychoanalysis: Max Eastman, "Exploring the Soul and Healing the Body," *Everybody's Magazine* June 1915, 741–50.

"fleet of limousines." Laura Fermi,

Illustrious Immigrants: The Intellectual Immigration from Europe 1930–41 (Chicago: University of Chicago Press, 1971), 144.

Freud charged . . . $20 per hour. Roazen, *Freud and His Followers*, 424.

fee Brill was charging Mabel Dodge. Christopher Lasch, *The New Radicalism in America: The Intellectual as a Social Type* (New York: Alfred A. Knopf, 1965), 140.

Zilboorg . . . was charging $100 per hour. Paul Johnson, *Intellectuals* (New York: Harper and Row, 1988), 297.

"in our actual work . . . to the world." Lawrence J. Friedman, *Menninger: The Family and the Clinic* (New York: Alfred A. Knopf, 1990), 142.

"If the Freudian doctorine is . . . the world." W. Beran Wolfe, "Twilight of Psychoanalysis," *American Mercury* Aug. 1935, 385–94.

1966 survey of psychoanalysts. Arnold A. Rogow, *The Psychiatrists* (New York: G. P. Putnam's Sons, 1970), 62.

Among patients seen in psychoanalysis. H. Aronson and Walter Weintraub, "Social Background of the Patient in Classical Psychoanalysis," *Journal of Nervous and Mental Disease* 146 (1968): 98–102.

"Brilliant results . . . are folk-lore and song." J. W. Courtney, "The View of Plato and Freud on the Etiology and Treatment of Hysteria," *Boston Medical Surgical Journal* 168 (1913): 649–52.

"If we regard it as a . . . private university." Robert Michels, "Psychoanalysis: The Second Century," *Harvard Mental Health Letter* Dec. 1990, 5–7.

"Whatever we . . . place of sex in life." Havelock Ellis, "Freud's Influence in the Changed Attitude Toward Sex," *American Journal of Sociology* 45 (1939): 309–17.

"Freud found sex . . . an honored guest." Wolfe, "Twilight of Psychoanalysis," 385–94.

"a speck afloat on a sea of feeling." Reuben Fine, *A History of Psychoanalysis* (New York: Columbia University Press, 1979), 345.

"a grand vision . . . of what people might be." Ibid., 539–40.

"in America it was . . . psychoanalysis." Fermi, *The Illustrious Immigrants*, 172.

"Like no man before . . . you and I." Walter Kaufmann, "Freud and the Tragic Virtues," *American Scholar* 29 (1960): 469–81.

Michael Harrington. Harrington underwent psychoanalysis and called Freud "one of the most profound thinkers in Western history." At the same time he acknowledged that "psychoanalysis as I have undergone it cannot possibly help the millions of people who desperately need aid . . . Psychoanalysis is an aristocratic, perhaps even an ascetic, discipline." See Michael Harrington, *Fragments of the Century*, 183, 191.

Robert Coles. Coles underwent psychoanalysis during his psychiatric training and subsequently utilized Freudian theory in some of his work including *Erik H.*

Erikson: The Growth of His Work.

"there is very little evidence . . . given other labels." Fisher and Greenberg, *Freud's Theories and Therapy*, 324.

Dr. Jerome Frank. Jerome Frank, *Persuasion and Healing: A Comparative Study of Psychotherapy* (Baltimore: Johns Hopkins University Press, 1961); recently published in 3rd edition, 1991.

principle of Rumpelstiltskin. E. Fuller Torrey, *Witchdoctors and Psychiatrists: The Common Roots of Psychotherapy and Its Future* (New York: Harper and Row, 1986).

a stew using one moose and one rabbit. Harry L. Senger, "The 'Placebo' Effect of Psychotherapy: A Moose in the Rabbit Stew," *American Journal of Psychotherapy* 41 (1987): 68–81. An alternate formulation of this is a horse and canary pie in which the generic aspects of psychotherapy are the horse; for this recipe see Lester Luborsky, Barton Singer, and Lisa Luborsky, "Comparative Studies of Psychotherapies," *Archives of General Psychiatry* 32 (1975): 995–1008.

"The overwhelming success of Freudianism . . . in modern times." Alfred Kazin, "The Freudian Revolution Analyzed," *New York Times Magazine*, 6 May 1956, 22–40.

"Gestalt Prayer." Fritz S. Perls, *In and Out of The Garbage Pail* (Lafayette, Cal.: Real People Press, 1969), no page numbers.

"In a dying culture . . . spiritual enlightenment." Christopher Lasch, *The Culture of Narcissism*

(New York: W. W. Norton, 1979), 396.

"Love thyself . . . the search continues." Charles Krauthammer, "An Answer for Patricia Godley," *Washington Post,* 5 May 1989, A:27.

"the lap-dog psychology . . . is really a poet." Alfred Kazin, "Psychoanalysis and Literary Culture Today," *Partisan Review* 26 (1959): 45–55.

"the one most outstanding . . . makes him overtly obnoxious." Edward R. Pinckney and Cathey Pinckney, *The Fallacy of Freud and Psychoanalysis* (Englewood Cliffs, N.J.: Prentice-Hall, 1965), 157.

"the golden age of . . . addiction." John Leo, "The It's-Not-My-Fault Syndrome," *U.S. News & World Report,* 18 June 1990, 16.

"Mental hygienists are stressing . . . the *parents.*" Margo Horn, *Before It's Too Late: The Child Guidance Movement in the United States, 1922–1945* (Philadelphia: Temple University Press, 1989), 41, quoting Dr. George Pratt, a psychiatrist.

[Women are] "a dark continent." Gay, *Freud: A Life for Our Time,* 501.

[Freud's wife put] toothpaste on his toothbrush. Paul Roazen, *Freud: Political and Social Thought* (New York: Alfred A. Knopf, 1968), 57.

[Freud's] failure to attend the funeral of his mother. Gay, *Freud: A Life for Our Time,* 573.

[Women] "little sense of justice." Freeman and Strean, *Freud and Women,* 227.

"the little creature without a penis." Ibid., 214.

"Freud . . . motivated to find substitutes." Seymour Fisher and Roger P. Greenberg, *The Scientific Credibility of Freud's Theories and Therapy* (New York: Basic Books, 1977), 199.

"have come . . . to become wise." Gay, *Freud: A Life for Our Time,* 522.

"as their main function . . . of men." Jones, *Freud,* 2:421.

"The deepest hurt . . . Mother failed me." Karl A. Menninger, "Men, Women and Hate," *Atlantic Monthly* Feb. 1939, 158–68.

"mothers were held responsible . . . to be healing." Paula J. Caplan, "Take the Blame Off Mother," *Psychology Today,* Oct. 1986, 70–71.

"we took it for granted . . . it must be maternal." Janna M. Smith, "Mothers: Tired of Taking the Rap," *New York Times Magazine,* 10 June 1990, 32–38.

13 percent of assaults: "Felony Defendants in Large Urban Counties, 1988." Bureau of Justice Statistics, U. S. Department of Justice, April 1990.

"in most of the cases": "16 Women Await the Death Penalty in U.S.," *New York Times,* 3 Nov. 1984, B:12.

"the average bookstore": Nancy McGrath, "By the Book," *New York Times Magazine,* 27 June 1976, 26–28.

"psychoanalysis is . . . itself the remedy." Percival Bailey, *Sigmund the Unserene* (Springfield: Charles Thomas, 1965), 86.

"I do not like these patients." Paul Roazen, *Freud: Political and Social*

Thought (New York: Alfred A. Knopf, 1968), 141.

nation's homeless population: See E. Fuller Torrey, *Nowhere to Go: The Tragic Odyssey of the Homeless Mentally Ill* (New York: Harper and Row, 1988).

"When we sever . . . the poor will forgive us." Howard J. Karger, Letter, "Private Practice and Social Work: A Response," *Social Work* 35 (1990): 479. Reprinted with the kind permission of *Social Work* and the National Association of Social Workers.

"being bled white by . . . objectivity." Arthur Miller, *Timebends: A Life* (New York: Grove Press, 1987), 320.

"In practice, psychoanalysis has by now become . . . less creative." Norman Mailer, *Advertisements for Myself* (New York: G. P. Putnam's Sons, 1959), 346.

"experienced Est . . . in New Consciousness." Lasch, 44.

"indifference to politics . . . by Freudianism." Rieff, 256.

"psychoanalysis in the hands . . . of the Catholic priest." Jeffrey Berman, *The Talking Cure: Literary Representations of Psychoanalysis* (New York: New York University Press, 1987), 4, quoting E. M. Jensen, "Anna O—A Study of Her Later Life," *Psychoanalytic Quarterly* 39 (1970): 269–93.

"There was an atmosphere of . . . his apostles." Max Graf, "Reminiscences of Professor Sigmund Freud," *Psychoanalytic Quarterly* 11 (1942): 467–76.

"Freud . . . development of a church history." Ibid.

"the apostle of Freud who was my Christ." Frank J. Sulloway, *Freud, Biologist of the Mind: Beyond the Psychoanalytic Legend* (New York: Basic Books, 1979), 481, quoting Steckel.

followers . . . a secret committee. Roazen, *Freud: Political and Social Thought*, 323.

all 17 members of the Wednesday Society were Jewish. Dennis Klein, *Jewish Origins of the Psychoanalytic Movement* (Chicago: University of Chicago Press, 1981), xi.

Viennese lodge of B'nai B'rith. Ibid., 74.

"startlingly close." David Bakan, *Sigmund Freud and the Jewish Mystical Tradition* (Princeton: D. van Nostrand, 1958), 19.

"special taxes on . . . a religious body." Morton Prince, "The Demand for Unifying Views," *Journal of Abnormal Psychology* 12 (1917): 270–71.

"I had found the one . . . I could live by." Roazen, *Freud: Political and Social Thought*, 323.

"Religions have . . . the novitiate of the Church." Ibid.

"the interpenetration . . . for the eradication of neurosis." Klein, 139.

"Some of us believed . . . like great men." Fritz Wittels, "Brill—The Pioneer," *Psychoanalytic Review* 35 (1948): 394–98.

"found in 'mental health' . . . scientific values." Barbara Sicherman, "The Quest for Mental Health in America, 1880–1917," (Ph.D. diss., Columbia University, 1967), 407.

"accept Freudism as . . . its tenents." Robert S. Woodworth, "Followers of Freud and Jung," *Nation* Oct. 26, 1916, 396.

"They were all . . . the masculine protest." Max Eastman, *Enjoyment of Living* (New York: Harper and Brothers, 1948), 491.

[Mabel Dodge's] faith healers. Fred H. Matthews, "Freud Comes to America: The Influence of Freudian Ideas on American Thought, 1909–1917," (M.A. thesis, University of California at Berkeley, 1957), 98.

"psychic tea." Catherine L. Covert, "Freud on the Front Page: Transmission of Freudian Ideas in the American Newspaper of the 1920's," (Ph.D. diss., Syracuse University, 1975), 264.

"the most stupendous . . . design with no posterity." Peter B. Medawar, "Victims of Psychiatry," *New York Review of Books*, 23 Jan. 1975, 17.

"our grandsons no doubt will regard . . . phrenology." Vladimir Nabokov, *Strong Opinions* (New York: Vintage Books, 1973), 47.

Appendix A.
An Analysis of Freudian Influence on America's Intellectual Elite

"on cultural or socio-political issues." Charles Kadushin, *The American Intellectual Elite* (Boston: Little, Brown, 1974), 30–31.

"read extensively . . . with some care." Personal communication from Daniel Bell, 5 Sept. 1990.

"delivered . . . while still a teenager." Alexander Bloom, *Prodigal Sons: The New York Intellectuals and Their World* (New York: Oxford University Press, 1986), 275.

"socialist in economics . . . conservative in culture." Bell, personal communication.

"*Henderson the Rain King* . . . with its Reichianism." Eusebio L. Rodrigues, "Reichianism in *Henderson the Rain King*," *Criticism* 15 (1973): 212–33.

"saturated with Reichianism." Ibid.

"One can assume . . . through the fifties and sixties." Ibid.

"a choice between . . . the human situation." Robert F. Kiernan, *Saul Bellow* (New York: Continuum, 1989), 83.

"Bellow sees the world . . . in *Civilization and Its Discontents*." Jonathan Wilson, *On Bellow's Planet* (Rutherford, N.J.: Fairleigh Dickinson University Press, 1985), 13.

"refused to fall in with . . . right and left." Wilson, *On Bellow's Planet*, 10.

"read Freud . . . fascinating." Personal communication from Noam Chomsky, 27 Aug. 1990.

"one of the leading liberal economists of the postwar era." Richard H. Pells, *The Liberal Mind in a Conservative Age: American Intellectuals in the 1940s and 1950s* (New York: Harper and Row, 1985), 164.

"has been extraordinarily haphazard and casual." Personal communication from John K. Galbraith, 28 Aug. 1990.

"chief spokesman for . . . western radicalism." Richard King, *The*

Party of Eros: Radical Social Thought and the Realm of Freedom (Chapel Hill: University of North Carolina Press, 1972), 78.

"read extensively the writings of Freud." Personal communication from Irving Kristol, 17 Oct. 1990.

"abounded with psychological and psychiatric terminology." Stephen J. Whitfield, A Critical American: The Politics of Dwight Macdonald (Hamden, Conn.: Shoe String Press, 1984), 48.

he was very interested in . . . Reich. Personal communication from Daniel Bell, 5 Sept. 1990.

"Mailer began as an ardent Freudian." Andrew Gordon, An American Dream: A Psychoanalytic Study of the Fiction of Norman Mailer (Rutherford, N.J.: Fairleigh Dickinson University Press, 1980), 32.

"a genius . . . and new questions." Ibid., 34.

"profoundly influenced by Wilhelm Reich." King, The Party of Eros, 5.

Mailer even built his own Reichian orgone. Hilary Mills, Mailer: A Biography (New York: Empire Books, 1982), 189–90.

"of the good orgasm . . . ailments." Gordon, An American Dream, 40.

"read most of Freud's works but . . . spotty." Personal communication from Norman Podhoretz, 1 Sept. 1990.

"a great book by a major thinker . . . in town." Norman Podhoretz, Breaking Ranks: A Political Memoir (New York: Harper and Row, 1979), 48.

"tilted toward the Left." Philip Nobile, Intellectual Skywriting: Literary Politics and the New York Review of Books (New York: Charterhouse, 1974), 6.

"the neoconservative brain trust." Bloom, Prodigal Sons, 369.

"an early and ardent admirer of Freud." Daniel Bell, Individualism Reconsidered and Other Essays (Glencoe: Free Press, 1954), 306.

[Mother] analyzed by Karen Horney. Personal communication from David Riesman, 5 Sept. 1990.

"There was a time . . . Freud wrote." Ibid.

"one of the . . . heroes of all time." King, The Party of Eros, 49–50.

"psychoanalytic ideas have shaped . . . general way." Personal communication from David Riesman, 5 Sept. 1990.

"perhaps the . . . field of sociology." Podhoretz, Breaking Ranks, 33.

"the manifesto of postwar liberalism." Allen J. Matusow, The Unraveling of America: The History of Liberalism in the 1960s (New York: Harper and Row, 1984), 4.

"a fair amount of . . . the psychoanalytic school." Personal communication from Arthur Schlesinger, 29 Aug. 1990.

"the chief theoretical organ of radical chic." Nobile, Intellectual Skywriting, 7.

"I have tried . . . unsympathetic." Personal communication from Robert Silvers, 2 Jan. 1991.

"platform of the radical Left." Nobile, Intellectual Skywriting, 4.

public support for liberal causes. Bloom, Prodigal Sons, 336.

"the most influential . . . revolutionary mind." Susan Sontag, Against Interpretation and Other

Essays (New York: Farrar, Straus Giroux, 1966), 256, 260.

married . . . Philip Rieff. Sohnya Sayre, *Susan Sontag: The Elegaic Modernist* (New York: Routledge, 1990), 27.

"Sontag helped . . . Rieff." Ibid., 7.

Appendix B.
Is Toilet Training Related to "Anal" Personality Traits? A Summary of the Research <u>Studies</u>

"The way in which . . . fixations result." Seymour Fisher and Roger P. Greenberg, *The Scientific Credibility of Freud's Theories and Therapy* (New York: Basic Books, 1977), 145; quoting Freud.

Index